Targeting of Drugs
With Synthetic Systems

NATO ASI Series
Advanced Science Institutes Series

A series presenting the results of activities sponsored by the NATO Science Committee, which aims at the dissemination of advanced scientific and technological knowledge, with a view to strengthening links between scientific communities.

The series is published by an international board of publishers in conjunction with the NATO Scientific Affairs Division

A	**Life Sciences**	Plenum Publishing Corporation
B	**Physics**	New York and London
C	**Mathematical and Physical Sciences**	D. Reidel Publishing Company Dordrecht, Boston, and Lancaster
D	**Behavioral and Social Sciences**	Martinus Nijhoff Publishers
E	**Engineering and Materials Sciences**	The Hague, Boston, and Lancaster
F	**Computer and Systems Sciences**	Springer-Verlag
G	**Ecological Sciences**	Berlin, Heidelberg, New York, and Tokyo

Recent Volumes in this Series

Volume 105—The Physiology of Thirst and Sodium Appetite
edited by G. de Caro, A. N. Epstein, and M. Massi

Volume 106—The Molecular Biology of *Physarum polycephalum*
edited by William F. Dove, Jennifer Dee, Sadashi Hatano,
Finn B. Haugli, and Karl-Ernst Wohlfarth-Bottermann

Volume 107—NMR in the Life Sciences
edited by E. Morton Bradbury and Claudio Nicolini

Volume 108—Grazing Research at Northern Latitudes
edited by Olafur Gudmundsson

Volume 109—Central and Peripheral Mechanisms of Cardiovascular Regulation
edited by A. Magro, W. Osswald, D. Reiss, and P. Vanhoutte

Volume 110—Structure and Dynamics of RNA
edited by P. H. van Knippenberg and C. W. Hilbers

Volume 111—Basic and Applied Aspects of Noise-Induced Hearing Loss
edited by Richard J. Salvi, D. Henderson, R. P. Hamernik,
and V. Colletti

Volume 112—Human Apolipoprotein Mutants: Impact on Atherosclerosis
and Longevity
edited by C. R. Sirtori, A. V. Nichols, and G. Franceschini

Volume 113—Targeting of Drugs with Synthetic Systems
edited by Gregory Gregoriadis, Judith Senior, and George Poste

Series A: Life Sciences

Targeting of Drugs
With Synthetic Systems

Edited by
Gregory Gregoriadis
and
Judith Senior

Royal Free Hospital School of Medicine
London, England

and
George Poste

Smith Kline & French Laboratories
Philadelphia, Pennsylvania

Plenum Press
New York and London
Published in cooperation with NATO Scientific Affairs Division

Proceedings of a NATO Advanced Study Institute on
Targeting of Drugs with Synthetic Systems,
held June 24–July 5, 1985,
at Cape Sounion Beach, Greece

Library of Congress Cataloging in Publication Data

NATO Advanced Study Institute on Targeting of Drugs with Synthetic Systems
(1985: Ákra Soúnion, Greece)
    Targeting of drugs with synthetic systems.

    (NATO ASI series. Series A, Life sciences; v. 113)
    "Proceedings of a NATO Advanced Study Institute on Targeting of Drugs with
Synthetic Systems, held June 24–July 5, 1985, at Cape Sounion Beach Greece"—
T.p. verso.
    "Published in cooperation with NATO Scientific Affairs Division."
    Includes bibliographies and index.
    1. Drugs—Dosage forms—Congresses. 2. Capsules (Pharmacy)—Congresses.
3. Microencapsulation—Congresses. I. Gregoriadis, Gregory. II. Senior, Judith.
III. Poste, George. IV. North Atlantic Treaty Organization. Scientific Affairs Divi-
sion. V. Title. VI. Series. [DNLM: 1. Drugs—administration & dosage—congresses. WB 340 N279t 1985]
RS200.N37   1985   615'.19   86-16891
    ISBN 978-1-4684-5187-0   ISBN 978-1-4684-5185-6 (eBook)
    DOI 10.1007/978-1-4684-5185-6

PREFACE

Targeting of drugs via carrier systems to sites in the body in need of pharmacologic intervention is a rapidly growing area of research in the treatment or prevention of disease. It has evolved from the need to preferentially deliver drugs, enzymes, vitamins, hormones, antigens, etc. to target cells and organs so as to avoid toxicity, waste of drugs through premature secretion or inactivation and at the same time render treatment more convenient and cost-effective. A wide assortment of naturally occurring or semi-synthetic drug carriers (e.g. antibodies, glycoproteins, lectins, peptide hormones, cells and liposomes), their interaction with relevant receptors and mediation of optimal pharmacological action were discussed in the two previous NATO Advanced Studies Institutes (ASI) of this series, "Targeting of Drugs" and "Receptor-Mediated Targeting of Drugs", the proceedings of which were published by Plenum in 1982 and 1984 respectively.

This book contains the proceedings of the 3rd NATO ASI "Targeting of Drugs with Synthetic Systems" held as before at Cape Sounion, Greece during 24 June-5 July 1985. It deals mostly with man-made carriers such as a variety of polymers, matrices, liposomes and other colloidal microparticles. The twenty chapters discuss the interaction of such carriers with the biological milieu, approaches to bypass the reticuloendothelial system (or, when needed, take advantage of its interception of carriers to optimally deliver drugs to phagocytes) and ways to improve delivery to specific cells, often with the help of carrier-linked ligands. Each of these are dealt with by leading authorities in terms of applications in pharmacology, immunology and medicine and related methodologies.

We express our appreciation to Drs. D. Chapman, G. Deliconstantinos, H. Kotsifaki, J.B. Lloyd, M. Maragoudakis and A. Trouet who as members of the international or local committees gave valuable advice throughout the planning stages of the Institute. We are also grateful to Miss O. Dansie for her enthusiastic management of many of the practical aspects of the ASI and to Mrs. B. Thompson for immaculate secretarial assistance. The ASI was held under the sponsorship of NATO and was cosponsored and generously financed by Smith Kline and French Laboratories, Philadelphia, U.S.A. Generous financial assistance was also provided by Pharmacia (Health Care), Uppsala, Sweden, CIBA Geigy, Horsham, U.K. and Wellcome Research Laboratories, Burroughs Wellcome Co., Research Triangle Park, N.C., U.S.A.

April 1986

Gregory Gregoriadis
Judith Senior
George Poste

CONTENTS

MANNOSE BINDING PROTEINS IN THE LIVER AND BLOOD

John A. Summerfield

Department of Medicine, Royal Free Hospital School of
Medicine, Rowland Hill Street, London NW3 2PF, U.K.

INTRODUCTION

The survival of glycoproteins in the circulation is determined by the
nature of the terminal non-reducing sugar on their oligosaccharides. Glycoproteins terminating in sialic acid have plasma half-lives measured in
days whereas those terminating in other sugars are cleared rapidly from
the blood (a few minutes) mainly by the liver. The first mechanism to be
defined was the hepatocyte galactose receptor which mediated the clearance
of galactose-terminated glycoproteins from the blood (Ashwell & Morell,
1974). Around this time, groups interested in enzyme replacement for lysosomal storage disorders noted that various rat lysosomal enzymes (e.g. β -
glucuronidase) administered intravenously were also rapidly cleared by the
liver (Achord et al., 1977a; Stahl et al., 1976a,b; Schlesinger et al, 1976).
Uptake was carbohydrate mediated since it was abolished by periodate treatment of the lysosomal enzymes. However in the case of lysosomal enzymes
hepatic uptake was inhibited, not by galactose, but both by mannan (a
mannose-terminated proteoglycan from yeast cell walls) and agalactooroso-
mucoid (N-acetylglucosamine-terminated) (Stahl et al., 1976c; Achord et al.,
1977b). Thus this novel receptor (in this paper referred to as the mannose
receptor) recognised glycoproteins terminating in two sugars, either mannose
(Man) or N-acetylglucosamine (GlcNAc). Furthermore the receptors were
present, not on hepatocytes, but on the sinusoidal lining cells of the liver
(Achord et al., 1978; Schlesinger et al., 1978; Steer & Clarenburg, 1979).

Mannose receptors have subsequently been shown to be present on alveolar (Stahl et al., 1978) and peritoneal (Ezekowitz et al., 1980; Imber
et al., 1982) macrophages and cultured (but not fresh) circulating mononuclear cells and bone marrow cells (Shepherd et al., 1982; Kataoka &
Tavassoli, 1985). However in the clearance of circulating glycoproteins
these cells play a minor role, the great majority being cleared by the
hepatic mannose receptors (Schlesinger et al., 1980; Parise et al., 1984).
Finally, an hepatic mannose binding protein isolated from whole rabbit, rat
and human liver, which was originally considered to be the sinusoidal cell
mannose receptor (Kawasaki et al., 1978; Townsend & Stahl, 1981; Mizuno
et al., 1981) has now been shown to be a distinct intracellular hepatocyte
protein (Maynard & Baenziger, 1982; Mori et al., 1983; Wild et al., 1983)
located within the cisternae of the rough endoplasmic reticulum (Maynard
& Baenziger, 1982; Mori et al., 1984). This paper will review recent data
on mammalian mannose binding proteins with particular emphasis on those

areas which are of potential relevance to the problems of drug targeting.

CELL SURFACE MANNOSE RECEPTROS

Cellular Localization of Sinusoidal Mannose Receptors

The cellular localisation of the hepatic mannose receptor has been the subject of controversy. The sinusoidal cells of the liver comprise a mixture of Kupffer cells, endothelial cells, fat-storing cells (Ito cells), and (at least in the rat) pit cells. The endothelial cells of the liver sinusoids are specialised cells which differ both structurally and funct-ionally from the endothelial cells lining the rest of the vasculature. Initially the mannose receptor was believed to be located on Kupffer cells (Achord et al., 1978; Schlesinger et al., 1978, 1980), largely because ex-trahepatic (e.g. alveolar) macrophages also possessed the receptor. A subsequent study, using isolated hepatic endothelial cells and Kupffer cells separated by centrifugal elutriation indicated that the mannose receptor was located principally on endothelial cells (Summerfield et al., 1982).

This discrepancy has been resolved by three other studies. In an electron microscopic autoradiographic study radiolabelled Man- and GlcNAc-terminated glycoproteins accumulated in Kupffer cells and entothelial cells but not in hepatocytes. However, the internalisation of these ligands by endothelial cells was two to six times greater than that by Kupffer cells (on a cell volume basis) (Hubbard et al., 1979). Similar results were ob-tained in an in-vitro study, using fractions of isolated Kupffer and endo-thelial cells prepared by centrifugal elutriation. Both cell types intern-alised Man- and GlcNAc-terminated glycoproteins, but uptake was greater by endothelial cells (on a cell number basis) (Praaning van Dalen et al., 1982). Finally in a study using a modification of the standard centrifugal elutriation technique for separating hepatic sinusoidal cells a series of fractions were collected and characterized by histochemistry and electron microscopy. These revealed that elutriation yielded cell populations ranging from almost homogeneous Kupffer cell fractions. The uptake of a radiolabelled GlcNAc-terminated glycoprotein (^{125}I-agalactoorosomucoid) was greatest (53% of total uptake) by elutriator fractions containing equal proportions of endothelial and Kupffer cells (Fig. 1) (Parise et al., 1984). Taken together these data indicate that mannose receptors are present on both Kupffer cells and endothelial cells in the liver. It remains to be determined whether these mannose receptors are structurally identical and whether they share a similar spectrum of carbohydrate specificities. Each mannose receptor participates in many internalisation events indicating that it is probably recycled (Stahl et al., 1980).

Effect of Diabetes Mellitus on the Sinusoidal Mannose Receptor

The uptake of Man- and GlcNAc-terminated glycoproteins by isolated hepatic sinusoidal cells is inhibited not only by mannose and N-acetyl-glucosamine but also by glucose, fructose and a glucose-albumin conjugate (Fig. 2). Inhibition by glucose is competitive over a wide range of con-centrations and is almost 100% at a glucose concentration of 56mM (Summer-field et al., 1982). Furthermore the rates of plasma clearance, hepatic accumulation and catabolism of GlcNAc-terminated glycoproteins are decreased in diabetic mice or rats (Pizzo et al., 1981; Summerfield et al., 1982). This phenomenon also occures if hyperglycaemia is induced by infusing glucose in the absence of diabetes mellitus (Pizzo et al., 1981). The number of cell surface mannose receptors on rat peritoneal macrophages are reduced when they are cultured in media containing high glucose concentrations (Weiel & Pizzo, 1983). These data indicate that the impaired clearance of glycoproteins by the mannose receptor in diabetes is due to hyperglycaemia having both a direct inhibitory effect on uptake and down-regulating the surface expression of mannose receptors.

However the effects of diabetes on glycoprotein metabolism by hepatic sinusoidal cells appear to be more complex than simply decreasing uptake.

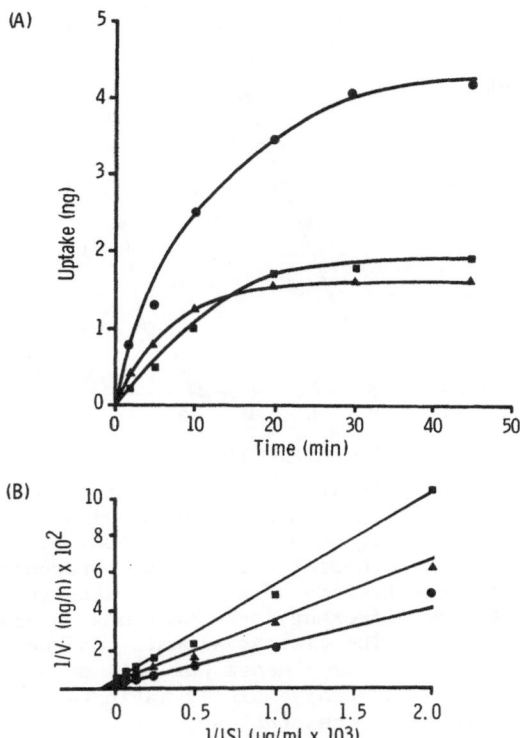

Fig. 1. The specific uptake of radioiodinated agalactoroso-mucoid (AGOR) by fractions of rat hepatic sinus-oidal cells separated by centrifugal elutriation: endothelial cell fraction (▲), mixed cell fract-ion (●), and Kupffer cell fraction (■). A, the time dependence of specific uptake of AGOR (14 nmol/l). B, the concentration dependence of spec-ific uptake of AGOR. Reproduced with permission from Parise et al., (1984).

Compartmental analysis of the metabolism of intravenously administered ^{125}I-agalactoorosomucoid (^{125}I-AGOR) in rats revealed another effect of diabetes. Not only was ^{125}I-AGOR uptake decreased but the ligand catabolic rate was greatly diminished indicating that diabetes also affected intracellular processing. These effects could be reversed by insulin treatment. Unexpectedly insulin also prolonged the intracellular transport time, the period between ligand internalisation and delivery to the lysosomes (Tanaka et al., 1985).

The functional consequences of impaired mannose receptor function in diabetes are unknown. The inhibition of AGOR uptake by fructose (Summerfield

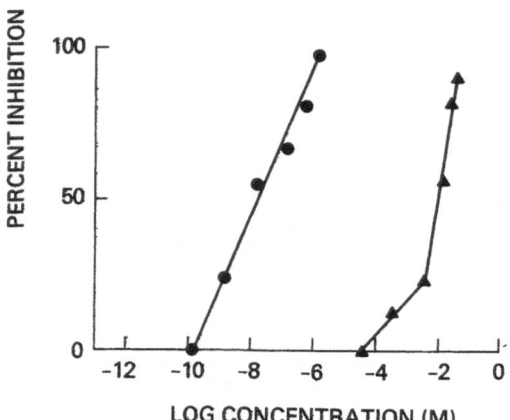

Fig. 2. Glucose-mediated inhibition of the uptake of radioiodinated AGOR by hepatic endothelial cells. Data on the inhibition of uptake of AGOR (14 nmoles/l) by endothelial cells in the presence of different concentraions of a glucose-albumin conjugate (●) or different concentrations of glucose (▲) are shown. The glucose-albumin conjugate contained 35 moles of glucose per mole of albumin. Uptake of AGOR was progressively inhibited by increasing glucose concentrations. The glucose-albumin conjugate was six orders of magnitude more potent an inhibitor (on a molar basis) than glucose. Reproduced with permission from Summerfield et al., (1982).

et al., 1982) is of particular interest because a fructose residue is the terminal sugar of spontaneously glycosylated proteins such as haemoglobin Alc and glycosylated albumin. The terminal fructose residue of such glycoproteins is generated by an Amadori rearrangement of the Schiff base formed between the free amino groups of the protein and glucose (Koenig et al., 1977). However for a glycosylated protein to be internalized by the mannose receptor depends largely on the degree of substitution of the protein with sugar residues (Krantz et al., 1976). At present, there is no evidence that sufficient glycosylation of proteins occurs in diabetes mellitus to permit their uptake by the mannose receptor. Conversely, the accumulation of glycosylated proteins in plasma in diabetes mellitus (Bunn, 1981) could be due, at least in part, to hyperglycaemia inhibiting their uptake by the mannose recpetor. Yeasts (Warr, 1980) and some bacteria (Perry & Ofek, 1984) are also cleared by the mannose receptor. Thus it is possible that inhibition of the normal function of the hepatic mannose receptor by hyperglycaemia may contribute to abnormal metabolism of glycoproteins and the increased incidence of fungal and bacterial infections in diabetes mellitus.

Effects of Other Agents on the Sinusoidal Mannose Receptor

Several other agents have been shown to influence the hepatic mannose receptor. Some of these studies have been performed on peritoneal or alveolar macrophages and so the evidence that the agents affect the hepatic mannose receptor is indirect. A further problem is that some agents alter the population of local macrophages by causing recruitment of cells from other sites (usually bone marrow). This phenomenon probably accounts for the differing reports of the effect of Bacillus Calmette Guerin (BCG) on the mannose receptor. In studies with isolated peritoneal macrophages, pretreatment with BCG caused a selective diminution of ligand binding to the mannose receptor on these cells (Imber et al., 1982; Ezekowitz et al., 1980). It was concluded that BCG down-regulated the macrophage mannose receptor. However, in another study, BCG was found to cause a great increase in the number of hepatic sinusoidal cells due to the accumulation of mononuclear cells morphologically distinct from the resident Kupffer and endothelial cells. Furthermore although the uptake of glycoprotein per cell was reduced (Fig. 3), in vivo the total hepatic uptake of glycoprotein by BCG treated rats was not different from control animals (Table 1) (Parise et al., 1984). Thus it appeared that, in the liver, BCG caused the recruitment and accumulation of extrahepatic mononuclear cells in the liver but did not alter mannose receptor expression on the resident sinusoidal cells.

In contrast, iron loading of hepatic sinusoidal cells by injections of iron sorbitol profoundly reduced hepatic glycoprotein uptake and caused a twofold increase in the proportion of ligand remaining in the circulation (Table 1). Iron loading caused a selective suppression of glycoprotein uptake by the fraction of cells containing equal numbers of Kupffer and endothelial cells (Fig. 3), that fraction containing the greatest density of mannose receptors (Parise et al., 1984). The mechanism by which iron appears to down-regulate the expression of the mannose receptor is unclear, however it is a specific effect since other agents, such as latex particles (Fig. 3), which are also phagocytosed by Kupffer cells have no influence (Parise et al., 1984).

Two manoeuvres have been described which up-regulate the cell surface expression of the mannose receptor. A 24h fast (Summerfield et al., 1982) or pretreatment with dexamethasone (Shepherd et al., 1985) increase the amount of ligand bound by isolated sinusoidal cells or cultured bone marrow cells about threefold without altering the apparent uptake (the affinity of the receptor for the ligand).

The consequences of the alterations of glycoprotein metabolism that can be induced by external factors acting on the mannose receptor remain to be defined. Dexamethasone treatment, as well as up-regulating the expression of the mannose receptor, also reduced the extracellular levels (in the culture medium) of hexosaminidase (a lysosomal enzyme) suggesting that the mannose receptor may play a role in regulating the extracellular levels of lysosomal enzymes (Shepherd et al., 1985). The demonstration that iron overload has a potent depressant effect on the hepatic sinusoidal

5

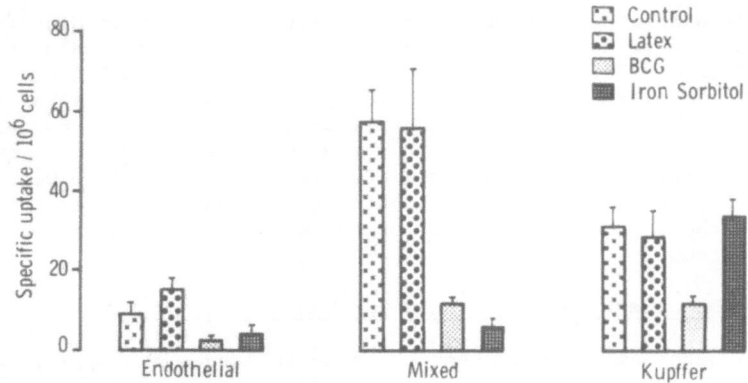

Fig. 3. The specific uptake of radioiodinated AGOR by rat
hepatic sinusoidal cells separated by centrifugal
elutriation into endothelial cell, mixed cell and
Kupffer cell fractions. The uptake by sinusoidal
cells from control rats and rats after latex, bacillus Calmette-Guerin and iron sorbitol are shown.
Reproduced with permission from Parise et al.
(1984).

mannose receptor raises the possibility that marked accumulation of lysosomal enzymes may occur in clinical states of iron overload such as haemo-
chromatosis and haemosiderosis in which the cellular iron burdens are much
greater.

Effects of Glycoprotein Structure on Uptake and Processing by the Hepatic
Mannose Receptor.

    There are several stages in the metabolism of a glycoprotein that is
targeted for the hepatic mannose receptor. The glycoprotein must first bind
to the mannose receptor. The ligand-receptor complex is internalized and
transported into the cell in endosomes. In the endosome compartment the
ligand-receptor complex is disrupted by acidification of the endosome
(Forgac et al., 1983). The endosomes then form another specialized organelle, the compartment for uncoupling the ligand-receptor complex (CURL)
where the ligand is segregated from the receptor (Geuze et al., 1983). The
receptor is recycled to the cell surface for reutilization and the glycoprotein is transported to the lysosomes to be catabolized.

    Surprisingly little data exists on the influence of glycoprotein struc-
ture on their uptake and processing by the hepatic mannose receptor despite
its potential importance for drug and enzyme targeting. However there are
observations that indicate that glycoprotein structure exerts an important
effect. In contrast to glycoproteins such as AGOR, which has an hepatic
half-life of approximately 30 min (Taylor et al., 1985), several mannose-
terminated lysosomal enzymes are not rapidly catabolized after their uptake
by hepatic sinusoidal cells (Schlesinger et al., 1978; Steer et al., 1979)
even though they have been transported to lysosomes (Schlesinger et al.,

Table 1. In-vivo metabolism of ^{125}I-AGOR

	n	Liver	Blood	Lung	Spleen	Recovery**
		Distribution*				
Control	4	60+1.4	12.3+1.4	0.5+0.1	0.9+0.1	82+5.0
Latex	3	58+0.7	8.0+0.8	0.3+0.0	1.0+0.2	74+0.5
BCG	3	61.5±3.8	11.2±2.3	0.6±0.1	1.4±0.1	86±5.0
Iron sorbitol	3	39.3±0.4‡	24.7±0.6	0.9±0.1	1.9±0.4	82±2.0
Sorbitol	5	81.2±7.2	6.9±1.3	0.2±0.004	0.4±0.04	93±6.0

AGOR = Agalacto-orosomucoid; BCG = Bacillus Calmette-Guerin
Values represent mean ± SEM for each group
*Distribution is expressed as percent injected radioactivity
 accumulating in an organ by 10 min.
**Recovery is expressed as percent of injected radioactivity
 that was recovered in the organs examined (including the
 kidneys).
‡Significantly lower than control (p < 0.05)
Reproduced with permission from Parise et al., (1984).

1976; Brown et al., 1978). Undoubtedly structural features that protect
from lysosomal degradation are critical properties of lysosomal enzymes.

   The binding rate of neoglycoproteins to the mannose receptor is dependent on the density of substituted sugars (Schlesinger et al., 1980;
Hoppe & Lee, 1983), the type of substituted sugar (the receptor has a
greater affinity for mannose than glucose) and the geometry of the oligo-
saccharide (the receptor has greater affinity for complex oligosaccharides
than a similar number of monosaccharides) (Taylor et al., 1985). In contrast the rate of ligand internalisation is independent of the receptor
binding rate constants (Hoppe & Lee, 1983; Taylor et al., 1985). Intracellular transport time, which is a composite of the events from ligand
internalization to its delivery to the lysosomes and catabolism have been
studied by compartmental analysis of in-vivo glycoprotein metabolism in
the rat (Taylor et al., 1985). Transport was significantly slower for the
Man-BSA (a mannose-terminated neoglycoprotein) than for the other ligands
(e.g. AGOR) suggesting that receptor-ligand uncoupling was slower for Man-
BSA (for which the receptor had the highest affinity in these studies),
or that extra-lysosomal catabolism of the other ligands occurred.

   Many studies have shown that deglycosylated forms of glycoproteins
are catabolized more rapidly than the original glycoprotein (Wang & Hirs,
1977; Olden et al., 1977; Loh & Gainer, 1978; Brown et al., 1979; Chu &
Maley, 1980; Olden et al., 1982; Bernard et al., 1982). Furthermore,
swainsonine, an inhibitor of lysosomal α-mannosidase, has been shown to
cause total inhibition of proteolytic degradation of endocytosed asialoglycoproteins (Winkler & Segal, 1984). These data indicate that deglycosylation of glycoproteins is the rate-limiting step in their catabolism.
But catabolism also seems to be determined by the nature of the carbohydrate on the glycoprotein. It was greater for glucose-terminated than

7

mannose-terminated BSA and slowest for AGOR which bears a complex oligo-saccharide (Taylor et al., 1985).

Thus it is evident that the structure of a glycoprotein has a critical effect on its processing by the mannose receptor. An understanding of the structural features that influence hepatic uptake, transport and catabolism will be of value in drug targeting and for enzyme replacement in lysosomal storage disorders.

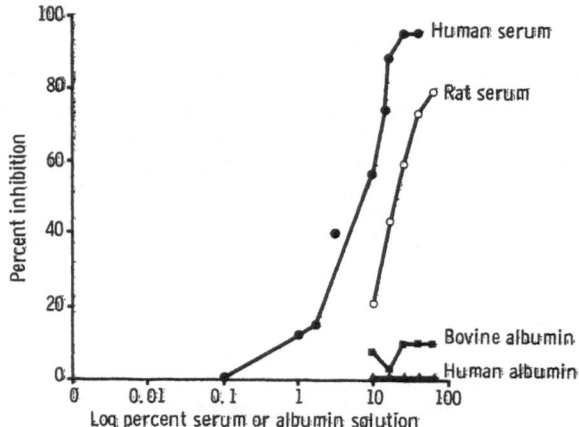

Fig. 4. Inhibition by dialyzed serum of the specific uptake of radioiodinated AGOR by rat hepatic sinusoidal cells. Sinusoidal cells were incubated with AGOR (14nmoles/l) with increasing amounts of dialyzed serum or albumin (5% w/v in glucose-free buffer). Serum, but not albumin, inhibits glycoprotein uptake by hepatic sinusoidal cells. Reproduced with permission from Taylor and Summerfield (1984).

MANNOSE BINDING PROTEINS IN THE BLOOD

It is now recognized that the metabolism of circulating Man- or GlcNAc-terminated glycoproteins is further complicated by the presence of mannose binding proteins in the blood. Human or rat serum inhibit the specific uptake of AGOR by isolated rat hepatic sinusoidal cells (Fig. 4). The serum inhibitors are not glycoproteins that bind to the hepatic mannose receptor but have the properties of lectins that bind Man- or GlcNAc-terminated glycoproteins. They can be isolated from serum by affinity chromatography and the isolated proteins will inhibit the sinusoidal uptake of glycoproteins in vitro (Taylor & Summerfield, 1984).

The nature of these serum mannose binding proteins is becoming clearer. Human serum contains calcium independent mannose binding proteins, the principal one is mannose specific immunoglobulin (IgG) (Summerfield & Taylor, 1985). A calcium dependent mannose binding protein has been isolated from rabbit and human serum (Kozutsumi et al., 1980; Kawasaki et al., 1983). This calcium dependent serum mannose binding protein, like the hepatocyte mannose binding protein (Townsend & Stahl, 1981), has specificities for GlcNAc and fucose as well as mannose (Summerfield & Taylor, 1985) Furthermore in Western blots, antibodies to the 30k subunit of this serum protein bind the 30k subunits of the hepatocyte mannose binding protein isolated by Wild et al. (1983) (Summerfield & Taylor, 1985). These data indicate that the calcium dependent serum mannose binding protein may originate by secretion from hepatocytes.

Thus serum contains at least two classes of mannose binding proteins; mannose specific, calcium independent immunoglobulins and a calcium dependent protein of broader carbohydrate specificity which is probably secreted by hepatocytes. The central questions that remain are the modes of action and functions of these serum mannose binding proteins. The factors that determine which serum mannose binding protein binds a glycoprotein that enters the circulation and the ways in which the glycoprotein-binding protein complex interacts with the sinusoidal mannose receptor are unknown. Since the binding of glycoprotein to both serum and sinusoidal mannose binding proteins is carbohydrate mediated it seems likely that the glycoprotein dissociates from the serum mannose binding protein prior to binding to the sinusoidal mannose receptor. As to the function of serum mannose binding proteins, they might bind and inactivate noxious glycoproteins (such as lysosomal enzymes, yeasts and bacteria) that enter the circulation prior to their removal by the sinusoidal mannose receptor. Serum mannose binding proteins may function in an analogous way to α 1-antitrypsin and the other components of the serum antiprotease system. The consequences of this system of serum mannose binding proteins for targeting drugs to the sinusoidal mannose receptor remains to be defined.

CONCLUSIONS

Mannose- and N-acetylglucosamine-terminated glycoproteins that enter the circulation are first bound by carbohydrate specific serum binding proteins and then cleared by carbohydrate specific receptors on hepatic sinusoidal cells. Both the structure of the saccharide moiety of the glycoprotein and external factors such as diabetes mellitus or treatment with glucocorticoids influence uptake and processing of the glycoprotein by the sinusoidal mannose receptor. The mannose receptor should be a potent means of targeting drugs to hepatic sinusoidal cells.

ACKNOWLEDGEMENT

JAS is a Wellcome Trust Senior Fellow.

REFERENCES

Achord, D., Brot, F., Gonzalez-Noriega, A., Sly, W., and Stahl, F., 1977a, Human β-glucuronidase. II. Fate of infused human placental β-glucuronidase in the rat, Pediatr. Res., 11:816.
Achord, D.T., Brot, F.E., and Sly, W.S., 1977b, Inhibition of the rat

clearance system for agalacto-orosomucoid by yeast mannans and by mannose, Biochem. Biophys. Res. Commun., 77:410.

Achord, D.T., Brot, F.E., Bell, C.E., and Sly, W.S., 1978, Human β-glucuronidase: In-vivo clearance and in-vitro uptake by a glycoprotein recognition system on reticuloendothelial cells, Cell, 15:269.

Ashwell, G., and Morell, A.G., 1974, The role of surface carbohydrates in the hepatic recognition and transport of circulating glycoproteins, Adv. Enzymol., 41:99.

Bernard, B.A., Yamada, K.M., and Olden, K., 1982, Carbohydrates selectively protect a specific domain of fibronectin against proteases, J. Biol. Chem., 257:8549.

Brown, J.A., Segal, H.L., Maley, F., Trimble, R.B., and Chu. F., 1979, Effect of deglycosylation of yeast invertase on its uptake and digestion in rat yolk sacs, J. Biol. Chem., 254:3689.

Brown, T.L., Henderson, L.A., Thorpe, S.R., and Baynes, J.W., 1978, The effect of α-mannose terminal oligosaccharides on the survival of glycoproteins in the circulation. Rapid uptake and catabolism of bovine pancreatic ribonuclease B by non-parenchymal cells of rat liver, Arch. Biochem. Biophys., 188:418.

Bunn, H.F., 1981, Nonenzymatic glycosylation of protein: relevance to diabetes, Am. J. Med., 70:325.

Chu, F.K., and Maley, F., 1980, The effect of glucose on the synthesis and glycosylation of the polypeptide moiety of yeast external invertase, J. Biol. Chem., 255:6392.

Ezekowitz, R.A.B., Austyn, J., Stahl, P.D., and Gordon, S., 1980, Surface properties of bacillus Calmette-Guerin activated mouse macrophages, J. Exp. Med., 154:60.

Forgac, M., Cantley, L., Weidenmann, B., and Altstiel, L., 1983, Clathrin coated vesicles contain an ATP-dependent proton pump, Proc. Natl. Acad. Sci. USA, 80:1300.

Geuze, H.J., Slot, J.W., Strous, G.J.A.M., Lodish, H.F., and Schwartz, A.L., 1983, Intracellular site of asialoglycoprotein receptor-ligand uncoupling-double label immunoelectronmicroscopy during receptor mediated endocytosis, Cell, 32:277.

Hoppe, C.A., and Lee, Y.C., 1983, The binding and processing of mannose-bovine serum albumin derivatives by rabbit alveolar macrophages. Effect of sugar density, J. Biol. Chem., 258:14193.

Hubbard, A.L., Wilson, G., Ashwell, G., and Stukenbrok, H., 1979, An electron microscopic autoradiographic study of the carbohydrate recognition systems in the rat liver, J. Cell. Biol., 83:47.

Imber, M.J., Pizzo, S.V., Johnson, W.J., and Adams, D.O., 1982, Selective diminution of the binding of mannose by murine macrophages in the late stages of activation, J. Biol. Chem., 257:5129.

Kataoka, M., and Tavassoli, M., 1985, Development of specific surface receptors recognizing mannose-terminal glycoconjugates in cultured monocytes: a possible early marker for differentiation of monocyte into macrophage, Exp. Hematol., 13:44.

Kawasaki, T., Etoh, R., and Yamashina, I., 1978, Isolation and characterization of a mannan-binding protein from rat liver, Biochem. Biophys. Res. Commun., 81:1018.

Kawasaki, N., Kawasaki, T., and Yamashina, I., 1983, Isolation and characterization of a mannan-binding protein from human serum, J. Biochem., 94:937.

Koenig, R.J., Blobstein, S.H., and Cerami, A., 1977, Structure of carbohydrate of hemoglobin Alc, J. Biol. Chem., 252:2992.

Kozutzumi, Y., Kawasaki, T., and Yamashina, I., 1980, Isolation and characterization of a mannan-binding protein from rabbit serum, Biochem. Biophys. Res. Commun., 95:658.

Krantz, M.J., Holtzman, N.A., Stowell, C.P., and Lee, Y.C., 1976, Attachment of thioglycosides to proteins. Enhancement of liver membrane binding, Biochemistry, 15:3963.

Loh, Y.P., and Gainer, H., 1978, The role of glycosylation on the biosynthesis, degradation and secretion of the ACTH-B-lipotropin common
precursor and its peptide products, FEBS Lett., 96:269.

Maynard, Y., and Baenziger, J.U., 1982, Characterization of a mannose and
N-acetylglucosamine specific lectin present in rat hepatocytes,
J. Biol. Chem., 257:3788.

Mizuno, Y., Kozutsumi, Y., Kawasaki, T., and Yamashina, I., 1981, Isolation
and characterization of a mannan-binding protein from rabbit liver,
J. Biol. Chem., 256:4247.

Mori, K., Kawasaki, T., and Yamashina, I., 1983, Identification of the
mannan-binding proteins from rat livers as a hepatocyte protein
distinct from the mannan receptor on sinusoidal cells, Arch. Biochem.
Biophys., 222:542.

Mori, K., Kawasaki, T., and Yamashina, I., 1984, Subcellular distribution
of the mannan-binding protein and its endogenous inhibitors in rat
liver, Arch. Biochem. Biophys., 232:223.

Olden, K., Pratt, R.M., and Yamuda, K.M., 1977, Role of carbohydrates in
protein secretion and turnover: effects of tunicamycin on the major
cell surface glycoproteins of chick embryo, Cell, 13:461.

Olden, K., Bernard, B.A., White, S.L., and Parent, J.B., 1982, Carbohydrate
moieties of proteins. A re-evaluation of their function, J. Cell.
Biochem., 18:313.

Parise, E.R., Taylor, M.E., and Summerfield, J.A., 1984, Effects of iron
loading and bacillus Calmette-Guerin on a glycoprotein recognition
system on rat hepatic sinusoidal cells, J. Lab. Clin. Med., 104:
908.

Perry, A., and Ofek, I., 1984, Inhibition of blood clearance and hepatic
tissue binding of Escherichia coli by liver lectin-specific sugars
and glycoproteins, Infect. Immun., 43:257.

Pizzo, S.V., Lehrman, M.A., Imber, M.J., and Guthrow, C.E., 1981, The
clearance of glycoproteins in diabetic mice, Biochem. Biophys. Res.
Commun., 101:704.

Praaning van Dalen, M., and Knook, D.L., 1982, Quantitative determination
of in-vivo endocytosis by rat liver Kupffer and endothelial cells
facilitated by an improved cell isolation method, FEBS Lett., 141:
229.

Schlesinger, P., Rodman, J.S., Frey, M., Lang, S., and Stahl, P., 1976,
Clearance of lysosomal hydrolases following intravenous infusion.
the role of liver in the clearance of β-glucuronidase and N-acetyl-
β-D-glucosaminidase, Arch. Biochem. Biophys., 177:606.

Schlesinger, P.H., Doebber, T.W., Mandell, B.F., White, R., de Schriyver,
C., Rodman, J.S., Miller, M.J., and Stahl, P., 1978, Plasma clearance of glycoprotein with terminal mannose and N-acetylglucosamine
by liver non-parenchymal cells, Biochem. J., 176:103.

Schlesinger, P.H., Rodman, J.S., Doebber, T.W., Stahl, P.D., Lee, Y.C.,
Stowell, C.P., and Kuhlenschmidt, T.B., 1980, The role of extrahepatic tissues in the receptor-mediated plasma clearance of glycoproteins terminated by mannose or N-acetylglucosamine, Biochem. J.,
192:597.

Shepherd, V.L., Campbell, T.J., Senior, R.M., and Stahl, P.D., 1982, Characterization of the mannose/fucose receptor on human mononuclear
phagocytes, J. Reticuloendothel. Soc., 32:423.

Shepherd, V.L., Konish, M.G., and Stahl, P., 1985, Dexamethasone increases
expression of mannose receptors and decreases extracellular lysosomal
enzyme accumulation in macrophages, J. Biol. Chem., 260:160.

Stahl, P., Rodman, J.S., and Schlesinger, P., 1976a, Clearance of lysosomal
enzymes following intravenous infusion. Kinetics and competition
experiments with β-glucuronidase and N-acetyl-β-D-glucosaminidase,
Arch. Biochem. Biophys., 177:594.

Stahl, P., Six, H., Rodman, J.S., Schlesinger, P., Tulsiani, T.R.P., and
Touster, O., 1976b, Evidence for specific recognition sites mediat-

ing clearance of lysosomal enzymes in vivo, <u>Proc. Natl. Acad. Sci. USA</u>, 73:4045.

Stahl, P., Schlesinger, P.H., Rodman, J.S., and Doebber, T., 1976c, Recognition of lysosomal glycosidases in vivo inhibited by modified glycoproteins, Nature, 264:86.

Stahl, P.D., Rodman, J.S., Miller, M.J., and Schlesinger, P.H., 1978, Evidence for receptor mediated binding of glycoproteins glycoconjugates, and lysosomal glycosidases by alveolar macrophages, <u>Proc. Natl. Acad. Sci. USA</u>, 75:1399.

Stahl, P., Schlesinger, P.H., Sigardson, E., Rodman, J.S., and Lee, Y.C., 1980, Receptor-mediated pinocytosis of mannose glycoconjugates by macrophages: characterization and evidence for receptor recycling, <u>Cell</u>, 19:207.

Steer, C.J., and Clarenburg, R., 1979, Unique distribution of glycoprotein receptors on parenchymal and sinusoidal cells of rat liver, <u>J. Biol. Chem.</u>, 254:4457.

Steer, C.J., Kusiak, J.W., Brady, R.O., and Jones, E.A., 1979, Selective hepatic uptake of human β-hexosaminidase A by a specific glycoprotein recognition system on sinusoidal cells, <u>Proc. Natl. Acad. Sci. USA</u>, 76:2774.

Summerfield, J.A., Vergalla, J., and Jones, E.A., 1982, Modulation of a glycoprotein recognition system on rat hepatic endothelial cells by glucose and diabetes mellitus, <u>J. Clin. Invest.</u>, 69:1337.

Summerfield, J.A., and Taylor, M.E., 1985, Mannose-binding proteins in human serum: identification of mannose specific immunoglobulins and a calcium dependent lectin secreted by hepatocytes, Submitted.

Tanaka, N., Leaning, Ml, Taylor, M., and Summerfield, J.A., 1985, The effects of diabetes and insulin on glycoprotein metabolism by rat liver, <u>J. Hepatol.</u>, in press.

Taylor, M.E., Leaning, M.S., and Summerfield, J.A., 1985, The influence of ligand structure on uptake and processing of glycoproteins by the rat hepatic mannose receptor, Submitted.

Taylor, M.E., and Summerfield, J.A., 1984, Human serum contains a lectin which inhibits hepatic uptake of glycoproteins, <u>FEBS Lett.</u>, 173:63.

Townsend, R., and Stahl, P., 1981, Isolation and characterization of a mannose/N-acetylglucosamine/fucose-binding protein from rat liver, <u>Biochem. J.</u>, 194:209.

Wang, F.F.C., and Hirs, C.H.W., 1977, Influence of the heterosaccharides in porcine pancreatic ribonuclease on the conformation and stability of the protein, <u>J. Biol. Chem.</u>, 252:8358.

Warr, S.A., 1980, A macrophage receptor for (mannose/glucosamine)-glycoproteins of potential importance in phagocytic activity, <u>Biochem. Biophys. Res. Commun.</u>, 93:737.

Weiel, J.E., and Pizzo, S.V., 1983, Down-regulation of macrophage mannose/N-acetylglucosamine receptors by elevated glucose concentration, <u>Biochim. Biophys. Acta.</u>, 759:170.

Wild, J., Robinson, D., and Winchester, B., 1983, Isolation of mannose-binding proteins from human and rat liver, <u>Biochem. J.</u>, 210:167.

Winkler, J.R., and Segal, H.L., 1984, Inhibition by swainsonine of the degradation of endocytosed glycoproteins in isolated rat liver parenchymal cells, <u>J. Biol. Chem.</u>, 259:1958.

CELL MEMBRANE MOLECULES ON NEOPLASTIC GELLS: THEIR ROLE IN MALIGNANT CELL

TRANSFORMATION AND DISSEMINATION

Lennart Olsson and Claus Due

Cancer Biology Laboratory
State University Hospital (Rigshospitalet)
Dk-2100 Copenhagen, Denmark

INTRODUCTION

Malignant, neoplastic cell populations are characterized by their autonomous growth, invasiveness and metastatic activity. These features, like normal cellular activities, are all dependant on the interaction between the cells and their microenvironment. The cell membrane therefore constitutes a crucial biological structure in neoplastic cell behaviour.

The cell membrane contains a number of molecules with different biological functions of which many are essential for the normal biological functions of the cell. Receptors like the insulin receptor[1,2] are typical examples, but also components like major histocompatibility antigens (MHC) are of vital importance for the normal interplay between the cells in a multicellular organism.[3,4] Normal cellular differentiation is associated with pronounced alterations in some cell membrane features as exemplified in bone marrow cell differentiation. In fact, expression of certain cell surface constituents, e.g. Ia-molecules, can be used as marker molecules for certain stages of differentiation and have, in several cell systems, been used to identify the cellular origin of malignant cell populations.[5-8]

Malignant transformation is associated with abnormal cellular differentiation that may lead to expression of membrane molecules that would not be expressed concomitantly in normal cells. On the other hand, the highly complicated biological processes such as autonomous growth, invasiveness and metastatic activity indicate that the molecules in the malignant cell membrane are able to be coordinated functionally. Indeed, this process of coordination seems to be so important and complicated that a high amount of malignant variants can be expected to be eliminated either because the coordination does not take place or it takes place in a way that does not result in functional molecules in the membrane.

Experimental investigations on the interaction of molecules in the membrane of malignant cells and the pertubation in the biological activities of such molecules as related to malignancy, are scarce. It is only recently that methods have been developed to dissect concomitantly the expression of a multitude of cell membrane molecules. However, more important seems to be the fact that the very recent elucidation of factors that may be of relevance in tumorgenesis and metastatic behaviour permits identification of cell membrane molecules that with a high probability are of direct importance with regards to the malignant state of the cells.

The cell membrane fluidity and the high concentration of molecules in the membrane's outer layer indicate that the molecules may associate into biologically active complexes.[9] We have elsewhere suggested that the so-called "compound" receptors may consitute a very substantial part of the biologically active constituents in the cell membrane.[10,11] Consequently, it becomes pertinent to investigate whether malignant transformation affects the fluidity in the membrane, as this consequently can be expected to affect the construction of compound receptors and, thus, the metabolic condition of the cells. Moreover, as malignant transformation may result in expression of membrane components that under normal conditions are not present in the same cells, formation of new types of compound receptor molecules may take place.[12] This also includes construction of molecular complexes that may be important for the progression and dissemination of malignant cells.

CELL MEMBRANE MOLECULES

The molecules in the membrane of malignant cells may conveniently be divided into three categories: (i) molecules expressed on virtually all eukaryotic cells, (ii) molecules only expressed in certain stages of cellular differentiation, and (iii) molecules expressed particularly or exclusively on malignant cells.[13]

Only a few molecules are known to belong to the first category. The class I molecules of the major histocompatibility complex (MHC I) is found on virtually all nucleated eukaryotic cells. The MHC I consists of a 45,000 d glycoprotein (heavy chain) that is bound non-covalently to $2-\beta$ microglobulin (12,000 d). The biological function of the MHC I molecule has primarily been related to its role in restricting the target cell spectrum of cytolytic T-lymphocytes. However, the MHC heavy chain is probably the most polymorphic molecule in the cell, and it seems difficult to accept this polymorphism and its phylogenetical preservation as solely based on its role in MHC-restriction of cytolytic T-lymphocytes. Several investigators, including ourselves, have recently suggested that MHC may also have a number of functions that do not directly relate to the immune system. Ohno thus suggested several years ago that MCH I was an anchorage protein of importance for organogenesis.[14] Our own hypothesis on compound receptors has specifically focused on the role of MHC I as a subunit in hormone receptors, and in particular, the insulin receptor. It has been suggested that the MHC I molecules may undergo significant changes as related to malignant transformation, and has in rare cases been found not to be expressed. The biological significance of this is unknown. Similarly, the insulin receptor (IR) is known to be expressed on virtually all nucleated cells. However, recent experiments on neoplastic cells have revealed that some cells may lose the IR, and are thus unable to use insulin-mediated metabolism.

The second category of membrane molecules are only expressed at certain steps of differentiation and/or certain cell types. The class II molecules of MHC have already been mentioned. Transferrin receptor is another molecule that is expressed only in proliferating cells.[15] These molecules may thus be used to identify the cellular origin of a malignant cell population. However, malignant transformation may result in abnormal expression of otherwise normal cell membrane compounds. Accordingly, a given differentiation antigen may be expressed irrespectively of the stage of differentiation from which the malignant cells have been derived. The biological function of MHC II and the transferrin receptor are known. However, for the majority of differentiation antigens, no specific biological functions have been described.

Analysis of tumor-associated antigens have shown that these antigens fall into the following categories: present only on autologous tumor cells, present on autologous and allogeneic tumor cells and a restricted range of normal cells, and finally some that are found on a broad range of both normal and malignant cells. The introduction of monoclonal antibodies (Mabs) as produced by hybridoma technology has enabled a more careful dissection of the antigenic repertoire on malignant cells.[12] These investigations have revealed that a very significant part (about one third) of tumor-associated antigens (TAA) is determined by carbohydrate epitopes.[15] The processing of carbohydrate structures seems thus very frequently to be abnormal in malignant cells.[16,17] This may imply that the genetic basis for such carbohydrate structures is closely linked to the set of genetic events that eventually result in malignant transformation. The biological function of TAA is virtually unknown, and such antigens have therefore mainly been used as tumor marker molecules. It is only very recently that some experimental indication has been obtained to suggest that some TAA may be of direct importance for the malignant behaviour of the cells, and that blockage down-regulation of such antigens therefore may affect very severely the malignant behaviour of the cells.

CELL MEMBRANE DIVERSITY

Individual malignant tumor cell populations have been shown to be highly diversified in respect to a number of phenotypic attributes such as growth properties in vitro and in athymic nude mice, metastatic activity, sensitivity to irradiation and chemotherapeutic drug and antigenicity.[18-20] The latter feature is directly related to the expression of cell membrane molecules. Intratumoral antigenic heterogeneity (IAH) has thus been demonstrated in several types of malignant tumors in relation to MHC molecules. as well as a variety of TAA.[21]

The biological basis for the generation of IAH has been discussed in detail previously.[20] The genetic instability in cancer cells as well as epigenetic factors seem to be of importance. Recent experiments have thus shown that changes in the genomic content of 5-methylcytosin as obtained by treating the cells with 5-azacytidine may result in pronounced changes not only in morphology and growth behaviour of the cells, but also in membrane density of MHI, IR, and TAA.[22-24] The intratumoral diversity is thus not only a result of abnormal differentiation processes within the tumor cells, but also a result of an interplay between microenvironmental factors and the cellular response to such factors.[25,26]

Antigenic modulation is another process that very significantly may contribute to IAH.[27] The modulation phenomenon has only been studied in a few systems. Both shedding and internalization can be expected to be involved in antibody-mediated antigen modulation. Hormones like insulin are known to result in internalization of the corresponding receptor, and it is conceivable that such modulation processes may occur continously and thereby contribute to IAH, if the modulation pattern differs among the different stages of differentiation. It is currently unknown whether carbohydrate epitopes modulate differently from protein epitopes upon exposure to specific antibody. It has become particularly important to elucidate this aspect since it was realized that a number of TAA are determined by carbohydrate epitopes.

THE ROLE OF CELL SURFACE CONSTITUENTS IN TUMORIGENICITY

The process of tumor progression results eventually in an autonomous growth pattern of the malignant cell population that, before this stage,

may be influenced by hormones and other growth regulating factors.[18]   The
progressive loss of regulatory interaction between the tumor cells and their
microenvironment implies that the cell membrane loses its ability to receive
and/or translate signals from the microenvironment or to process orderly
signals into an adequate response. The former deficiencies are clearly related to alterations of the cell membrane consitituents. However, as the
metabolic activity is high in most malignant cells, such cells clearly
have the ability to utilize factors in the microenvironment for their growth
activity.

The biochemical nature of the processes that are elicited by normal
growth factors and result in cell proliferation is poorly understood. Some
knowledge has in recent years been obtained for insulin and insulin-like
hormones, but knowledge is very scarce for most other growth factors, their
equivalent receptor molecules and the events upon interaction of these molecules with the related ligands. Only little attention has been paid to
these growth factors in relation to malignant neoplastic cell population,
even though it was realized that the proliferation and expansion of the
tumors must depend on stimulation of cells through appropriate growth stim-
ulatory signals.

The identification of extensive homology between certain onc-gene encoded products and growth factors have however resulted in bringing the
role of growth factors for tumorigenicity in focus. Thus, two groups have
reported concomitantly and independently that the v-sis proteins of Simian
sarcoma virus had extensive homology to platelet-derived growth factors.[28,29]
Subsequent investigations have clearly suggested that other growth factors
also may have extensive homology to onc-gene encoded substances.[30,31]

It is currently virtually unknown what the relationship is between the
onc-gene encoded product and tumorigenesis and tumor progression.[30,31] As
for the relevance to the growth and dissemination of in vivo growing tumors,
an important reservation should still be kept in mind. The assay system
for onc-genes is almost exclusively based on cessation of contact inhibition of growth of cultured fibroblasts resulting in focus formation in
vitro and in tumorigenicity in athymic nude mice. These assays only focus
on proliferative properties and it may therefore not be highly surprising
that onc-gene encoded products have high homology to growth factors.

The interaction between growth factors and neoplastic cells can be
assumed to take place through specific receptors in the cell membrane. A
malignant cell may thus produce high amounts of growth factors without
effect on proliferation if cells lack the corresponding receptor. The dependence of expression of appropriate receptor molecules on tumor-associated
growth factors has become linked to the observation that the transforming
product of the Rous sarcoma virus, pp60[src], is highly homologous to the
epidermal growth factor receptor.[32] Thus, onc-gene encoded products may
thus be both homologous to growth factors and receptors for such factors.
It is consequently tempting to speculate that by activation of the appropriate onc-genes the cells may be provided with the growth factors and the
corresponding receptors. Finally, tumor progression may result in activation and expression of a number of normal genes, including those for
growth factors and ligand receptors that in normal cells would have otherwise been silent. During progression, it can be assumed that a selective
advantage will be obtained by those cells which, by change, both produces
growth factors and express the corresponding receptors. This also implies
that blockage production of either growth factors or in particular, the
corresponding receptor molecules may impair the tumor growth significantly.
Accordingly, it seems of crucial importance to elucidate the structural and
functional mechanisms of such receptors.

METASTATIC ACTIVITY AND THE CELL MEMBRANE

The metastatic process is often characterized by a high degree of organ selectivity. The mechanisms involved in this organ selectivity are virtually unknown, but can on the other hand be expected to be of direct importance for metastatic activity. The organ selectivity thus suggests that recognition mechanisms between the metastatic cells and the target organ are involved. Such recognition could conceivably take place through receptor molecules on the tumor cells and/or on the cells in the target organ. Another mechanism may be that the microenvironmental conditions in the target organ provide support for tumor cell growth. However, this mechanism can also be expected to be based on the expression of specific receptor molecules on the cell surface.

The role of cell membrane molecules in the metastatic process is consequently a central issue in the analysis of the biological basis of metastasis. Several molecules have recently been found to affect metastatic activity. Our own laboratory identified a metastasis-associated antigen in the murine Lewis lung carcinoma (3LL) system. The antigen was only expressed on metastatic cells and the corresponding gene(s) were found to be regulated in their expression by m^5Cyt.[33] The antigen was characterized as a glycoprotein with a M_r of about 45,000 d, but the molecule did not coprecipitate other molecules. We initially did not consider the antigen as a possible abnormal HMC heavy chain, because β 2-microglobulin did not coprecipitate. However, subsequent analysis revealed that the molecule has high homology to MHC antigens. These observations are in line with recent findings that MHC molecules are of importance in metastatic activity.[34]

The impact of MHC on metastatic activity is also in line with our hypothesis on compound receptors. The MHC molecules may associate with other membrane components to form receptors that are involved in the establishment and expansive growth of metastatic lesions. It is highly intriguing that MHC molecules may be involved in determining the selection of organs for metastatic activity, because it implies crucial non-immunological roles of these molecules in organ identification, and thereby also non-immune functions of the MHC complex. These molecules may thus provide the first clues to the complex of different molecules that together can be expected to be required for the metastatic process. Both tumorigenicity and metastatic activity seem thus highly dependent on the interaction between cell membrane molecules (receptors) and factors in the cellular microenvironment. Identification of these molecules is therefore of central importance and the use of anti-idiotypic antibodies (anti-id) against antibodies to the corresponding growth factor seems to be a very promising approach.

THE ANTI-IDIOTYPE APPROACH

The antigenic determinants located in the variable portion of Ig-molecules (idiotypes) are usually formed by interaction of both the heavy and light chains of the molecule. The idiotype is specific for the individual clone of B-lymphocytes, a fact that has been explored in attempts to treat B-cell lymphomas.[35,36] The efficiency of this procedure for therapy seems low, and indeed somewhat disappointing relative to the promises of the initial reports.

The idiotypic specificity was recently used as a basis of a very intriguing and highly attractive idea: Mouse monoclonal antibodies against a highly specific tumor-associated antigen for gastrointestinal cancers[37] or melanoma[38] were produced, and monoclonal mouse anti-idiotype antibodies generated against such tumor-specific Mabs. The idea was that the variable region of the anti-idiotype antibody has a high structural homology to the tumor associated antigen, and hybridoma technology would therefore permit production of a molecule with high similarity to the tumor associated antigen. Therefore, the antiidiotype antibody should be usable as antigen for active immunization against the tumor. The method has only been examined to a very limited extent. However, it seems likely that human anti-idiotype Mabs will have a significantly higher chance to elicit a specific immune response against the idiotype compared to a similar mouse or rat Mabs. The latter antibodies can be expected primarily to elicit a murine response against epitopes specific for all rodent Ig-molecules and only to an insignificant extent against the idiotypic fragment. In contrast, human Mabs (HMabs) used as antigens in human beings can be expected primarily to elicit an immune response against the idiotype. The anti-idiotype approach can therefore first be fully evaluated, when HMabs are used as immunogen for active immunization. This also implies that experimental evaluation of the hypothesis can be carried out in mice (e.g. athymic nude mice carrying human tumors) only with difficulty: the tumor-bearing animals will react primarily against epitopes common for all human Ig-molecules.

It has become increasingly recognized that carbohydrate and carbohydrate lipid compounds may be of importance in the construction of both receptor structures and ligands. However, studies on such structures are often prohibited by the obstacles and limitations in carbohydrate biochemistry as compared to protein biochemistry. It would therefore be highly desirable to have the possibility to convert carbohydrate epitopes into protein epitopes. The anti-idiotype approach offers the possibility to convert a carbohydrate epitope into a protein epitope and thereby largely facilitate identification sites on cells of the carbohydrate structure. Thus, carbohydrates with effects on growth activity of eukaryotic cells presumably bind to receptors on the cells. Anti-id antibodies that mimic such carbohydrate molecules can consequently be used to identify and characterize the corresponding receptors by conventional antibody-dependent protein biochemistry techniques.

HUMAN SQUAMOUS LUNG CARCINOMAS (SLC): AN EXAMPLE

The strategy of our laboratory in establishing in vitro cloned cell lines and some of the experimental data of analysis of these cell lines have been reported recently.[39] In addition to these lines, two other lines (designated SLC-L12 and SLC-L13) have been established. The latter two lines originate from the primary tumor lesion and a metastatic foci respectively of the same patient. The morphological features, including electron microscopic, the growth pattern in liquid and semisolid culture medium, the tumorigenicity in athymic nude mice, karyotypes, and cell surface antigenic attributes were determined (Table 1). The cellular expression of most of these phenotypic features were related to the stage of differentiation of SLC cells. Treatment of the cells with drugs like phorbol ester and retinoid acid resulted thus in terminal differentiation of the cells with concomitant loss of autonomous growth activity both in vitro and in vivo.[40]

Mabs with high specificity for SLC were derived. One of these, designated 43-9F, was found to be highly specific to SLC (Table 2). The epitope for 43-9F Mab was found on a set of glycoproteins released into the culture medium of SLC lines.[41] Approximately 20% of the glycoproteins released from the SLC-L11 cells were found to express the 43-9F epitope. Immunofluorescence analysis with FITC-conjugated 43-9F Mab (FITC-43-9F) on viable and acetone fixed SLC cells showed that the antigens are both on the cell membrane and in the cytoplasm. Western-blots with the 43-9F Mab of both SLC-L11 cellular extracts and of conditioned SLC-L11 culture medium showed smear-like staining in the molecular weight range of about $5x10^4 \rightarrow 30x10^4$ d. Subsequent analysis demonstrated that the epitope could be found on a div-

Table 1. Phenotypic attributes of cloned in vitro cell lines established from human squamous lung cancer lesions

Line desig nation	Population doubling time / Cloning effi- ciency in agar	Morphology Modal chromosome number	Tumori genicity in nude mice	Cell Surface antigens		
				MHC class I	insulin receptors	43.9F Mab binding
SLC-L11	37 h / 5.6%	Pleomorphic 50	++	++	++	++
8 subclones	28.49 h / 4.0-6.6%	Pleomorphic 48-52	++	++	++	+-++
3 subclones	80-160 h / 0	Pleomorphic 49-51	Neg	++	++	Neg
SLC-L12	37 h / 1.6%	Pleomorphic 51	++	++	++	++
SLC-L13	51 h / 1.1%	Pleomorphic 50	++	++	++	Neg

Table 2. Binding of the 43-9F Mab to cell lines, tissue samples and to serum from patients with various diseases.

Origin of human biopsy material	No specimens examined	Reactivity with 43-9F		
		Dot blot	ELISA	% positive
Carcinomas				
Squamous lung carcinomas	117	96	87	82
Small cell carcinomas	81	0	0	0
Adenocarcinomas	56	3	3	5
Others	29	0	0	0
Other carcinomas				
Stomach	45	0	0	0
Colo-rectal	48	0	0	0
Breast	34	0	0	0
Ovarian	29	1	0	4
Bladder	34	0	0	0
Sarcomas				
Fibrosarcomas	19	0	0	0
Leukemias	69	0	0	0
Normal bone marrow	74	1	0	1

Human cell lines positive for 43-9F: Squamous lung carcinoma (designated RH-SLC-L11: RH-SLC-L12). Human cell lines negative for 43-9F: B-lymphoma, T-lymphoma, Myeloid leukemia, Small cell lung carcinoma, melanoma, ovarian cancer, firbroblasts.
Murine cell lines negative for 43-9F: Lewis lung carcinoma, B16 melanoma, T-cell leukemia, normal bone marrow cells.

erse set of glycoproteins with M_r $5x10^4$->$2x10^6$ d. The outcome of the Western-blots is in accordance with this. Although the SLC lines established in our laboratory with one exception all produce and release the 43-9F antigen, they produce and release different amounts of the antigen.

The 43-9F antigenic glycoproteins could be retained on wheat-germ agglutinin columns and also to some extent on concanavalin A columns. Moreover, treatment of the 43-90F positive antigens with endoglycosidase F (cleaves complex and high mannose type glycans near their asparagine attachment sites in glycoproteins) showed that the epitope was on the carbohydrate moiety of the glycoprotein molecules.

Radioimmune assays were established to examine binding of ^{125}I-labeled 43-9F to 43-9F antigens in the presence of competitors. SLC-L11 glycoproteins were labeled with ^{3}H-glycosamine. The glycoproteins were treated extensively with proteinase K and the remaining ^{3}H-labeled carbohydrates that eluted in void volume on a S200 column were found to compete to the same extent as intact glycoproteins with the binding of [125]I-43-9F Mab to purified 43-9F antignes. The 43-9F Mab did not bind to blood group antigens as well prior to as after the neuraminidase treatment. The SLC-L11 line has a cloning efficiency of about 5.6% in semisolid agar medium and a population doubling time of about 32h in liquid culture medium. Eleven subclones (eight 43-9F positive, three negative) of the SLC-L11 line were established in vitro. All positive clones could be cloned in semisolid medium and were tumorigenic in nude mice, whereas the three 43-9F antigen negative sublines were non-clonable in vitro and non-tumorigenic in nude mice (Table 1). One of the negative clones became 43-9F antigen positive after eight months in culture and became concomitantly clonable in agar and tumorigenic in nude mice. Moreover FACS separation of SLC-L11 cells with high and low 43-9F antigen density revealed that the cells with high antigen density contained almost all the clonogenic cells. This implies indirectly that the 43-9F epitope may be associated with the growth activity of the SLC-L11 cells. A number of glycoproteins have been reported as having a growth promoting effect either by their direct effect on target cells or because they act as carrier molecules for small growth hormones. The 43-9F positiye glycoproteins are a very diverse set of glycoproteins and do not permit determination of the effect of the 43-9F epitope on cellular growth, before the exact chemical nature of the epitope is known and we were therefore urged to use other experimental procedures to elucidate these effects.

ANTI-IDIOTYPIC MABS AGAINST THE 43-9F MAB

Our general strategy for the production of anti-idiotypic and anti-anti-idiotypic antibodies against Mabs with specificity for TAA is as follows: Mice are immunized with the 43-9F Mab and Mabs generated against 43-9F. Only Mabs that bind to the 43-9F Mab, but not to other murine Ig molecules are analysed further for their ability to (i) compete with the interaction between the 43-9F Mab and the corresponding epitope and (ii) to result in the generation of polyclonal antibodies against the 43-9F epitope, when mice are immunized with the anti-idiotypic Mab. Anti-id Mabs against the 43-9F epitope have been generated by these procedures and may consequently be applied in the analysis for specific receptors for the carbohydrate structure defining the 43-9F epitope. These anti-id antibodies should therefore be highly useful in the identification the cell membrane molecules on SLL cells that are intermediate steps in the processes by which the 43-9F epitope has growth promoting effects on various malignant cell lines.

CONCLUSION

The cell membrane constitutes the crucial barrier between the cellular microenvironment and the intracellular processes and it eventually determines the phenotypic attributes of the cell and thereby its interaction with surrounding cells and tissues. The membrane constituents on neoplastic cells are therefore of crucial importance for the expression of malignancy stigmata such as tumorigenicity and metastatic activity. Tumor progression, including expansive growth, is dependent on appropriate growth stimuli. The identification of onc-gene encoded products as having substantial homology to growth factors have suggested that production of such growth factors is essential for tumorigenesis. Similarly, very recent experiments have elucidated some of the membrane molecules that may be involved in the metastatic process. Detailed analysis of the interaction between growth factors and malignant neoplastic cells require a careful dissection of the receptor molecules for such factors. However, the receptor molecules may be very difficult to identify for some epitopes, e.g. carbohydrate, and the anti-idiotypic approach seems to have very promising features as a tool in this type of work. Finally, identification of such receptor molecules should provide some clues to the possibilities for manipulatory manoeuvres with the receptor molecule(s) and thereby obtain growth control and perhaps even decay of the malignant tumor cell population. These principles thus suggest the development of a new membrane-based therapeutic approach which can be highly efficient in interfering with tumor cell growth without resulting in substantial destruction of normal cells and tissues.

ACKNOWLEDGEMENTS

Work reported in this paper was in part supported by grants from U.S. Public Health Service, grant No CA-35227, The Danish Medical Research Council, The Danish Cancer Society, The Novo Foundation and The Boel Foundation.

REFERENCES

1. J.R. Gabin, P. Gordon, J. Ruth, J.A. Archer and D.N. Buell, Characteristics of the human lymphocytes insulin receptor, J. Biol. Chem. 248:2202 (1973).
2. S. Jacobs and P. Cuatrecasas, Insulin receptor structure and function, Endocrine Rev. 2:251 (1981).
3. H.O. McDevitt, Current concepts in immunology: Regulation of the immune response by the major histocombatibility system, N. Engl. J. Med. 303:1514 (1980).
4. L. Hood, M. Steinmetz and B. Mulissen, Genes of the major histocompatibility complex of the mouse, Ann. Rev. Immunol. 1:529 (1983).
5. J. Lotem and L. Sachs, Regulation of normal differentiation in mouse and human myeloid leukemia cells by phorbol ester and the mechanism of tumor promotion, Carcinog. Compr. Surv. 7:385 (1982).
6. R.W. Schroff, K.A. Foon, R.J. Billing and J.L. Fahey, Immunologie classification by lymphocytic leukemias based on monoclonal antibody-defined cell surface antigens, Blood 59:207 (1982).
7. A.J. McMichal and J.W. Fabre (eds), "Monoclonal antibodies in clinical medicine." Academic Press, London (1982).
8. M.S. Mitchell and H.F. Oettgen, Hybridomas in cancer diagnosis and treatment, Prog. Cancer Res. Ther. 21:1 (1982).
9. R.J. Cohen and H.N. Eisen, Hypothesis: Interaction of macromolecules on cell membranes and restriction of T-cell specificity by products of the major histocompatibility comples, Cell Immunol. 32:1 (1977).
10. M. Simonsen and L. Olsson, Possible roles of compound receptors in the immune system, Ann. Immunol. 134D:85 (1983).
11. M. Simonsen, C. Skjødt and M. Crone, Compound receptors in the cell membrane: Ruminations from the borderland of immunology and physiology, Prog. Allergy 36:151 (1985)
12. L. Olsson, Human monoclonal antibodies in experimental cancer research, J. Natl. Cancer Inst., in press.
13. L. Old, Cancer immunology: The search for specificity, Cancer Res. 41:361 (1981).
14. S. Ohno, The original function of MHC antigen as the general plasma membrane anchorage site of organogenesis-directing proteins. Immunol. Rev., 33:59 (1977).
15. V. Ginsburg, P. Fredman and J. Magnani, Cancer associated carbohydrated antigens detected by monoclonal antibodies, Contrib. Oncol. 19:51 (1984).
16. S. Hakamori and R. Kannagi, Glycosphingolipids as tumor-associated and differentiation markers, J. Natl. Cancer Inst. 71:231 (1983).
17. T. Feizi, Demonstration and glycolipids are onco-developmental antigens, Nature 314:53 (1985).
18. P.C. Nowell, The clonal evolution to tumor cell populations, Science 194:23 (1976).
19. G. Poste and I.J. Fidler, The pathogenesis of cancer metastasis, Nature 283:139 (1980).
20. G.H. Heppner, Tumor heterogeneity, Cancer Res. 44:2259 (1984).

21. L. Olsson, Phenotypic diversity in leukemia cell population, Cancer Metastasis Rev. 2:153 (1983).
22. A.D. Riggs and P.A. Jones, 5-methylcytosine, gene regulation and cancer, Adv. Cancer Res. 40:1 (1983).
23. P.A. Jones, Altering gene expression with 5-azacytidine, Cell 40:485 (1985).
24. L. Olsson, C. Due and M. Diamant, Treatment of human cell lines with non-mutagenic, non-toxic doses of 5-azacytidine may result in profound alterations in clonogenicity and growth rate, J. Cell. Biol. 100:108 (1985).
25. L. Ossowski and E. Reich, Changes in malignant phenotype of a human carcinoma conditioned by growth environment, Cell 33:323 (1983).
26. C. Honsik, M. Diamant and L. Olsson, Generation of stable cellular phenotypes in a human malignant cell line conditioned by alterations in the cellular microenvironment, Submitted.
27. L. Chatenaud and J.F. Bach, Antigenic modulation - a major mechanism of antibody action, Immunol. Today 5:20 (1984).
28. M.D. Waterfield, G.T. Scarce and N. Whittle, Platelet-derived growth factor is structurally related to the putative transforming protein p28^{sis} of Simian sarcoma virus, Nature 304:35 (1983).
29. R.F. Doolittle, M.W. Hunkapilar and L. Hood, Simian sarcoma virus oncogene, v-sis, is derived from the gene (or genes) encoding a platelet-derived growth factor, Science 221:275 (1983).
30. T. Hunter, Oncogenes and proto-oncogenes: How do they differ? J. Natl. Cancer Inst. 73:773 (1984).
31. T. Burgess, Growth factors and oncogenes, Immunol. Today 6:107 (1985).
32. A. Ullrich, L. Cousseus and J.S. Hayflink, Human epidermal growth factor receptor. DNA sequence and oberrant expression of the amplified gene in A431 epidernoid carcinoma cells, Nature 309:18 (1984).
33. L. Olsson and J. Forchhammer, Induction of the metastatic phenotype in a mouse tumor model by 5-azacytidine and characterization of an antigen associated with metastatic activity, Proc. Natl. Acad. Sci USA 81:7897 (1984).
34. I.R. Hart, Molecular basis of tumour spreads, Nature 315:274 (1985).
35. R.A. Miller, D.G. Maloney, R. Warnke and R. Levy, Treatment of B-cell lymphoma with monoclonal anti-idiotype antibody, N. Engl. J. Med. 306:517 (1982).
36. R. Levy and R.A. Miller, Biological and clinical implications of lymphocyte hybridomas: Tumor therapy with monoclonal antibodies, Ann. Rev. Med. 34:107 (1983).
37. H. Koprowski, D. Herlyn and M. Lubeck, Human anti-idiotype antibodies in cancer patients: Is the modulation of the immune response beneficial for the patient?, Proc. Natl. Acad. Sci. USA 81:216 (1984).
38. G.T. Nepom, U.T. Nelson and S.L. Holbeck, Induction of immunity to a human tumor marker by in vivo administration of anti-idiotypic antibodies in mice, Proc. Natl. Acad. Sci USA 81:2864 (1984).
39. L. Olsson, H.R. Sørensen and O. Behnke, Intratumoral phenotypic diversity of cloned human lung tumor cell lines and consequences for analysis with monoclonal antibodies, Cancer 54:1757 (1984).
40. L Olsson, O. Behnke and H.R. Sørensen, Modulatory effects of 5-azacytidine, phorbol ester, and retinoid acid on the malignant phenotype of human lung cancer cells, Int. J. Cancer 35:189 (1985).
41. D. Pettijohn, C. Due and E. Rønne, Specific glycoproteins released by human squamous lung carcinoma cells: Characterization and diagnostic applications, J. Natl. Cancer Inst., in press.

TUMOR-ASSOCIATED GLYCOLIPID MARKERS: POSSIBLE TARGETS FOR DRUG AND IMMUNO-TOXIN DELIVERY

Sen-itiroh Hakomori

Program of Biochemical Oncology and Membrane Research
Fred Hutchinson Cancer Research Center, Departments of
Pathobiology, Microbiology, and Immunology, University
of Washington, Seattle, WA 98104

INTRODUCTION

In order to achieve effective targeting of antibody-drug conjugates to specific types of cells, a comprehensive understanding of the chemical, physical, and dynamic properties of cell surface structures is essential. Glycosphingolipids (briefly, glycolipids), as discussed here, are potentially useful to achieve effective targeting for the following reasons: (i) They are an integral part of the lipid bilayer, and the majority are assumed to be inserted at the outer leaflet of the plasma membrane; (ii) their structure and organization in membranes differ from one type of cell to another and constitute characteristic cell surface specificity of each type of cell; and (iii) they may be more readily internalized than other membrane components, since receptor carbohydrates are directly inserted in membranes.

Three classes of glycolipids with different carbohydrate core structures have been characterized. Molecular models of these classes are presented in Fig. 1. They are lactoseries, ganglioseries, and globoseries, respectively. Structural variation in their peripheral regions results in over 100 molecular species, and combinations of these glycolipids are characteristic of specific types of cells.[1] A possible organizational framework of glycolipids in membranes is shown in Fig. 2. In recent years, a great deal of interest has been aroused by the presence of tumor-associated glycolipid markers, first demonstrated in experimental animal cancer, and subsequently in a large variety of human cancers as well. I will present first the general structural features of human cancer-associated glycolipids. Subsequently, the results of a few experiments using polyclonal or monoclonal antibodies directed to one type of mouse lymphoma-associated antigen, gangliotriaoxylceramide (Gg$_3$; GalNAcβ1→4Galβ1→4Glcβ1→1Cer), for targeting neocarcinostatin (NCS), liposome-encapsulated actinomycin D,

---

Abbreviations: Glycolipids are abbreviated according to the recommendations of the Nomenclature Committee of the IUPAC, but the suffix OseCer is omitted (IUPAC-IUB Commission on Biochemical Nomenclature, Lipids 12, 455-463; 1977). Gg$_3$ is, therefore, Gg$_3$OseCer (GalNAcβ1→4Galβ1→4Glcβ1→1Cer). Ganglioseries gangliosides are abbreviated according to Svennerholm (J. Lipid Res. 5, 145-162; 1964).

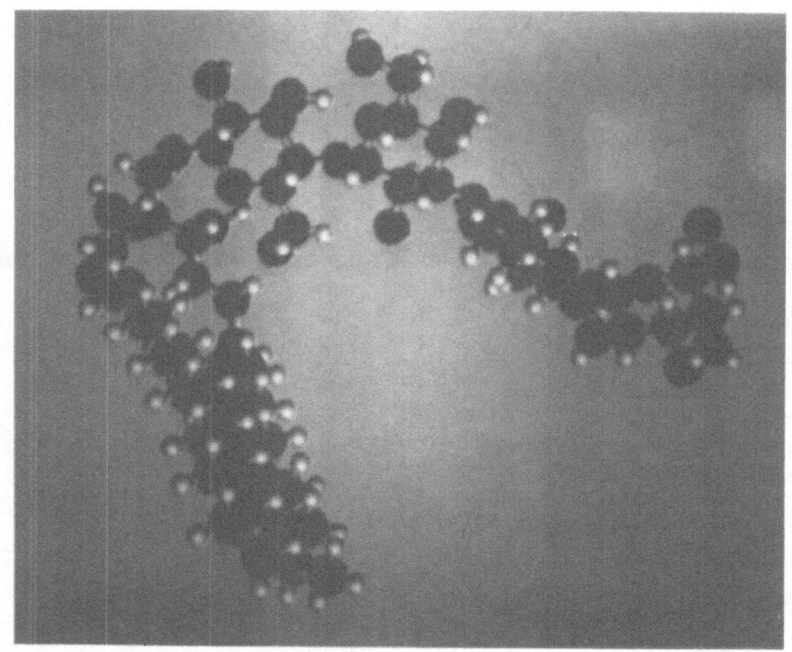

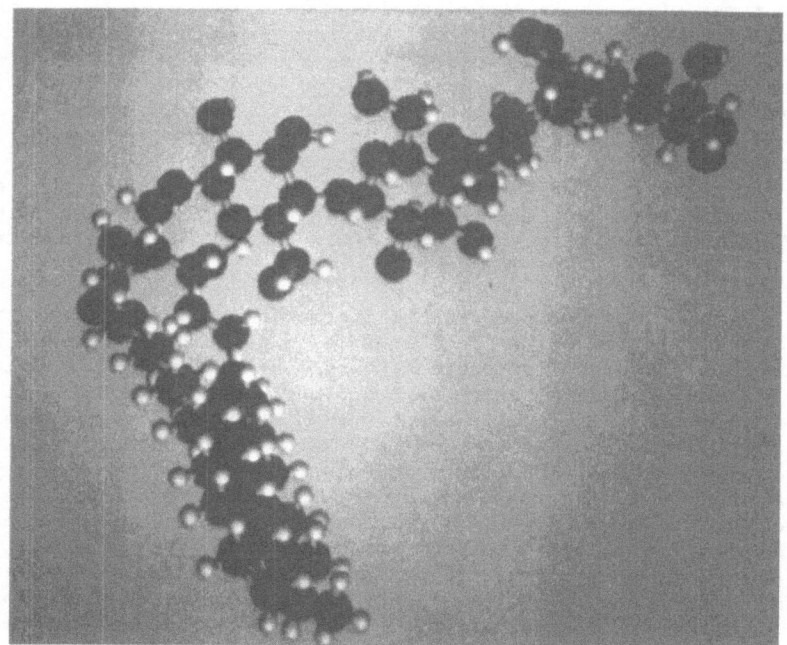

and ricin A chain conjugate will also be presented, although none of these approaches have been entirely satisfactory, particularly for in-vivo targeting. Finally, various drawbacks and possibilities for improvement in targeting efficiency will be discussed.

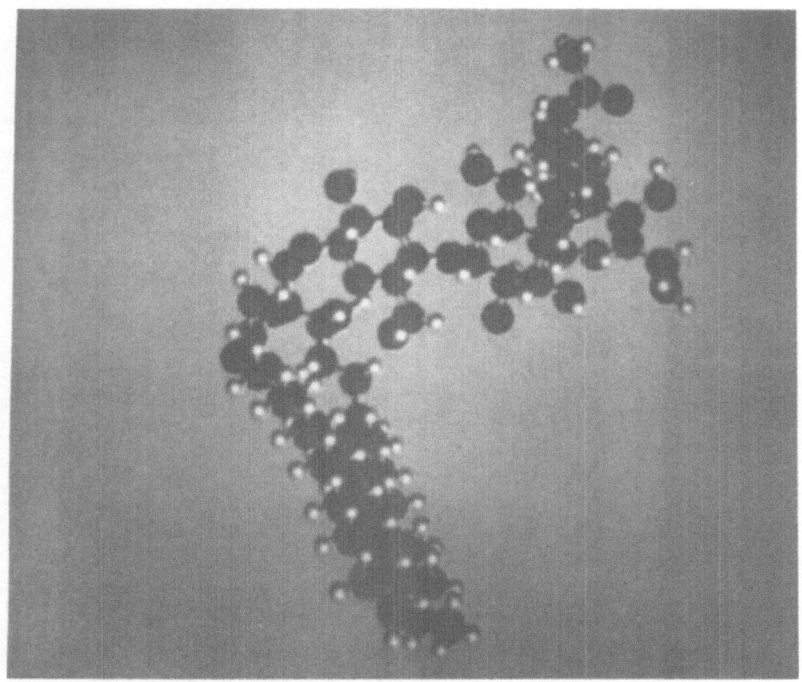

Fig. 1. The minimum energy conformation models of the core
         structures of glycosphingolipids. Upper and middle
         panels on previous page and panel on this page rep-
         resent, respectively, lactoseries, ganglioseries,
         and globoseries core structures. The models were
         constructed by Drs. R. Stenkamp and K. Watenpaugh of
         the Department of Biological Structure, University
         of Washington and by Mr. Steven B. Levery of this
         laboratory by computer-based calculations of bond
         angles, interatomic distances, and interactions.
         Glc, glucose; Gal, galactose; GlcNAc, N-acetylgluco-
         samine; GalNAc, N-acetylgalactosamine. It should
         be noted that the axis of ceramide is perpendicular
         to that of the carbohydrate.

CHARACTERIZATION OF TUMOR-ASSOCIATED GLYCOLIPID MARKERS

     Oncogenic transformation accompanies dramatic changes in chemical comp-
osition, metabolism, and organization of cell-surface glycoconjugates.[1,2]
Three types of transformation-dependent changes in glycolipids have been
observed: (i) blocked synthesis of complex-type carbohydrates with or without accumulation of precursors; (ii) synthesis of "neoglycolipids", which
are essentially absent in progenitor cells or tissues but may be present
in small quantities in unrelated normal cells; and (iii) organizational
changes of glycolipids in membranes, including loss of crypticity and possible aggregation of glycolipids in high density. Any of these changes or
a combination of changes may lead to the formation of tumor-associated
antigens. The presence of tumor-associated glycolipid antigens was established in experimental tumors based solely on chemical analysis and polyclonal antibodies.[3-5] With the development of the monoclonal antibody
approach, it has become increasingly apparent that antibodies selected by

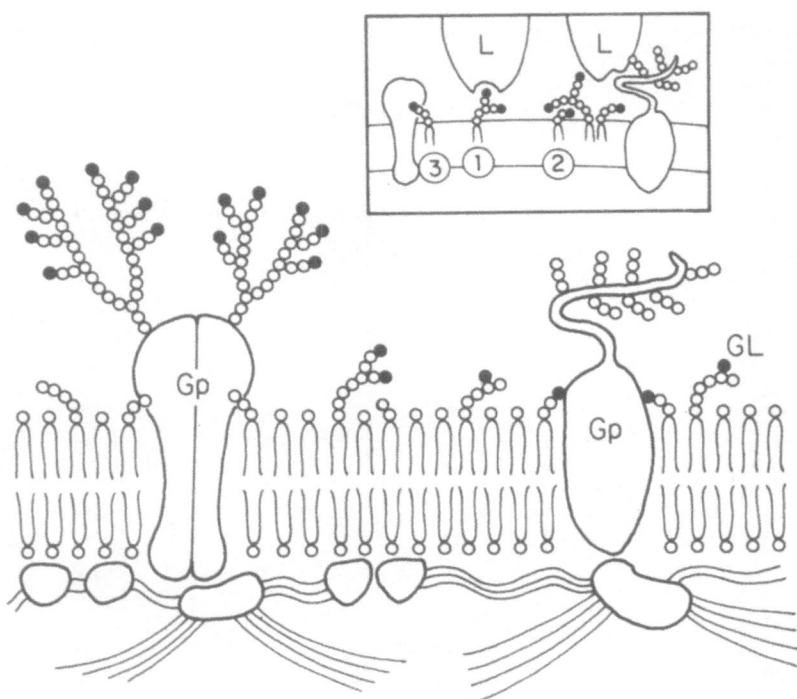

Fig. 2. Plasma membrane model including glycosphingolipids
and glycoproteins. Glycoproteins (Gp) are floating
among the lipid bilayer. Some of them could be surrounded by glycolipids, which are closely associated
with glycoproteins. The lipid moiety (ceramide) of
glycosphingolipids is inserted in the outer leaflet
of the lipid bilayer, and the carbohydrate moiety
lies on the lipid bilayer, since the axis of the
carbohydrate is perpendicular to the axis of ceramide,
as shown in Fig. 1. Some of the peripheral regions of
glycosphingolipids show the same structure as the peripheral regions of the carbohydrate chains attached to
proteins. Not all glycosphingolipids in membranes are
exposed at the cell surface, as schematically shown in
the insert. Glycolipids that are closely associated
with other glycolipids having longer carbohydrate
chains or with membrane glycoproteins are cryptic. The
glycolipid in location 1 is accessible to ligand (L),
while the glycolipid in location 2, associated with
a long chain glycolipid, and that in location 3, associated with a membrane protein, are not accessible to
ligand (L); they are cryptic.

specific or preferential reactivity with tumor cells or tissues are indeed
directed to glycolipids. Monosialosyl Le[a],[6] monosialosyl Le[x],[7] monosialosyl
dimeric Le[x],[8] and di- or trimeric Le[x] [9] and Le[y] [10-12] have been chemically
identified as tumor-associated antigens defined by specific monoclonal antibodies. Essentially all of these antigens have been found to be develop-
mentally regulated and are highly expressed at specific stages of embryogenesis or organogenesis;[13,14] therefore, these are typical oncofetal antigens.[15] These antigens as described above are related to blood group
carriers, either type 1 or type 2 chain. On the other hand, some types of

Table 1. Novel fucolipids and fucogangliosides as human tumor-associated markers defined by specific monoclonal antibodies.

	Association	Structure	Antibody	Ref
A. Lacto-series type 1 chain				
Sialosyl Le[a]	gastro-intestinal pancreas cancer	$Gal\beta1 \rightarrow 3GlcNAc\beta1 \rightarrow 3Gal\beta1 \rightarrow R$ with $3 \leftarrow NeuAc\alpha2$ and $4 \leftarrow Fuc\alpha1$	N-19-9	6
B. Lacto-series type 2 chain				
Difucosyl Y2 III3Fuc2nLc6	gastro-intestinal/ lung/breast cancer	$Gal\beta1 \rightarrow 4GlcNAc\beta1 \rightarrow 3Gal\beta1 \rightarrow 4Glc\beta1 \rightarrow 1Cer$ with $3 \leftarrow Fuc\alpha1$ (and $3 \leftarrow Fuc\alpha1$)	FH4	9,13
Sialosyldifucosyl Y2	as above	$Gal\beta1 \rightarrow 4GlcNAc\beta1 \rightarrow 3Gal\beta1 \rightarrow 4Glc\beta1 \rightarrow 1Cer$ with $3 \leftarrow NeuAc\alpha2$, $3 \leftarrow Fuc\alpha1$ (and $3 \leftarrow Fuc\alpha1$)	FH6	8
III3V3FucVI3NeuAc	as above	$Gal\beta1 \rightarrow 4GlcNAc\beta1 \rightarrow 3Gal\beta1 \rightarrow R$ with $3 \leftarrow NeuAc\alpha2$, $3 \leftarrow Fuc\alpha1$	CSLEX1	7
Sialosyl Le[X]				
C. Globo-series	breast	$Gal\beta1 \rightarrow 3GalNAc\beta1 \rightarrow 3Gal\alpha1 \rightarrow 4Gal\beta1 \rightarrow 4Glc\beta1 \rightarrow 1Cer$ with $2 \leftarrow Fuc\alpha1$	MBr1	21
D. Ganglio-series				
GD3	melanoma	$NeuAc\alpha2 \rightarrow 8NeuAc\alpha2 \rightarrow 3Gal\beta1 \rightarrow 4Glc\beta1 \rightarrow 1Cer$		16,17
GD2	melanoma	$NeuAc\alpha2 \rightarrow 8NeuAc\alpha2 \rightarrow 3Gal\beta1 \rightarrow 4Glc\beta1 \rightarrow 1Cer$ with $4 \leftarrow GalNAc\beta1$		18.19
Fuc-GM1	small cell lung carcinoma	$Gal\beta1 \rightarrow 3GalNAc\beta1 \rightarrow 4Gal\beta1 \rightarrow 4Glc\beta1 \rightarrow 1Cer$ with $2 \leftarrow Fuc\alpha1$ and $3 \leftarrow NeuAc\alpha2$		20

tumors express ganglio- or globoseries glycolipids, which are also defined by specific monoclonal antibodies. GD$_3$ ganglioside in melanoma,[16,17] GD$_2$ ganglioside in melanoma and neuroblastoma,[18,19] fucosyl GM$_1$ ganglioside in small-cell lung carcinoma,[20] and fucosylgalactosylgloboside (globo H)

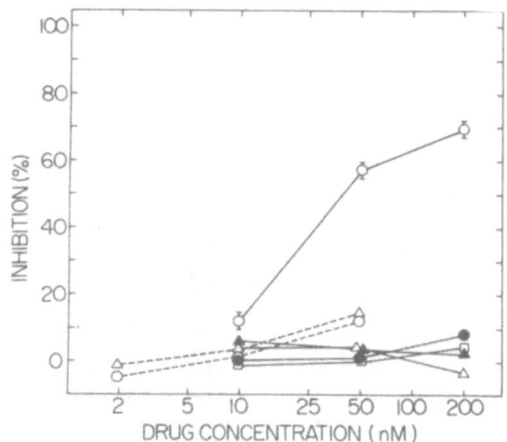

Fig. 3A. Targeting of biotinyl-neocarcinostatin (NCS) to mouse lymphoma L5178 cl 27 cells: Requirement for biotin on antibody and drug. O—O , biotinyl-anti-Gg$_3$ + avidin + biotinyl neocarcinostatin; □——□ , biotinyl-anti-Gg$_3$ + avidin in 1 mM biotin + biotinyl neocarcinostatin; △—△ , non-immune serum + avidin + biotinyl neocarcinostatin; ●—● , unsubstituted anti-Gg$_3$ + avidin + biotinyl neocarcinostatin; ▲—▲ , biotinyl-anti-Gg$_3$ + biotinyl neocarcinostatin (no avidin); O·····O , biotinyl-anti-Gg$_3$ + avidin + unsubstituted neocarcinostatin; △----△ , nonimmune serum + avidin + unsubstituted neocarcinostatin. Cell proliferation (increase in cell number) in all cultures without drug but with different antibody and avidin treatments was more than 600%. No difference was seen between control cultures. The reproducibility of the assay is reflected by the standard error brackets (n = 5) on the specific inhibition curve (O—O).

in breast cancer[21] are typical examples. Structures of these antigens are shown in Table 1, of which lactoseries structures are shared with glycoproteins. All of these glycolipid markers are possible targets for immunotherapy and for drug and immunotoxin delivery, which is the major topic of this symposium.

Antibody inactivation by drug coupling has been one of the technical problems in preparation of antibody-conjugates. In order to effectively avoid antibody inactivation, a technique using a biotin/avidin system has been developed, i.e., cells are first treated with biotinyl-antibody directed

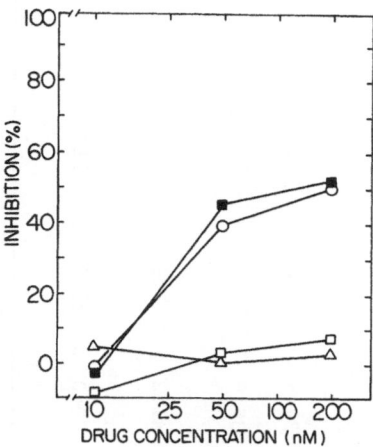

Fig. 3B. Targeting of biotinyl-neocarcinostatin (NCS) to mouse lymphoma L5178 cl 27 cells: Antigen specificity. O , biotinyl-anti-Gg_3 (not absorbed) + avidin + biotinyl neocarcinostatin; △ , nonimmune serum + avidin + biotinyl neocarcinostatin; □ , biotinyl anti-Gg_3 (absorbed on guinea pig erythrocytes) + avidin + biotinyl neocarcinostatin; ■ , biotinyl anti-Gg_3 (absorbed on sheep erythrocytes) + avidin + biotinyl neocarcinostatin. Absorption of antibody was accomplished by incubating 125 μl of antibody with 200 μl of packed red blood cells for 1 h at room temperature followed by centrifugation to remove red blood cells.

to a cell-surface glycolipid, followed by successive treatments with avidin and biotinyl drug. As shown in Fig. 3A, only a complete system (three successive treatments) effectively inhibited tumor cell growth. If one treatment was omitted, cell growth was not inhibited at all. Such inhibition of tumor cell growth due to targeting of avidin-dependent biotinyl-drug to tumor cells is highly specific for the target structure, as shown in Fig. 3B, i.e., cell growth was not inhibited when biotinyl-antibodies were pre-

Table 2. In-vivo targeting with monoclonal IgM anti-gangliotriaosylceramide.

Group	Therapy[a]			Days survived			Survivors (>80 days)	Therapeutic effect
	1	2	3	mean	SD	median		
1	NaCl/Pi	NaCl/Pi	NaCl/Pi	29	6.2	27	0/5	control
2	Avidin alone			34.8	1	35	0/4	−
3	Biotinyl IgM	NaCl/Pi	NaCl/Pi	32.4	6.3	35	0/5	−
4	IgM	Avidin	NaCl/Pi	33.4	7.0	35	0/5	−
5	IgM	Avidin	Biotinyl-NCS (2.4 µg)	36.8	13.3	35	0/5	−
6	Biotinyl-avidin IgM	NaCl/Pi		41.2	5.9	44	0/5	−
7	Biotinyl-avidin		Biotinyl-NCS (2.4 µg)	70.0	22.0	51	2/5	+

[a] Animals received 10^6 L5178cl27Al cells i.p. on day 0. Therapy was given on days 1 and 4 and consisted of 3 i.p. injections: 1) IgM monoclonal anti-gangliotriaosylceramide (256 H.A. units); 2) avidin (150 µg), and 3) biotinyl drug (2.4 µg) with 30 min intervals between injections. Group 2 received avidin (150 µg) alone in a single injection on days 1 and 4.

treated with the antigen (gangliotriaosylceramide) or absorbed with cells containing such antigen. Pretreatment with other glycolipids or with cells containing other antigens did not inhibit cell growth. These results clearly indicate that the approach with biotinyl-antibody and biotinyl-drug

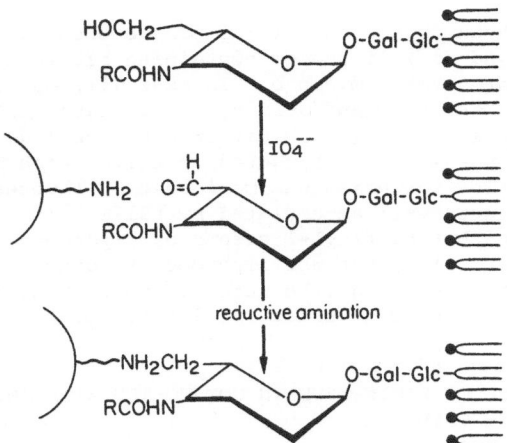

Fig. 4. Covalent attachment of protein ligands to liposome surface. GM_3-containing liposome is oxidized by periodate (IO_4^{-2}) to create an aldehyde group located on the pyranose ring of sialic acid, to which ligands are attached through reductive amination with sodium cyanoborohydride.

is a useful method for targeting drugs to tumor cells. Furthermore, biotinyl-antibody has been proven not only to be active in its antigen-binding activity but also to maintain activity in complement fixation.[22] The approach has been extensively studied in vitro; however, in-vivo studies are still limited. One example is described in the subsequent section.

IN VIVO TARGETING OF BIOTINYL-NEOCARCINOSTATIN TO MOUSE LYMPHOMA L5178 CELLS

The system as described above has been applied in a limited number of in-vivo studies. The results of one study are reproduced in Table 2. Although the numbers are limited, only a complete system, i.e. injection of biotinyl-anti-Gg_3 (IgM) followed by successive injection of avidin and biotinyl-neocarcinostatin, showed a clear effect.[23]

TARGETING OF ANTIBODY-LIPOSOME COMPLEXES

A liposome attached to an antibody encapsulating drugs or enzymes is an ideal assembly to delivery drugs or enzymes to target cells. A number of studies along these lines have been made in the past.[24-27] However, covalent attachment of antibodies to liposome surfaces has been one of the crucial technical problems involved in this approach.[24-29] The efficiency of antibody attachment to the liposome surface should be high and reproducible. We have developed a method using a ganglioside liposome treated with periodate, followed by coupling with immunoglobulin by reductive amination, as shown in Fig. 4. A similar method using galactosylcerebroside has been independently developed.[30] However, the efficiency of ganglioside liposomes is higher than that of galactosylcerebroside liposomes because sialic acid is highly sensitive to periodate oxidation, and the resulting aldehyde group on the stable pyranose ring is highly reactive to the amino group of proteins (Fig. 4).

This method has been applied for (i) liposome-encapsulated actinomycin D coupled to avidin, and (ii) liposomes containing biotinyl-phosphatidyl-ethanolamine-encapsulated actinomycin D. In case (i), cells were pretreated with biotinyl-anti-Gg_3 antibody and avidin, and in case (ii), cells were pretreated with biotinyl-anti-Gg_3. In either case, cell growth was not inhibited, although liposomes were targeted to L5178 lymphoma cells that express Gg_3 antigen. The targeting was monitored by fluorescein-encapsulated liposomes, which were well accumulated by L5178 lymphoma cells.[22] These results indicate that biotinyl-liposome or liposome-avidin complexes can successfully target cells, but the liposomes targeted to cells may not fuse with cell membranes or may not be internalized. Some technical improvements, described in the discussion, will avoid such a situation.

TARGETING EFFECTIVENESS OF RICIN A CHAIN CONJUGATED WITH ANTI-Gg_3 ANTIBODY TO MOUSE LYMPHOMA L5178 CELLS

Since we have demonstrated that the glycolipid antigen Gg_3 is a typical tumor-associated antigen in L5178 lymphoma cells, and IgG3 monoclonal antibody but not IgM monoclonal antibody effectively inhibits lymphoma growth in vivo,[31] we have been trying to use IgM antibody conjugated with ricin A chain for selective growth inhibition of lymphoma cells expressing Gg_3. In-vitro experiments consistently showed a very efficient and highly specific inhibition of lymphoma cell growth by the ricin A-IgM antibody conjugate. Such an inhibition was not demonstrated by ricin A chain alone nor by the IgM antibody alone. Furthermore, the growth of lymphoma cells that did not express Gg_3 was not inhibited, as shown in Fig. 5. However, when the system was applied in vivo, the results showed extensive variation. In some experiments, survival of mice inoculated with lymphoma cells highly expressing Gg_3 antigen became much higher than that of mice inoculated with lymphoma cells not expressing Gg_3 antigen. In other experiments, however, such differences were minimal. Such variation may well be due to subtle differences in injection site or viability of cells, but a major reason could be the use of IgM antibody conjugates. A typical positive result for the in-vivo inhibition of L5178 lymphoma through administration of IgM-ricin A conjugate is shown in Table 3. It should be noted, however, that such a positive result could only be obtained when tumor cells and IgM-ricin A conjugate were both inoculated intraperitoneally and initial therapy was performed not later than 24 h after inoculation of tumor cells.

DISCUSSION

Since the monoclonal antibody approach has been introduced, our knowledge of the chemical properties of cell type-specific antigens has been greatly enriched. That many tumor-associated antigens are carbohydrates bound to either proteins (glycoproteins), lipids (glycolipids), or both (epitopes shared by glycoproteins and glycolipids) has been well-established.[2,15] On the other hand, some monoclonal antibodies directed to tumor-associated antigens define specific epitopes along a polypeptide chain of integral membrane proteins,[32] pericellular matrix proteins, or macromolecules classified as proteoglycans.[33] These antigens differ in

Table 3. Effect of monoclonal anti-gangliotriaosylceramide (IgM)-ricin A chain conjugate on the survival rate of mice after lymphoma L5178 inoculation.

Group	Tumor Cells Inoculated	Therapy	Days Survived (Mean)	Survivors >150 (Mean)	Therapeutic Effect
1	L51784 27AV, 5 x 10^5	NaCl/Pi	26.5	0/10	–
2	L51784 27AV, 5 x 10^5	IgM anti-Gg$_3$	32.5	0/10	–
3	L51784 27AV, 5 x 10^5	IgM anti-Gg$_3$-ricin A chain conjugate	31.5	0/10	–
4	L51784 1C2, 5 x 10^5	NaCl/Pi	32.5	0/10	–
5	L51784 1C2, 5 x 10^5	IgM-anti-Gg$_3$		1/10	+
6	L51784 1C2, 5 x 10^5	IgM anti-Gg$_3$ ricin A chain conjugate		8/10	++

DBA/2 mice received i.p. 5 x 10^5 lymphoma cells L51784 27AV (which do not express Gg$_3$ antigen) or L51784 1C2 (which express high levels of Gg$_3$ antigen) on day 0. Therapy was given on days 1 and 4. 100 μl of NaCl/Pi (for groups 1 and 4) or containing 50 μg of IgM (for groups 2 and 5) or 100 μg of IgM-ricin A conjugate (for groups 3 and 6) were applied.
Data in collaboration with H. Miyazaki and T. Osawa).

their degree of potential release from cells to the external environment (exocytotic property) and in their ability to be internalized (endocytotic property). These classes of tumor-associated antigens and their potential as targets are schematically shown in Fig. 6. In general, targeting to tumor-associated antigens depends on the following three conditions: (i) shedding or exocytosis should be minimal; (ii) the antigen should be an integral membrane component, not a pericellular or peripheral membrane component; and (iii) the antigen should be internalized when antibody or antibody-drug conjugate is bound to the antigen. Some carbohydrate antigens with mucin-type glycoproteins, such as those defined by antibodies N-19-9, CA-1, CA-50, DU-PAN-1, etc., have been detected in the plasma of patients with cancer. Those antigens are shed and circulate in blood and, therefore, are useful for diagnosis of human cancer, but they may not be useful for targeting because the antigens may not be efficiently internalized when antibodies are bound. As compared with such peripheral or secretory glycoproteins, integral membrane proteins or membrane glycolipids such as p97 antigen[32] or GD$_3$ of human melanoma[16] are not released into blood. The level of GD$_3$ in sera of patients with cancer remains at a normal level (Nudelman, E., Hakomori, S., Hellström, I., and Hellström, K.-E., unpublished data). These antigens may well be more suitable for targeting, although experiments have not been made in human cancer.

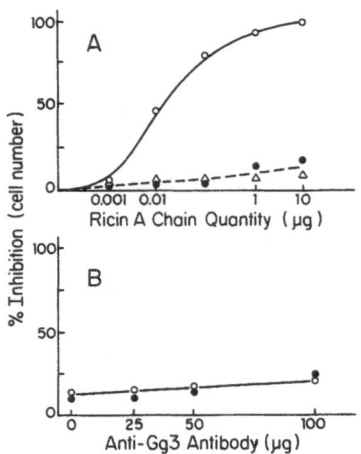

Fig. 5. Inhibition of cell growth by anti-Gg$_3$ IgM antibody
(2D4) conjugated with ricin A chain. <u>Panel A,</u>
specific effect of ricin A chain conjugated with
anti-Gg$_3$ IgM antibody (2D4). Aliquots of 1 x 10^5
mouse lymphoma cells L5178 variants were grown on
24-well plates in RPMI medium supplemented with
10% fetal calf serum. To each well was added
varying quantities (0.001 to 10 μ g) of ricin A chain
conjugated to anti-Gg$_3$-IgM antibody. The conjugation was performed by the SPDG method. As a control,
only ricin A chain was added. O—O, 1A1 clone expressing a high quantity of Gg$_3$ antigen; ●——●, AV27
clone not expressing Gg$_3$ antigen; △——△, 1A1 clone
in the presence of ricin A chain alone. <u>Panel B,</u>
effect of Gg$_3$ IgM antibody on cell growth. Aliquots of 1 x 10^5 cells were grown in 24-well plates
in the presence of various quantities of anti-Gg$_3$
IgM antibody alone. O——O, 1A1 clone expressing a
high quantity of Gg$_3$; ●——●, AV clone, not expressing Gg$_3$. Cell numbers were counted at time 0 and
at 48 h, and the ordinate expresses the % growth
inhibition. (H. Miyazaki, T. Osawa, N. Cochran,
and S. Hakomori, unpublished data).

a) Carbohydrate antigens with mucin-type glycoproteins

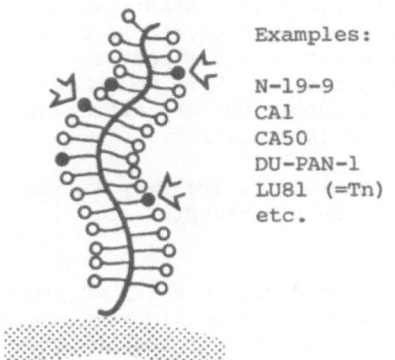

Examples:

N-19-9
CA1
CA50
DU-PAN-1
LU81 (=Tn)
etc.

Usefulness in serum diagnosis: Good,

b) Proteoglycan antigens

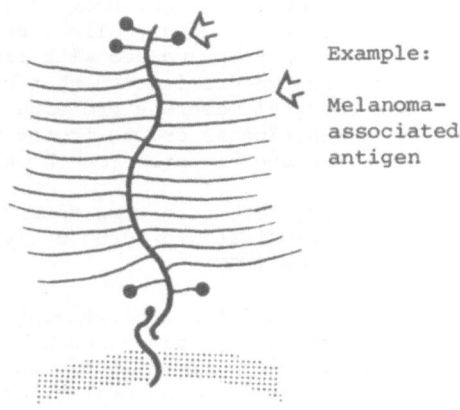

Example:

Melanoma-
associated
antigen

Targetability: Untested, but
questionable.

c) Protein antigens (integral protein antigens)

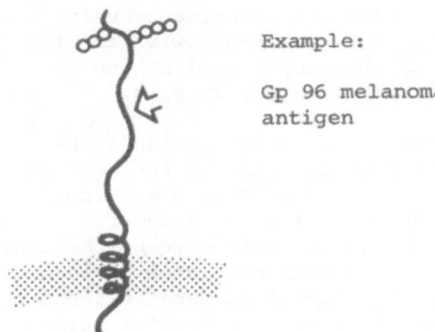

Example:

Gp 96 melanoma
antigen

Usefulness in serum diagnosis: Poor,

d) Carbohydrate epitopes carried by integral proteins shared with glycolipids

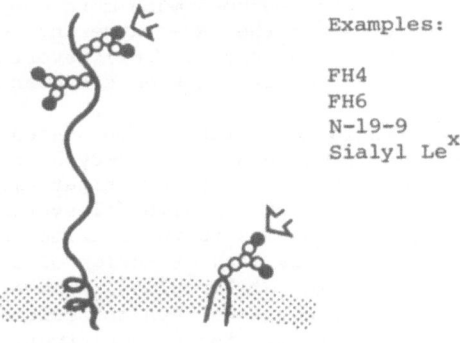

Examples:

FH4
FH6
N-19-9
Sialyl Le^x

Targetability: Untested

e) Carbohydrate epitopes carried by glycolipids only

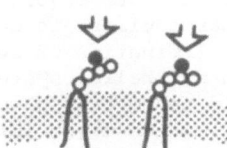

Examples:

Gb$_3$
GD$_3$
Gg$_3$
GM$_2$, etc.

Usefulness in diagnosis: Poor,
Targetability: Proved to be good
for Gb$_3$ and Gg$_3$

Fig. 6. Schematic presentation of various types of
tumor-associated antigens and their targetability. Arrows indicate the possible location of epitopes within the antigen macromolecules.

Among integral membrane antigens, some glycolipids could be very efficient for targeting of antibody conjugates and drug delivery. Wiels et al.[34] demonstrated that glycolipid antigen Gb_3 was highly sensitive to anti-Gb_3 IgM antibody conjugated with ricin A chain. However, it is premature to make the generalization that glycolipid antigens are better targets than integral membrane proteins or glycoproteins. Some membrane proteins that function as growth factor receptors should be very efficient for targeting, as the receptors can be efficiently internalized.

Nevertheless, such receptors specific for tumor cells are very limited. One interesting candidate for drug delivery could be the receptors for tumor growth factor (TGF).

Targetability and target-dependent cell killing depend greatly on properties of the antigen, as discussed above, and also depend on ligand characteristics. In the case of antibody conjugates, the isotype of the antibody and the affinity and gross morphology of the antibody conjugate, including hydrophobicity (interaction with the lipid bilayer), may be important in determining the rate of internalization of the antibody conjugate. Furthermore, each type of cell may have different endocytotic ability. Therefore, the same antigen targeted with the same antibody conjugate may not be treated equally by different types of cells. Considering all these factors, the targetability of antibody-drug conjugates to a specific type of cell is a highly complex matter. Success in a simple experimental system, as demonstrated for in-vitro targeting, can not easily be achieved in vivo. Tumor growth inhibition becomes much more complex in in-vivo experiments using the same cells and the same targeting system. Our experience with in-vivo targeting has been very limited; nevertheless, the simple application of an established in-vitro system has been difficult to apply in vivo.

Utilization of antibody-conjugated liposomes and other particulate-encapsulated drugs has been a very attractive approach and has been extensively documented by many investigators, as discussed in this symposium. Results from our two experimental systems have demonstrated successful targeting of liposomes to tumor cells; however, tumor cells were not killed, probably because of lack of fusion of liposomes to cell membranes or lack of internalization of liposomes. The size of the liposomes used in these experiments may not have been sufficiently small to allow effective internalization. Smaller, single compartment liposomes may well be more efficient in cell killing. Lack of internalization or cell fusion may be avoided by transplanting fusogenic material, such as the F protein of Sendai virus or a polypeptide region with fusogenic activity. If we include a radioactive element emitting a high dose of α-rays, efficient targeting per se is sufficient to kill the tumor cells. A similar approach can be applied to synthetic microspheres for efficient drug delivery, which I believe is extensively discussed by various speakers in this symposium. Another major drawback in performing targeting of drug-encapsulated particulates to cells, either liposomes or synthetic microspheres, is a rapid elimination of such particulates from blood by macrophages and the reticuloendothelial system. One possible way to avoid such a rapid elimination would be to transplant highly sialylated groups to the surface of particulates, which might interfere with the interaction with macrophages and reticuloendothelial cell system.

REFERENCES

1. S. Hakomori, Glycosphingolipids in cellular interaction, differentiation, and oncogenesis, Ann. Rev. Biochem. 50:733 (1981).
2. S. Hakomori, Aberrant glycosylation in cancer cell membranes as focused on glycolipids: Overview and perspectives, Cancer Res. 45:2405 (1985).

3. J. Sundsmo and S. Hakomori, Laco-N-neotetraosylceramide ("paraglaobo-side") as a possible tumor-associated surface antigen of hamster NILpy tumor, Biochem. Biophys. Res. Commun. 68:799 (1976).

4. G. Rosenfelder, W.W. Young, Jr., and S. Hakomori, Association of the glycolipid pattern with antigenic alterations in mouse fibroblasts transformed by murine sarcoma virus, Cancer Res. 37:1333 (1977).

5. S. Hakomori and W.W. Young, Jr., Tumor-associated glycolipid antigens and modified blood group antigens, Scand. J. Immunol. Supplement 6:97 (1978).

6. J.L. Magnani, B. Nilsson, M. Brockhaus, D. Zopf, Z. Steplewski, H. Koprowski, and V. Ginsburg, A monoclonal antibody-defined antigen associated with gastrointestinal cancer is a ganglioside containing sialylated lacto-N-fucopentaose II, J. Biol. Chem. 257:14365 (1982).

7. K. Fukushima, M. Hirota, P.I. Terasaki, A. Wakisaka, H. Togashi, D. Chia, N. Suyhama, Y. Fukushi, E. Nudelman, and S. Hakomori, Characterization of sialosylated Lewis^x as a new tumor-associated antigen, Cancer Res. 44:5279 (1984).

8. Y. Fukushi, E. Nudelman, S.B. Levery, H. Rauvala, and S. Hakomori, Novel fucolipids accumulating in human cancer. III. A hybridoma antibody (FH6) defining a human cancer-assoicated difucoganglioside (VI^{3}NeuAcV^{3}III^{3}Fuc$_2$nLc$_6$), J. Biol. Chem. 259:10511 (1984).

9. Y. Fukushi, S. Hakomori, E. Nudelman, and N. Cochran, Novel fucolipids accumulating in human adenocarcinoma. II. Selective isolation of hybridoma antibodies that differentially recognize mono-, di-, and trifucosylated type 2 chain, J. Biol. Chem. 259:4681 (1984).

10. K. Abe, J.M. McKibbin, and S. Hakomori, The monoclonal antibody direct-ed to difucosylated type 2 chain (Fucα1→2Galβ1→4[fucα1→3]GlcNAβ1→ R; Y determinant), J. Biol. Chem. 258:11793 (1983).

11. A. Brown, T. Feizi, H.C. Gooi, M.J. Emblton, J.K. Picard, and R.W. Baldwin, A monoclonal antibody against human colonic adenoma recog-nizes difucosylated type 2 blood group chains. Biosci. Rep. 3:163 (1983).

12. K.O. Lloyd, G. Larson, N. Stromberg, J. Thurin, and K.-A. Karlsson, Mouse monoclonal antibody F-3 recognizes difucosyl type 2 blood group structure, Immunogenetics 17:537 (1983).

13. Y. Fukushi, S. Hakomori, and T. Shepard, Localization and alteration of mono-, di-, and trifucosylα1→3 type 2 chain structures during embryogenesis and in human cancer, J. Exp. Med. 159:506 (1984).

14. Y. Fukushi, R. Kannagi, S. Hakomori, T. Shepard, B.G. Kulander, and J.W. Singer, Localization and distribution of difucoganglioside (VI^{3}NeuAcV^{3}III^{3}Fuc$_2$nLc$_6$) in normal and tumor tissues defined by its monoclonal antibody FH6, Cancer Res. 45:3711 (1985).

15. S. Hakomori, and R. Kannagi, Glycosphingolipids as tumor-associated and differentiation markers, J. Natl. Cancer Inst. 71:231 (1983).

16. E. Nudelman, S. Hakomori, R. Kannagi, S. Levery, M.-Y. Yeh, K.E. Hellström, and I. Hellström, Characterization of a human melanoma-associated ganglioside antigen defined by a monoclonal antibody, 4.2, J. Biol. Chem. 257:12752 (1982).

17. C.S. Pukel, K.O. Lloyd, L.R. Trabassos, W.G. Dippold, H.F. Oettgen, and L.J. Old, GD$_3$, a prominent ganglioside of human melanoma: Detection and characterization by mouse monoclonal antibody, J. Exp. Med. 155:1133 (1982).

18. L.D. Cahan, R. Irie, R. Singh, A. Cassidenti, and J.C. Paulson, Ident-ification of a human neuroectodermal tumor antigen (OFA-I-2) as ganglioside GD$_2$, Proc. Natl. Acad. Sci. USA 79:7629 (1982).

19. K. Watanabe, C.S. Pukel, H. Takeyama, K.O. Lloyd, H. Shiku, L.T.C. Li, L.R. Trabassos, H.F. Oettgen, and L.J. Old, Human melanoma antigen AH is an autoantigen ganglioside related to GD$_2$, J. Exp. Med. 156:1884 (1982).

20. O. Nillson, J.-E. Mansson, T. Brezicka, J. Holmgren, L. Lindholm, S. Sorenson, F. Yngvason, and L. Svennerholm, Fucosyl GM$_1$, a ganglio-

side associated with small cell lung carcinomas, Glycoconjugate J., 1:43 (1984).

21. E.G. Bremer, S.B. Levery, S. Sonnino, R. Ghidoni, S. Canevari, R. Kannagi, and S. Hakomori, Characterization of a glycosphingolipid antigen defined by the monoclonal antibody MBr1 expressed in normal and neoplastic epithelial cells of human mammary gland, J. Biol. Chem. 259:14773 (1984).

22. D.L. Urdal, and S. Hakomori, Tumor-associated ganglio-N-triosylcer-amide: Target for antibody dependent, avidin-mediated drug killing of tumor cells, J. Biol. Chem. 255:10509 (1980).

23. S. Hakomori, W.W. Young, Jr., and D. Urdal, Glycolipid tumor cell markers and their monoclonal antibodies: Drug targeting and immuno-suppression, in: "Monoclonal Antibodies in Drug Development", T. August, ed., Johns Hopkins University Press, Baltimore (1982).

24. G. Weissman, D. Bloomgarden, R. Kaplan, C. Cohen, S. Hoffstein, T. Collins, A. Gotlieb, and D. Nagle, A general method for the introduction of enzymes, by means of immunoglobulin-coated lipo-somes, into lysosomes of deficient cells, Proc. Natl. Acad. Sci. USA 72:88 (1975).

25. C.M. Cohen, G. Weissmann, S. Hoffstein, Y.C., Awasthi, and S.K. Srivastava, Introduction of purified hexosaminidase A into Tay-Sachs leukocytes by means of immunoglobulin-coated liposomes, Biochemistry 15:452 (1976).

26. G. Gregoriadis, and E.D. Neerunjun, Homing of liposomes to target cells, Biochem. Biophys. Res. Commun. 65:537 (1975).

27. G. Gregoriadis, E.D. Neerunjun, and R. Hunt, Fate of liposome-assoc-iated agent injected into normal and tumor-bearing rodents: Attempts to improve localization in tumor tissues, Life Sci. 21:374 (1977).

28. L. Huang, and S.J. Kennel, Binding of immunoglobulin G to phospholipid vesicles by sonication, Biochemistry 18:1702 (1979).

29. D. Sinha, and F. Karush, Attachment to membranes of exogenous immuno-globulin conjugated to a hydrophobic anchor, Biochem. Biophys. Res. Commun. 90:554 (1979).

30. T.D. Heath, R.T. Farley, and D. Papahadjopoulous, Antibody targeting of liposomes: Cell specificity obtained by conjugation of $F(ab')_2$ to vesicle surface, Science 210:539 (1980).

31. W.W. Young, Jr., and S. Hakomori, Therapy of mouse lymphoma with mono-clonal antibodies to glycolipid: Selection of low antigenic variants in-vivo, Science 211:487 (1981).

32. J. Brown, R.G. Woodbury, C.E. Hart, I. Hellstrom, and K. Hellstrom, Quantitative analysis of melanoma-assoicated antigen p97 in normal and neoplastic tissues, Proc. Natl. Acad. Sci. USA 78:539 (1981).

33. A.C. Morgan, D.R. Galloway, and R.A. Reisfeld, Production and charact-erization of monoclonal antibody to melanoma-specific glycoproteins, Hybridoma 1:27 (1981).

34. J. Wiels, S. Junqua, P. Dujardin, J.-B. LePecq, and T. Tursz, Prop-erties of immunotoxins against a glycolipid antigen associated with Burkitt's lymphoma, Cancer Res. 44:129 (1984).

INTERACTION OF MACROMOLECULAR DRUGS WITH RECEPTORS

Josef Pitha

Macromolecular Chemistry Section, National Institute on
Aging/GRC, National Institutes of Health, 4940 Eastern Ave
Baltimore, Maryland, 21224

Leonardo da Vinci, in that part of his writings which is currently
known as Hammer Codex, put forward a hypothesis to explain the circulation
of water in the world. Although this renaisance scientist correctly specified all the elements involved, he erroneously rejected rain as a serious
factor and suggested that water circulation was due to spontaneous seepage
of water from the seas through underground porous rocks to the mountains,
from where the rivers flow again. Obviously, we often overestimate the
adverse effects of an erroneous hypothesis; such a hypothesis may be temp-
orarily useful and definitely does not limit more rational analyses by
others. Appreciation of these aspects of research suitably prepares a
reader for the present chapter. Experimental observations described here
are hopefully as solid as the rocks, rivers, and rain of da Vinci. We cannot be as certain about the correctness of the hypotheses explaining the
observations, but at least the hypotheses are guaranteed harmless to anyone
who finds a better explanation.

In this chapter we will deal with receptors located on the outside surfaces of vertebrate cells. These receptors are glycoproteins which are
able to bind messengers such as hormones in blood. After the messenger is
bound, these receptors relay the signal to the interior of cells. Hormones
and their congeners thus affect the internal metabolism of the cells from
the outside. Since these compounds affect cells they also are termed agonists. Medicinal chemists eventually developed a number of compounds which
also bind to hormonal receptors but without eliciting any intracellular
response. Since binding of such compounds to receptors prevents the binding of agonists, these compounds are termed antagonists or blockers. Cell
surface receptors have been detected and partially characterized for a
great number of hormones and synthetic compounds; prominent among these are
peptidic hormones and catecholamines. We will deal mainly with the latter.
Steroid hormones, on the other hand, have receptors which are located
intracellularly and thus of less interest to polymer chemists.

Receptors for hormones or drugs have been the center of attention of
pharmacologists and medicinal chemists for nearly a century and thus, there
are well tested methods for the study of their effects. On the other hand,
up to quite recently, the molecular aspects of receptors could not be
effectively investigated. That has been mainly due to the great efficiency
of the action of receptors. Thus, for example, receptors for catecholamines which mediate all of the well appreciated dramatic effects of epi-

nephrine (adrenalin) are present in human tissues in less than parts per million amounts. Fortunately at present, mainly thanks to the studies of Dr. Snyder of Johns Hopkins University and his coworkers,[1] these receptors and their interaction with drugs can be well quantitated even in an amount as small as a few milligrams of tissue. The methods for such quantitation can be described as follows. The vertebrate tissue in question is minced and homogenized and an insoluble fraction, called "membrane preparation", is isolated by a simple centrifugation. This membrane preparation contains substantially all the cell-surface receptors present in the tissue. Such a membrane preparation, when suspended, can be pipeted without difficulty, but the membrane can be quantitatively recovered from the suspension either by centrifugation or by a filtration on glass filters. Thus, it is relatively easy to measure the binding of highly radioactive hormone to receptors located on these membranes or to measure how much a newly prepared compound interferes with such binding. Of course, the radioactive hormone binds to all components of the membrane preparation. That is, the hormone binds not only to receptors but also to other components (e.g., lipids) which are in the membrane preparation up to 10^5 times higher concentration than receptors. Fortunately, these two types of binding can be separated. Binding to receptors is very strong and to very few sites; consequently, the addition of a small amount of non-radioactive hormone can completely displace a very small amount of radioactive hormone which is bound to receptors in a specific manner. In this way the specific binding of a hormone or competition of new drug to that binding can be measured and the appropriate dissociation constants characterizing the equilibria in question calculated. The methods of study for such binding have been perfected and presently in several days, full binding characterization of a new compound to hormonal receptors can be obtained. It also has been established that the above binding data have relevance in therapeutic effects of drugs; this was accomplished through painstaking comparisons of data on binding of drugs to receptors and data on doses needed to achieve standard therapeutic effects. Direct correlations were found there.

Thus, rather fine experimental methods are available to measure the binding of hormones or drugs to their proper receptors. For chemists interested in direct-acting polymeric drugs, no further introductions or qualifications are necessary. New polymeric drugs can be evaluated within a few days and with considerably less uncertainty than when they were tested by methods involving isolated organs or whole animals. However, additional information is required to put these fine methods into the armatorium of chemists interested in the use of receptors to target polymers to a specific group of cells.

The first and obvious question is how many different receptors, in this context targets, can be differentiated. This question cannot be answered with precision. Receptors, since they were first recognized, keep multiplying and their number keeps increasing with the number of researchers bent on the discovery of new ones. Nevertheless, an illustrative sample can be provided. Table 1 presents some data on receptors frequently studied by practitioners of pharmacology in drug houses. Newly synthesized compounds are routinely put through the battery of binding assays presented in table 1 to find their pharmacological potential. Even this small sample of binding assays shows that there is a multitude of receptors/ targets and that these can be reproducibly differentiated. Some additional important features of drug-receptor interactions can be easily deduced from the numbers in Table 1. Obviously the associations of drugs with receptors are quite strong: K_D are in nanomolar range. Another important point concerning receptors can be deduced from the data on spiperone (Table 1). This is quite typical for butyrophenones, a class of drugs to which spiperone belongs. In the cortex, where a large number of serotonin $5HT_2$ receptors are located, a large portion of spiperone present is bound

Table 1. Some receptors and systems for their study

Receptor	Tissue used in membrane preparation	Commonly used radioligand	Affinity K_D(M)
α_1-adrenoceptor	brain, rat	prazosin	4×10^{-10}
α_2-adrenoceptor	forebrain, rat	clonidine	6×10^{-9}
β_1-adrenoceptor	heart, rat	dihydroalprenolol	3×10^{-9}
β_2-adrenoceptor	lungs, rat	dihydroalprenolol	5×10^{-9}
D_1-dopamine	caudate nucleus, calf	dopamine	2×10^{-8}
D_2-dopamine	caudate nucleus, calf	spiperone	3×10^{-10}
$5HT_1$ serotonin	brain, rat	serotonin	2×10^{-9}
$5HT_2$ serotonin	frontal cortex, rat	spiperone	7×10^{-10}
H_1 histamine	forebrain, rat	mepyramine	4×10^{-9}
muscarinic	forebrain, rat	quinuclidol benzilate	7×10^{-11}

Data in this table are from A. Closse et al., 1984[21] or the results of Dr. J.W. Kusiak (NIA/GRC).

to those receptors. In the caudate part of the brain, which is rich in dopamine D_2 receptors, a large part of spiperone is bound to those receptors.

To complete this elementary description of receptors a few data on receptors at the cellular and tissue levels should be given. Let us consider β-adrenoceptors on transformed C6-2B rat astrocytoma cells which are grown in vitro. There is an average of 4400 molecules of β-adrenoceptor on the surface of one cell, or 10 fmoles per mg of protein.[2] The probe often used for this receptor, iodopindolol, binds quite cleanly in these cells. At very low concentrations of iodopindolol about 95% of that probe is bound to cellular β-adrenoceptors, whereas the rest of the cell (about 99.99999%) binds only 5% of the probe. Binding to receptors is also quite strong, in this system the K_D being 4×10^{-11}M. On the cellular or tissue level there is another aspect of receptors worth mentioning. A cell may have several thousand receptors on its surface, but not all of these are constantly required. Compounds called alkylating β-blockers can be used for selective destruction of β-adrenoceptors without lethal effects on cells or animals, and rats with only 10% or 30% of their β-adrenoceptors left in their hearts or brains, respectively, survive such selective destruction.[3,4]

All of these features make drug receptors quite suitable targets for polymeric drugs. Few associations as strong and clean as the above one are known in biology. Of course, some caution is called for. Consider opiate receptors. These receptors are present in high concentrations in two tissues, brain and guts. That is rather to be expected because morphine not only affects the mind, but also stops diarrhoea. Any magic bullet shot at opiate receptors will thus hit a man not at one but at two locations and these particular locations should never be hit simultaneously; with both brains and guts gone, not much can be accomplished in life.

In our own work on polymeric drugs and receptors we studied mainly the β-adrenergic system. These receptors respond to catecholamines (epinephrine, norepinephrine and congeners). On the tissue level pharmacologists identify β-adrenoceptors by their ability, when activated, to cause vasodilation and inhibition of the uterine contraction and myocardial stimulation.[5] Three classes of drugs act on β-adrenoceptors. Cardiotonics are drugs which strenghten heart activity and are used, for example, after heart attacks; some drugs of that class are β-adrenoceptor agonists. Agonists acting on this receptor are used also as bronchodilators, i.e., as medication in relieving asthma. Drugs which act as antagonists on β-adrenoceptor are colloquially called β-blockers and form one of the largest groups of medications in use currently. β-Blockers are used to control hypertension, angina pectoris, and tachycardia. In the United States about one third of the people over 65 years old require antihypertensive treatment.

From a chemical point of view we work with two groups of compounds. β-Blockers are pleasant compounds to work with of the general structure below.

$$Aryl - O - CH_2 - CHOH - CH_2 - NH - Alkyl$$

In our work we have extensively used congeners and derivatives of a commercial drug, alprenolol, in which the aryl above is the o-allylphenyl group and the alkyl is the isopropyl group. Two main approaches for affixation of alprenolol pharmacophores to polymeric carriers have been used. The first approach follows the commercial synthesis of alprenolol. From o-allylphenol by alkylation with epichlorohydrin is prepared o-allylphenyl glycidyl ether (Fig. 1). When that compound is reacted with amines, whether of small or large molecular weight, alprenolol congeners are prepared (Fig. 2). There is a complicating factor in that reaction: two molecules of o-allylphenyl glycidyl ether may react with an amine and form fully substituted derivatives (Fig. 3) which have rather low biopotency.

Carrier macromolecules for the above described immobilizations were found in commercially available products. Jeffamines, which are manufactured in many ton quantities are used as additives in the preparation of polyurethane foams. They have the structures indicated in Fig. 4.

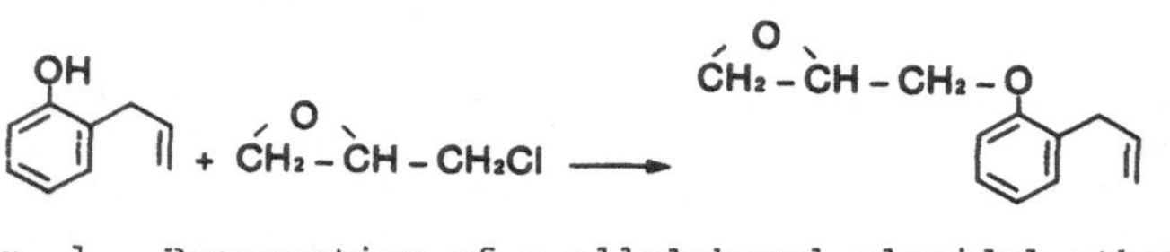

Fig. 1. Preparation of o-allylphenyl glycidyl ether.

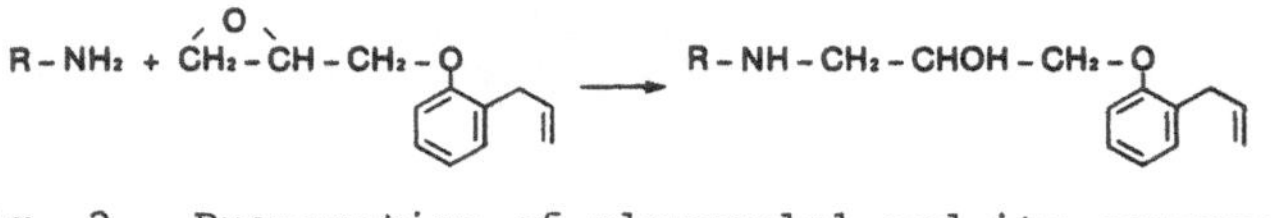

Fig. 2. Preparation of alprenolol and its congeners.

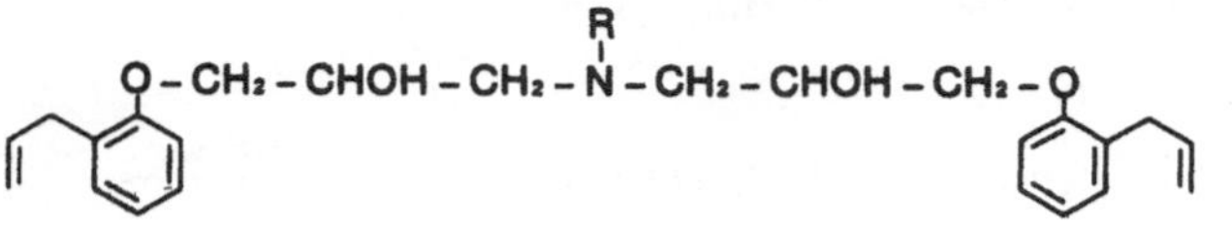

$$O-CH_2-CHOH-CH_2-\overset{\overset{\displaystyle R}{|}}{N}-CH_2-CHOH-CH_2-O$$

Fig. 3. Inactive by-products formed in the preparation of
alprenolol and its congeners.

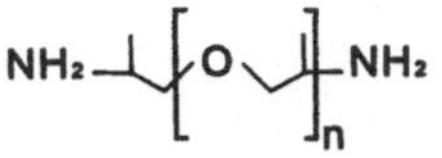

$$NH_2-\left[\right]_n-NH_2$$

Fig. 4. General structure of Jeffamine. Preparations with
average n varying between 2.6 and 49 are commercially
available.

Reaction of the above mentioned o-allylphenyl glycidyl ether yields
compounds with one or two full pharmacophores in one molecule.[6] Of course,
up to four times substituted Jeffamines can be prepared, but these do not
carry high affinity pharmacophores. Also, a reversed scheme may be used,
where epoxy groups are on the polymer and a reactive amino group is on the
drug residue.[7]

The alternative approach for affixing the alprenolol pharmacophore
to a polymer is based on the reactivity of allyl residue in the already
made drug. In our work we use polymers containing mercapto groups and
these, by free radical reactions, are added to the allyl group of alprenolol. Such an addition is an exceptionally clean reaction and is easy to
perform.[8-10] As a matter of fact, the chemicals and operations used are
quite similar to those used in the preparation of polyacrylamide gels and
thus well within the skills of practicing biochemists. The reaction is
illustrated in Fig. 5.

Macromolecules containing sulfhydryl groups were prepared from dextran,
a polysaccharide which is commercially available in the form of fractions
with well defined molecular weight. In preparation of these macromolecular
mercaptanes, bis-epoxides were used to form linkers and Bunte salts were
intermediates in the synthesis, as illustrated in Fig. 6.

Allyl groups of alprenolol are quite reactive in free radical reactions, a fact which has not been used much. Allyl derivatives may be copolymerized with vinyl polymers. Thus, when alprenolol is simply added to
acrylamide and free radical polymerization of the latter is initiated with
potassium persulfate, a substantial part of drug (up to 37%) may be incorporated into the polymer; probably macromolecules of the structures in
Fig. 7 are formed.[11]

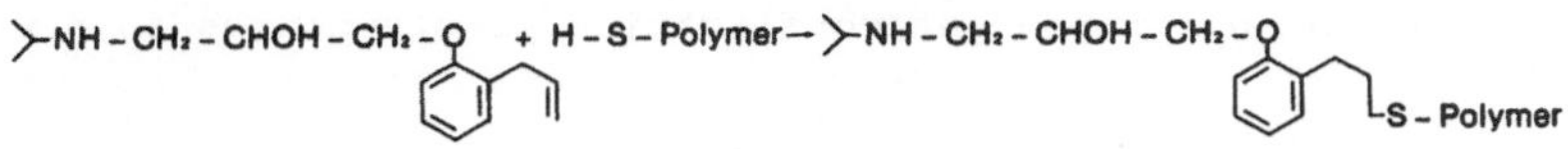

$$\rangle NH-CH_2-CHOH-CH_2-O \quad + \; H-S-Polymer \longrightarrow \rangle NH-CH_2-CHOH-CH_2-O$$
$$S-Polymer$$

Fig. 5. Affixation of alprenolol to a polymer containing
sulfhydryl groups.

45

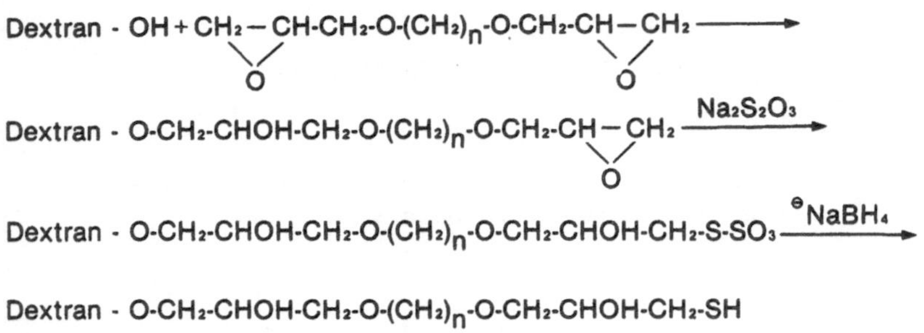

Dextran · OH + CH$_2$—CH-CH$_2$-O-(CH$_2$)$_n$-O-CH$_2$-CH—CH$_2$ ——————→

Dextran · O-CH$_2$-CHOH-CH$_2$-O-(CH$_2$)$_n$-O-CH$_2$-CH—CH$_2$ $\xrightarrow{\text{Na}_2\text{S}_2\text{O}_3}$

Dextran · O-CH$_2$-CHOH-CH$_2$-O-(CH$_2$)$_n$-O-CH$_2$-CHOH-CH$_2$-S·SO$_3$ $\xrightarrow{\ominus \text{NaBH}_4}$

Dextran · O-CH$_2$-CHOH-CH$_2$-O-(CH$_2$)$_n$-O-CH$_2$-CHOH-CH$_2$-SH

Fig. 6.   Preparation of dextrans substituted with sulfhydryl
          groups.

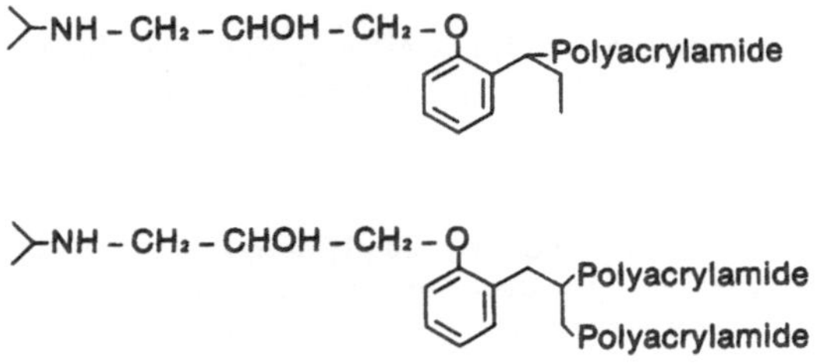

>—NH – CH$_2$ – CHOH – CH$_2$ – O $\longrightarrow$ Polyacrylamide

>—NH – CH$_2$ – CHOH – CH$_2$ – O $\longrightarrow$ Polyacrylamide / Polyacrylamide

Fig. 7.   Structures possibly present in alprenolol-acrylamide
          copolymer.

Using the above methods we prepared three groups of macromolecules
containing β-blocking pharmacophores.

(Blocker – S – Linker)$_n$ Dextran

Blocker – Oligopropylene oxide – Blocker

Blocker – Polyacrylamide

Chemical work in the field of β-adrenoceptor agonists is far from
pleasant. Pharmacological consideration demands that only pharmacophores
which are able to fully activate β-adrenoceptors are used and that means a
synthesis of catecholamines and congeners. These compounds are quite
unstable to oxidation by air and their colorless solutions easily turn
brown. Several months were spent on attempts to introduce agonist pharmacophores into macromolecules starting with 3,4-dihydroxyphenacyl chloride,
a compound used in the commercial synthesis of catecholamines, but no products of acceptable purity were obtained. Eventually we adapted for our
purposes the methods developed by Goodman and collaborators;[12] the reaction
sequence is illustrated in Fig. 8.

Even this approach is not free of difficulties. Catecholamines prepared by this method are contaminated by isomeric products of the structure
in Fig. 9, and often an independent confirmation of structures is necessary.

Using the above preparative method we obtained a group of macromolecules as shown below.

Catecholamine – Oligopropylene oxide – Catecholamine

Catecholamine – Oligopropylene oxide – Blocker

Blocker – Oligopropylene oxide – Blocker

Eventually, we collected a satisfactory number of polymeric ligands to β-adrenoceptor and could start their bio-evaluation. These studies, performed by Dr. J. Kusiak at NIA/GRC and by Drs. M.G. Caron and R.J. Lefkowitz and their collaborators at Duke University, were quite thorough and thus can be simply described. Binding of compounds to the following ligands were studied.

(a) Antibodies raised in rabbits to the β-adrenoceptor/hormone complex which was isolated from frog erythrocytes. This sounds complicated but this antibody is simply reactive towards any ligand of the β-adrenoceptor, including catecholamines or alprenolol.

(b) β-Adrenoceptors located on erythrocytes of frogs. This system has an advantage of being remarkably clean and well defined from a pharmacologist's point of view.

(c) β-Adrenoceptors located on rat erythrocytes. These cells were chosen as a mammalian counterpart to the system above.

(d) β-Adrenoceptors located on membrane preparations from rat heart. This preparation contains mainly β_1-subclass of receptors.

(e) β-Adrenoceptors located on membrane preparations from rat lungs. This preparation contains mainly β_2-subclass of receptors.

The results of binding studies of alprenolol-dextran conjugates[9,10] to β-adrenoceptors and to antibody are summarized in Fig. 10. Dissociation constants of binding are plotted versus the length of the linker expressed as the number of atoms (C,O,S) separating the alprenolol residue from the dextran carrier.

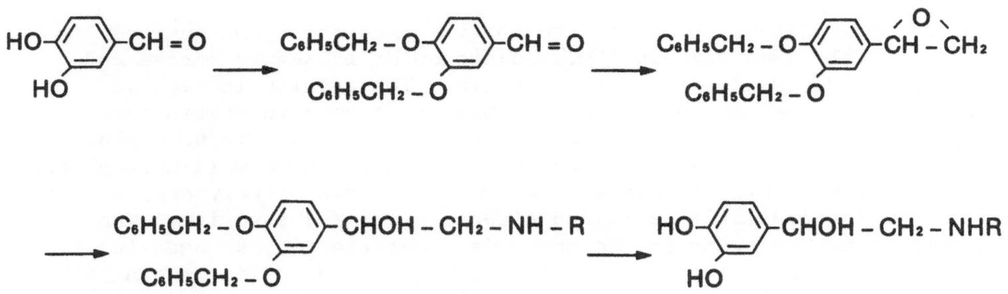

Fig. 8. Sequence of chemical reactions used to introduce catecholamine pharmacophores into macromolecules.

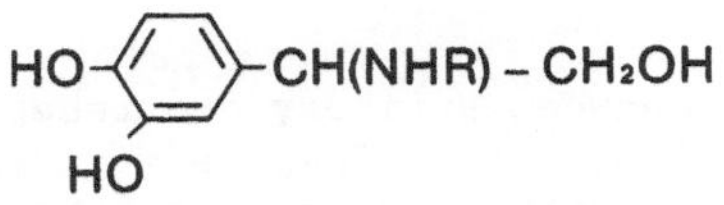

Fig. 9. Inactive by-product formed in the synthesis of catecholamines.

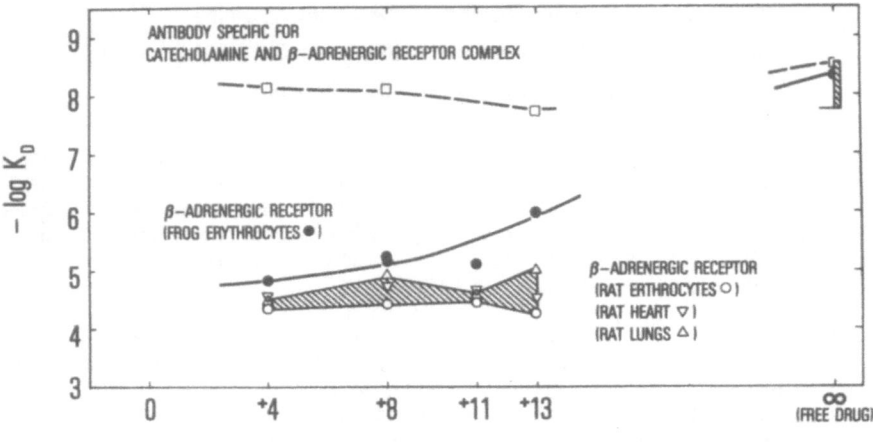

Fig. 10. Binding of dextrans to which alprenolol pharmacophores were attached by links of various
lengths to β-adrenoceptors and to antibodies.

    Obviously, all the macromolecules containing alprenolol residue bound
to the antibody with about the same affinity as the drug itself. With the
β-adrenoceptor the situation is dramatically different. When alprenolol
residue is close to the dextran, there is hardly any bonding at all.
Binding to β-adrenoceptor again increases only for compounds in which the
drug and macromolecule are separated by a long linker. About the same
results were obtained with copolymers of acrylamide and alprenolol. While
these compounds were strongly bound to antibody, their binding to β-adreno-
ceptor was very weak.[11]

    Quite obviously the binding site for the drug moiety on antibody is
exposed while the same site in the β-adrenoceptor is deeply buried and
thus, subject to considerable steric strain. The results in Fig. 10
furthermore show that there is a detectable difference in steric strain
when frog and rat β-adrenoceptors are compared, but there is no useful difference between rat β_1 and β_2 adrenoceptors. It has been well appreciated
that steric strains play a role in biological effects of polymers, but the
magnitude of the presently described effects came as a surprise. The
linker with thirteen atoms in the chain is about the longest used in affinity chromatography, another field in which the interaction of synthetic
macromolecule/natural macromolecule is of interest. The vast majority of
affinity chromatography work has been done with much shorter linkers.

    It is interesting to compare our data on alprenolol linked to dextran
with published data on alprenolol linked to the macromolecular complex of
biotin with avidin.[13] The linker of three atom lengths did not enable
binding of this probe to β-adrenoceptors on frog erythrocytes, whereas the
linkers of ten or sixteen atom lengths did allow the binding. In spite of
the different nature of macromolecules, both systems are in rough quantitative agreement.

    The differences found between antibody and receptors may have some
practical use. Some endocrinological diseases stem from immunodefects;
affected individuals produce antibodies against a hormone or against
hormone-receptor combinations and these antibodies grossly interfere with
hormonal regulation. Polymeric derivatives of the hormones involved in

such diseases may be of help; they would react only with the antibodies and not with receptors. Thus, it would be basically possible by use of polymeric drugs to neutralize the interferring antibodies without affecting the regulatory function of the receptor. No diseases of that type involving the β-adrenoceptor system have as yet been discovered and thus, there is no immediate practical use for alprenolol-dextran conjugates. But a similar approach can be used for diseases known to be of that type; it is plausible that the above scheme of steric hindrances is a general one.

The binding site of β-adrenoceptors also is very sensitive to polarity of the ligand. This circumstance makes it rather difficult to prepare a good polymeric ligand for this receptor. If the macromolecule is nonpolar it is not water-soluble and thus useless as a drug, and when the macromolecule is polar it again does not bind strongly to the receptor. The sensitivity of binding to polarity is best seen on binding data of a series of small molecular weight compounds, which were prepared by Dr. A. Liptak and evaluated by Dr. J.W. Kusiak, both at NIA/GRC. Structures and dissociation constants (membranes from rat tissues) are given in Fig. 11. It is obvious that any incorporation of hydroxy groups into a parent drug decreases the affinity to receptors.

As a next step of our studies we wanted to evaluate polymeric ligands of β-adrenoceptors which would have chains with low steric hindrance and high flexibility. Eventually we decided to eliminate the carrier macromolecules, i.e., dextran or polyacrylamide altogether, and to use only the linker moieties from the previous systems to hold the pharmacophores together. These linkers were derived from oligopropylene glycols. The resulting drugs are not really polymeric drugs, possibly they can be called oligomeric, but the series is well suited for structure/activity studies of multiple pharmacophores. The structures of the first series of compounds used are illustrated in Fig. 12. Binding of these compounds to β-adrenoceptors and their ability to activate the catecholamine system (i.e., activation of enzyme adenylate cyclase) was then carefully evaluated.[14,15] The results can be summarized as follows:

(A) Compounds with two pharmacophores were bound to receptors at best as strong as the compound with one pharmacophore; no strengthening of binding oligo-ligands was observed.

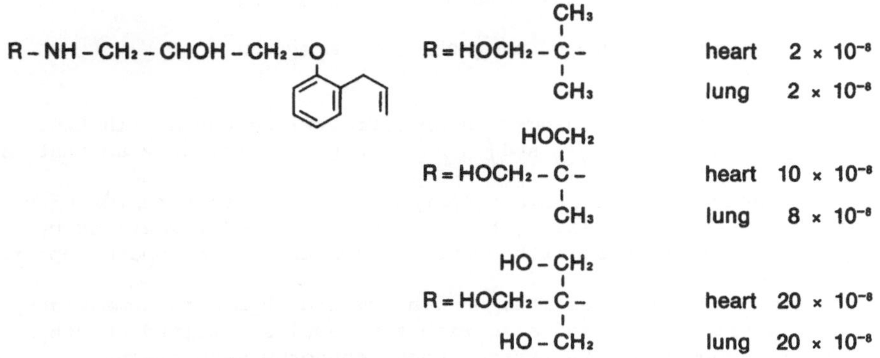

Fig. 11. Effect of polarity on binding (K_I in [M]) of alprenolol congeners to β-adrenoceptors; membrane preparations from rat tissues were used.

49

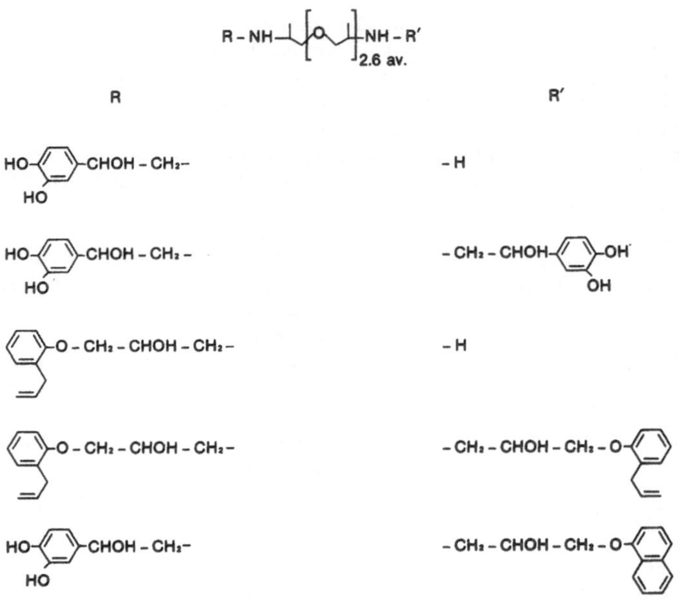

Fig. 12. Structure of drugs containing two pharmacophores.

(B)   There were no changes in the type of activity: compounds containing two antagonist pharmacophores functioned as antagonists, those with two agonist pharmacophores acted as partial agonists.

(C)   Antagonist-agonist combination acted as an antagonist. Findings sub(B) and sub(C) can be expressed as the series of equivalencies below.

agonist	– linker – NH_2	≡	partial agonist
agonist	– linker – agonist	≡	partial agonist
antagonist	– linker – NH_2	≡	antagonist
antagonist	– linker – antagonist	≡	antagonist
agonist	– linker – antagonist	≡	antagonist

(D)   There were no dramatic changes in specificity; compound with two pharmacophores bound to β_1 and β_2 adrenoceptors similarly as that with one.

(E)   Some compounds with two pharmacophores were bound to receptor with a remarkable persistency (i.e., binding could not be dissociated by washings, as is the case with compounds possessing one pharmacophore).

Having summarized the results, we can compare them with some widespread beliefs on polymeric drugs, and also with the results obtained on other receptors where compounds with several pharmacophores were evaluated. One of the recurring ideas in the field of polymeric drugs is that "many" is better than "one". This implies that polymeric drugs may bind more strongly than parent drugs. There are examples supporting these views and some of our own previous results on the interaction of polymer-cell surface also support such views. But in all these examples the interactions

GnRH-Ant ≡ ANTAGONIST

GnRH-Ant-CO-CH₂-CH₂-CO-O-CH₂-CH₂-O-CO-CH₂-CH₂-CO-Ant-GnRH ≡ ANTAGONIST

GnRH-Ant-CO-CH₂-CH₂-CO-O-CH₂-CH₂-O-CO-CH₂-CH₂-CO-Ant-GnRH
                    |
                    Antibody   ≡ AGONIST
                    |

GnRH-Ant-CO-CH₂-CH₂-CO-O-CH₂-CH₂-O-CO-CH₂-CH₂-CO-Ant-GnRH

Fig. 13. Pharmacological activity of gonadotropin congeners.

involved are considerably less specific (for example, coulombic interactions or organomercurial:sulfhydryl) than those presently studied and thus, on the surface of cells there is more "receptors" for such polyligand. With highly specific interactions, the steric match between the pharmacophores on the polymer and the binding receptors on cell surface is obviously of critical importance.

Another interesting aspect of compounds with two pharmacophores involves receptors of a peptide hormone, gonadotropin releasing hormone. It was shown that compounds containing two antagonist pharmacophores may act on this receptor as an agonist.[16] This situation is summarized in Fig. 13. To form bi-antagonist, a complex with antibody was used; presumed distance between the two pharmacophores is 120-150 Å.

In the above series of β-adrenoceptor ligands which we evaluated, this change from antagonistic to agonistic activity was not observed. But that series had had a small separation of pharmacophores, less than 20 Å. Eventually we synthesized an extended series of compounds with a larger separation of pharmacophores, as shown in Fig. 14.[15] The alprenolol pharmacophore (3-(o-allylphenoxy)-2-hydroxypropyl) is denoted as Alp.

The fully extended conformations of the two compounds in that series span about a distance which otherwise may be achieved with an antibody spacer. Nevertheless, all the compounds containing two pharmacophores with antagonistic activity on β-adrenoceptor which we prepared and studied had only antagonistic effects.

In the field of opiates it was found that drugs with two pharmacophores may have improved selectivity in the interaction with subclasses of opiate receptors.[17] This is illustrated by the results in Fig. 15, in which binding to opiate δ and μ receptors were compared.

We have not observed any change in the selectivity of binding to β_1 and β_2 receptors in the series of oligo-ligands of β-adrenoceptors we prepared and studied. The reason for the difference is not known.

The persistence of binding of some compounds with two or three pharmacophores to the β-adrenoceptor mentioned above is a remarkable phenomenon.[6,14] This persistence does not correlate with affinity of binding. This can be seen in Fig. 16 in which compounds with several pharmacophores, denoted as Alp, were compared. The abbreviation Jeff is used to denote the linker of the type shown in Fig. 17.

The strongest binding in this series is by the compound with only one pharmacophore (Alp₁Jeff) which washed off of the receptors very easily. Compounds with two or three pharmacophores, Alp₂Jeff or Alp₃Jeff, bind less

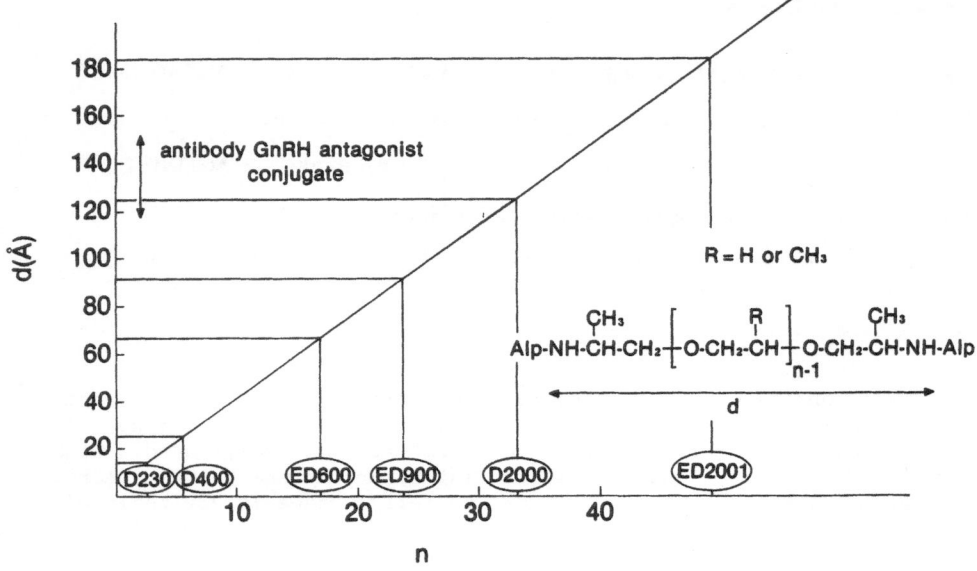

Fig. 14. Span between the pharmacophores in fully extended
conformation of divalent β-blockers; Alp denotes
3-(o-allylphenoxy)-2-hydroxypropyl-residue.

	relative δ: μ selectivity
[D-Ala²,des Leu⁵]-enkephalin amide	1
[D-Ala²,des Leu⁵]-enkephalin-NH-(CH₂)₁₂-NH-enkephalin[D-Ala²,des Leu⁵]	1000

Fig. 15. Ligands of opiate receptor

strongly to the β-adrenoceptors but, on the other hand, cannot be washed
off.

The probable reason for the persistent binding of these compounds is
their two-point attachment to membranes. These molecules can span hormone
binding sites of two receptors, or a receptor site and some secondary
binding site. Originally we thought that the compounds with two pharmacophores span two receptors, but presently we favor the other hypothesis.
This change in views was due to the observation that the ability to bind to
the membranes persistently is not greatly sensitive to the length of the
linker between the pharmacophores. Compounds in which two alprenolol pharmacophores were connected with oligopropylene oxide linkers of degree of
polymerization varying between 3-50, all were persistently bound.[15] This
finding suggests that the second pharmacophore of these ligands is bound to
a more common component of membrane than are β-adrenoceptors. The results
of Goodman and collaborators[18] and of Roche's group[19,20] suggest that on or
near β-adrenoceptors there may be a secondary binding site for suitably
substituted aromatic residues; perhaps our compounds with two pharmacophores bind to that secondary site as well.

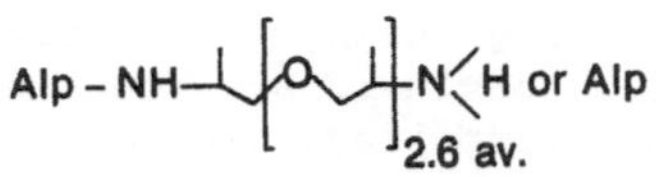

$$\text{Alp} - \text{NH} - \left[\underset{\text{2.6 av.}}{\bigvee} O \right] - N \Big\langle \begin{array}{c} \text{H or Alp} \end{array}$$

Fig. 16. Structure of compounds with multiple pharmacophores with β-blocker activity.

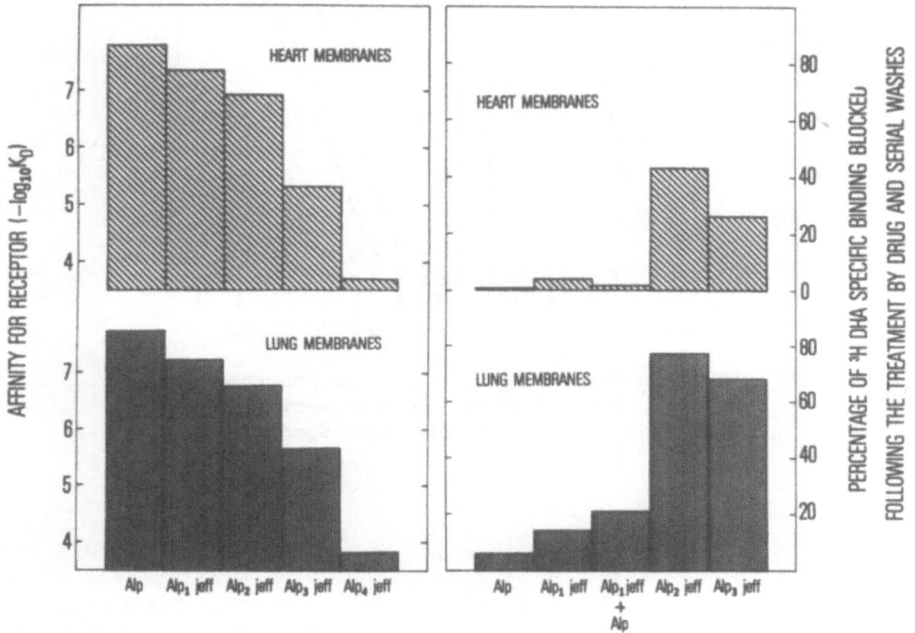

Fig. 17. Affinity and adhesiveness of drugs with mulitple
pharmacophores to β-adrenoceptor; membrane preparations from rat tissues were used.

With many medications a longer-lasting action is desired. One possible
way to obtain such action would be by persistently binding compounds with
two pharmacophores mentioned above. But this is not the usual way in which
the prolongation of drug affects have been achieved. This is usually done
by making drugs resistant to biodegradation. An example of such a drug is
nadolol which has been recently introduced into clinical practice as a
β-blocker; nadolol is degraded and eliminated from the organism only very
slowly and thus has long-lasting therapeutic action. This is an important
point to remember when polymeric drugs are evaluated; many of these also
resist biodegradation and excretion. This is illustrated in Fig. 18 in
which compounds with one or two pharmacophores mentioned above were
injected into a rat and β-adrenoceptor of rat heart and rat lung were
evaluated 18 hours later.[14] Both polymeric blockers were still acting upon
receptors, but of course, the parent drug, alprenolol, had already lost all
of its effects.

In conclusion it may be noted that the β-adrenoceptor system enabled us to study problems of polymeric drugs using highly simplified systems and to reach at least some unequivocal conclusions. These studies confirmed some expectations, but also provided surprises. The effect of steric strains on polymer interactions and the resistance of polymeric drugs to metabolic

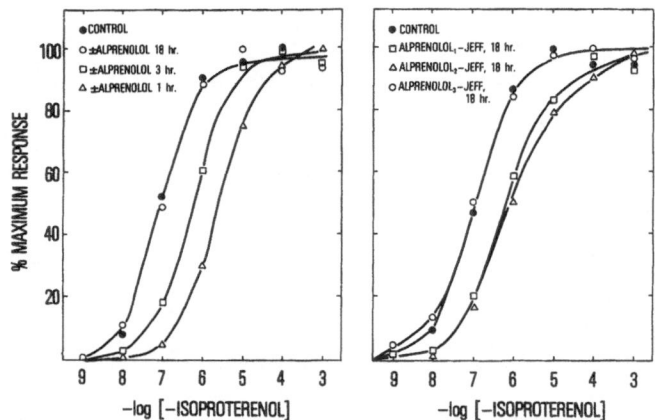

Fig. 18.   In vitro activation of the β-adrenoceptor system by isoproterenol after animals were pretreated with drugs (i.p.) at intervals specified.

degradation are not surprising results; however, persistent binding of drugs with two pharmacophores is. Equally surprising were the differences in properties of drugs with two pharmacophores of a peptide hormone class and of a β-adrenoceptor class. Hopefully these results may have practical application in the future.

REFERENCES

1. S.H. Snyder, Neurotransmitter receptor binding and drug delivery, J. Med. Chem., 26:1667 (1983).
2. K. Barovsky and G. Brooker, (-)-[125I]-Iodopindolol, a new highly selective radioiodinated β-adrenergic receptor antagonist: measurement of β-receptors on intact rat astrocytoma cells, J. Cyclic Nucleot. Res., 6:297 (1980).
3. J. Pitha, B.A. Hughes, J.W. Kusiak, E.M. Dax and S.P. Baker, Regeneration of β-adrenergic receptors in senescent rats: a study using an irreversible binding antagonist, Proc. Natl. Acad. Sci. USA, 79:4424 (1982).
4. B.L. Largent, A.L. Gundlach, J. Pitha and S.H. Snyder, Measurement of in-vivo β-adrenoceptor turnover in the rat striatum (abstract), Society for Neuroscience Meeting (1985).
5. P.B. Molinoff, α- and β-Adrenergic receptor subtypes: properties, distribution and regulation, Drugs, 28:1 (1984).

6. J. Pitha, J. Milecki, T. Czajkowska and J.W. Kusiak, β-Adrenergic antagonists with multiple pharmacophores: persistent blockage of receptors, J. Med. Chem., 26:7 (1983).

7. J. Pitha, S. Zawadzki and B.A. Hughes, Carriers for drugs and enzymes based on copolymers of allyl glycidyl ether with acrylamide, Makromol. Chem., 183:781 (1982).

8. M.G. Caron, Y. Srinivasan, J. Pitha, K. Kociolek and R.J. Lefkowitz, Affinity chromatography of the β-adrenergic receptor, J. Biol. Chem., 254:2923 (1979).

9. J. Pitha, J. Zjawiony, R.J. Lefkowitz and M.G. Caron, Macro-molecular β-adrenergic antagonists discriminating between receptor and anti-body, Proc. Natl. Acad. Sci. USA, 77:2219 (1980).

10. J.W. Kusiak and J. Pitha, Mapping of mammalian β-adrenoreceptors by use of macromolecular alprenolol derivatives, Biochem. Pharmacol., 31:2071 (1982).

11. J. Pitha, J. Zjawiony, R.J. Lefkowitz and M.G. Caron, Polymeric drugs by direct copolymerization: polymer of β-adrenergic antagonist alprenolol and its binding to receptors and antibodies, Makromol. Chem., 182:1945 (1981).

12. M.A. Avery, M.S. Verlander and M. Goodman, Synthesis of 6-aminoiso-proterenol, J. Org. Chem., 45:2750 (1980).

13. K.E. Meier and A.E. Ruoho, Formation of complexes between avidin and β-adrenergic receptors using biotinyl-alprenolol derivatives, Biochem. Biophys. Acta, 761:257 (1983).

14. J.W. Kusiak and J. Pitha, β-Adrenoreceptor antagonists with multiple pharmacophores: persistent inhibition of rat heart adenylate cyclase, J. Auton. Pharmacol., 3:195 (1983).

15. J.W. Kusiak, G. Heja and J. Pitha, Two β-adrenergic pharmacophores on the same molecule: complete set of agonist-antagonist combinations, submitted for publication.

16. P.M. Conn, D.C. Rogers, J.M. Stewart, J. Niedel and T. Sheffield, Conversion of a gonadotropin-releasing hormone antagonist to an agonist, Nature, 296:653 (1982).

17. Y. Shimohigashi, T. Costa, H.-C. Chen and D. Rodbard, Dimeric tetra-peptide enkephalins display extraordinary selectivity for the δ opiate receptor, Nature, 297:333 (1982).

18. K.A. Jacobson, D. Marr-Leisy, R.P. Rosenkranz, M.S. Verlander, K.L. Melmon and M. Goodman, Conjugates of catecholamines. 1. N-Alkyl-functionalized carboxylic acid congeners and amides related to isoproterenol, J. Med. Chem., 26:492 (1983).

19. R.W. Kierstead, A. Faraone, F. Mennona, J. Mullin, R.W. Guthrie, H. Crowley, B. Simko and L.C. Blaber, β_1-Selective adrenoceptor antagonists. 1. Synthesis and β-adrenergic blocking activity of a series of binary (aryloxy)propanolamines, J. Med. Chem., 26:1561 (1983).

20. P.J. Machin, D.N. Hurst, R.M. Bradshaw, L.C. Blaber, D.T. Burden, A.D. Fryer, R.A. Melarange and C. Shivdasani, β_1-Selective adreno-ceptor antagonists. 2. 4-Ether-linked phenoxypropanolamines, J. Med. Chem., 26:1570 (1983).

21. A. Closse, W. Frick, A. Dravid, G. Bolliger, D. Hauser, A. Saufer and H.-J. Tobler, Classification of drugs according to receptor binding profiles, Naunyn-Schmiedeberg's Arch. Pharmacol., 327:95 (1984).

ENDOCYTOSIS AND LYSOSOMES: RECENT PROGRESS IN INTRACELLULAR TRAFFIC

J.B. Lloyd

Biochemistry Research Laboratory, Department of
Biological Sciences, University of Keele, Staffordshire
England

INTRODUCTION

The cell membrane is a formidable barrier to macromolecules, and one
that is circumvented only by the phenomenon of endocytosis. Even then the
endocytosed material is segregated in a membrane-bounded vesicle and does
not have access to the general cytoplasm. Its ability to influence intracellular events depends on the routing of the vesicle (to the lysosome or
elsewhere), the effects of the intravesicular and intralysosomal environments, and the permeability properties of the vesicle and lysosome membranes.
My remit is to review the current understanding of this area of cell biology,
which is pivotal for attempts to target drugs with synthetic systems. Because of the focus of this symposium, I omit consideration of some related
issues, such as the biogenesis of lysosomal enzymes (but see references 1
and 2 for reviews). The area remaining is however both extensive and
active and, to avoid an unwieldy bibliography, I cite review articles where
available plus some selected recent primary research papers.

ENDOCYTOSIS

Endocytosis was the term introduced by de Duve[3] as a generic name sub-
suming the two processes of pinocytosis and phagocytosis. It is a useful
word when one is not sure which of these processes is responsible for a
particular uptake phenomenon but, in my view, is much overworked today.
Most examples of "receptor-mediated endocytosis" are quite clearly "receptor-mediated pinocytosis", and it is unhelpful and unnecessary to introduce an apparent ambiguity.

Phagocytosis,[4] the uptake of particulate matter, is not a general
property of mammalian cells. Its occurrence appears to be limited to the
so-called professional phagocytes of the body, those cells responsible for
patrolling the extracellular compartments and removing potentially harmful
particles such as bacteria. Neutrophils and cells of the macrophage family
are the principal cell-types involved.

Phagocytosis is initiated by the adherence of a particle to the cell,
and this in turn requires the particle to present a suitable surface. It
is found that many particles do not induce phagocytosis unless they are
coated with, for example, plasma proteins, a process known as opsonization.

Attachment is followed by major membrane movements, in which the phagocyte's plasma membrane flows around and engulfs particle. Subsequently a "phagosome" detaches from the plasma membrane, carrying its prey into the cell, normally to fuse with the lysosomes. Not surprisingly, in view of the extent of membrane movements involved, phagocytosis is strongly inhibited by lowered temperature or by metabolic inhibitors. Microfilament involvement is indicated by the sensitivity of phagocytosis to inhibition by cytochalasins.

The greater part of the phagocytic capacity of the mammalian body resides in the so-called reticuloendothelial system, a network of fixed macrophages in the tissues. The liver is quantitatively the most significant and accessible component of this system and, when particles are introduced into the bloodstream, most of them are trapped by the phagocytes of this organ, known as the Kupffer cells. Other contributions to this symposium (see chapters by S.S. Davies et al. and G. Gregoriadis et al.) describes how the distribution of particles to different sites within the reticuloendothelial system may be manipulated to some degree.

Pinocytosis, the uptake of liquid, is similar in many ways to phagocytosis. It too involves the formation of intracellular vesicles from the plasma membrane. Like phagosomes, these "pinosomes" can carry in extracellular material for digestion by the lysosomes. But there are important differences. First, pinocytosis occurs in most if not all nucleated cells. Although pinocytic activity varies with cell-type, the differences are surprisingly small.[5] Secondly, pinocytosis appears to be a constitutive phenomenon, and not substrate-triggered. The formation and inward migration of pinosomes is happening constantly, and solutes in the ambient fluid will inevitably find themselves engulfed and internalized. Pinocytosis has been compared to a paternoster-style elevator[6] or an escalator[7]; in contrast, a model for phagocytosis is the regular elevator, where the mechanism does not operate unless activated by a potential passenger. Thirdly, pinosomes are small (100-200 nm) and so they exclude most particles, capturing only liquid and any contained solutes. Very small particles, such as 30nm Percoll, can however enter pinosomes.[8]

The effect of inhibitors on pinocytosis is a confused area with currently no agreed concensus. Strikingly different results, following decreased temperature or the application of metabolic or cytoskeletal inhibitors, have been reported with a variety of experimental systems. It is not yet clear whether these apparent contradictions reflect real differences between cell-types or are merely methodological in origin.

Although, as already noted, all solutes present in the liquid surrounding a cell will be captured by pinocytosis, differential uptake of substrates is also a prominent feature. This is possible by "absorptive" pinocytosis, in which a solute binds to the external face of the plasma membrane and so is drawn into the forming pinosome in a concentration higher than that in the ambient liquid. This process can be highly efficient and lead to uptake rates two or three orders of magnitude greater than rates found with substrates captured only in the fluid phase. The latter category of substrates, which includes sucrose, polyvinylpyrrolidone and deglycosylated horseradish peroxidase, is useful however to measure the rate of pinocytic uptake of liquid. It is current practice to distinguish two categories of adsorptive pinocytosis:- receptor-mediated and non-specific. In the former the cell surface recognized and internalizes a ligand of narrowly defined structural characteristics, whereas in the latter substrate-specificity is much broader. It may be that this distinction is more apparent than real, and will evaporate as the physiological substrates of non-specific adsorptive pinocytosis are identified. Like enzymes, cell-surface binding-sites may exhibit a continuum of substrate specificity.

Substrate-specific adsorptive pinocytosis has been described for several classes of proteins, glycoproteins and lipoproteins and for a range of growth factors.[7] It is beyond the scope of this paper to discuss these systems in detail, but some general points should be made. One concerns the mode of pinocytosis. It was perhaps natural that those investigating some particular uptake system should suppose that the ligand in question was taken up by a pinocytic event specific for that one substrate, and distinct from fluid-phase pinocytosis. However, for cell-types that display several pinocytosis receptors (and this is the norm), that view appears rather implausible. Moreover it is hard to envisage how any near-spherical vesicle could fail to pick up liquid and so be a vehicle for fluid-phase pinocytosis. It is now increasingly recognized that each pinosome may carry into the cell both ambient fluid and a range of membrane-bound ligands. Several receptors may be present in the one pinosome,[9,10] and these may be occupied by their ligand or unoccupied.[11] Thus fluid-phase pinocytosis and receptor-mediated pinocytosis are not separate cellular events, but different facets of the same event.

Some readers may consider that the case I have just made is overstated. It may be so, and there may indeed be pinocytic events that capture one substrate only and exclude fluid-phase or other adsorptive substrates. I believe, however, that alleged instances should not be accepted without strong evidence.

Non-specific adsorptive pinocytosis[12,13] is responsible for the uptake of many non-glycosylated proteins, particularly following damage or denaturation. Cells appear to bind and pinocytose macromolecules displaying cationic or hydrophobic moieties; conversely proteins in which such groups are buried in the molecule's interior avoid capture except by fluid-phase pinocytosis.

Piggy-back pinocytosis[13] is a curious extension of adsorptive pinocytosis, and occurs when a substrate with little or no intrinsic affinity for the plasma membrane is nevertheless taken up efficiently because it binds to a simultaneously present adsorptive substrate. Examples of piggy-back pinocytosis have sometimes been mistakenly interpreted as evidence that a substrate was stimulating pinocytosis by increasing membrame internalization. In fact, pinocytosis appears to be extremely refractory to stimulation (in this sense): there are few well-authenticated examples of an evoked increase in the rate of pinosome formation.

I shall finally touch briefly on the question of coated pits, and this will lead naturally into my next section. When pinocytic invaginations form, they do so in regions of the plasma membrane whose cytoplasmic face bears a bristle-coat composed chiefly of the protein clathrin. Clathrin trimers can associate, with other proteins, in a basket-shape around the forming pinosome (thus the "coated pit"), and it is plausible that this phenomenon in some way transduces the energy needed for the membrane movements of pinosome formation and detachment. Some authors have rather uncritically supposed that coated pits were a feature of receptor-mediated pinocytosis and that fluid-phase uptake was into uncoated pinosomes. This view should be discarded along with the supposed separateness of the two phenomena (see above). It appears that in most cell-types (an exception is the capillary endothelial cell) all pinotyosis is into coated pits.[14]

THE ENDOSOME

Pinosomes, once they have detached from the plasma membrane, migrate centripetally towards a vacuolar compartment variously known as the endosome, the receptosome, or CURL (Compartment of uncoupling of receptors and ligands). This compartment, which is still far from fully characterized either morphologically or biochemically, contains many but not all of the elements found in the plasma membrane.[9] Notably clathrin is absent, and some authors hold that the coating of coated pits is left behind when the pinosome detaches from the plasma membrane. While others would contend that genuine coated vesicles (i.e. not artifacts of the plane of section) exist in the cytoplasm, there seems to be agreement that the coat has been lost by the endosome stage. The endosome is rich in pinocytosis receptors, and contains an active ATP-powered proton pump which decreases the pH inside the endosome relative to the surrounding cytoplasm. The endosome has a higher cholesterol:phospholipid ratio than the plasma membrane and fewer integral proteins. The endosome appears to have two important functions: the return of membrane to the plasma membrane, and the sorting of pinocytosed ligands between several possible destinations. The molecular basis of both these functions is at present obscure.

Pinocytosis, being constitutive, involves the internalization of a large amount of membrane,[7] and it is a logical necessity that this must either be degraded or reinserted in the plasma membrane. While some membrane may be degraded, turnover times indicate that most is not. Furthermore merely geometrical considerations suggest membrane withdrawal at an early stage following pinocytosis. Small pinosomes have a high surface-to-volume ratio and, when they fuse into larger endosomal structures, there is inevitably much surplus membrane. It was argued on this basis some years ago[15,16] that most recycling of membrane probably takes place soon after pinocytosis, and occurs by the budding-off of small vesicles from a rather flaccid vacuole. Rome[17] has recently amplified this notion by reference to the ultrastructure of the endosome. He proposes that the disproportionation of an endosomal vesicle into domains with very different membrane-to-volume ratios could lie behind the differential routing of receptors and ligands. The view that membrane recycling takes place at the level of the endosome has found recent support,[16] although direct experimental evidence for any membrane recycling following pinocytosis is sparse. The subject has been well reviewed in recent years.[7,14] It is now clear that the endosome is a sorting station both for the receptors carried into the cell by pinocytosis and for the ligands attached to them. There appear to be three major pathways leading from the endosome. One leads to the lysosomes and consequent exposure to that organelle's panoply of digestive enzymes. The other two both lead to the plasma membrane, but not to the same domain. I have already discussed the shorter of these two routes, which returns membrane in vesicular form to that aspect of the cell at which the incoming pinosomes arose. However, in some cells the endosome generates vesicles that travel through the cell and fuse with the plasma membrane on the far side. In cells that have a polarity in vivo, with opposite sides in contact with different extracellular compartments, this mechanism, known as diacytosis, can achieve a vectorial translocation of substances through an otherwise impenetrable barrier of cells. Diacytosis is a prominent phenomenon in vascular endothelium, and some more specialized examples are mentioned briefly below.

As implied by the name CURL, the endosome is often the site at which ligand detaches from receptor. Where this is the case, the lower pH in the endosome appears to be the cause. A consequence of ligand uncoupling is that receptor and ligand may take different routes out of the endosome. Nature has taken full advantage of the range of options inherent in the above scenario.[9] When insulin and its receptor are internalized by pinocytosis, both are channelled to the lysosome. Likewise for epidermal growth factor and its receptor. In contrast, the asialoglycoprotein receptor of mammalian hepatocytes releases its ligand in the endosome: the receptor returns to the plasma membrane while the asialoglycoprotein is delivered to the lysosome. A similar picture has emerged for uptake of low-density lipoprotein by fibroblasts. The transferrin receptor on the surface of cells also constantly cycles through the endosome, irrespective of whether transferrin is attached.[11] When a molecule of iron-loaded transferrin is captured, the lower pH of the endosome releases the iron from the transferrin but not the apotransferrin from the receptor. At the surface of the cell, however, the apotransferrin is released.

Receptors that retain their ligands in the endosome and then proceed into the diacytosis pathway include IgG in the neonate gut and the visceral yolk sac of many mammals, thyroglobulin, and polymeric IgA in the hepatocyte. This pathway is discussed more fully by G. Wilson in this volume. In addition to these vesicular pathways, the endosome apparently provides some macromolecules with an opportunity of direct entry into the cytoplasm. Bacterial and plant toxins enter the cell in this way, and the endosome is probably also the point of entry of DNA in the technique of transfection.

Finally, a word about nomenclature. In this section I have used the term endosome, preferring it to receptosome (a horrid hybrid of Latin and Greek) and CURL (attractive, but inaccurate for many ligands). Nevertheless endosome is etymologically a particularly empty word, meaning merely an inside-body. Unfortunately it is probably too late to replace endosome with a more meaningful word. May I also make a plea for de Duve's (recently restated[18]) usage, diacytosis, for transcellular vesicle movement? Inexplicably some authors are using diacytosis to denote the short route back to the plasma membrane and transcytosis (another hybrid formation) for translocations to the opposite pole.

THE LYSOSOME

The digestive capacity of the lysosomal enzymes has recently been reviewed.[19] The action of some fifty enzymes acting either singly, sequentially or in concert, can break down virtually any biopolymer to its monomer units. Rarely is lysosomal digestion rate-limiting: the enzymes are present in sufficient activity to process substrates arriving from the endosome without any build-up. Genetic diseases resulting in an absent or faulty lysosomal enzyme can, in contrast, lead to the accumulation of undigested material in the lysosomes.[20] Like the endosome, the lysosome has a low pH, maintained partly by the Donnan effect of impenetrant anions (probably the enzymes themselves), but chiefly by an ATP-driven proton pump.[21] The endosomal and lysosomal pumps are apparently not identical.[22]

The lysosome's membrane, like the plasma membrane, is impermeable to macromolecules. Even quite small molecules such as sucrose, mannitol and gluconate appear non-penetrating, although some other substances of similar molecular size can escape, particularly if they are uncharged and more hydrophobic. Most dipeptides appear to cross the membrane without difficulty, but amino acids cross more slowly. It has been assumed for many years that molecules that cross the lysosome membrane do so by passive diffusion.[23,24] This generalization has been seriously questioned in respect of amino acids[25] and there is now good evidence for membrane porters for cystine[26] and for lysine.[27]

The permeability properties of the lysosome membrane, combined with the pH difference between lysosomal matrix and cytoplasm, have some remarkable consequences. Cells exposed to weak bases, such as methylamine, concentrate these compounds in their lysosomes. This phenomenon is explained in terms of passage of the uncharged amine into the lysosomes from the cytoplasm, its protonation in the lysosome, and its inability to escape in the charged form.[28] If the amine has a long aliphatic chain, the protonated form has detergent properties and its accumulation in the lysosomes leads to their disruption and consequent cytolysis.[29]

Physiologically the limited permeability of the lysosome membrane serves the cell well. It prevents escape of the acid hydrolases into the cytoplasm, and it ensures that substances being digested do not escape in a partially digested state. The physiological macromolecules are probably retained in the lysosome until digested to the monomer level. Proteins will be a partial exception here, as dipeptides can also escape, and nucleotides must be further degraded to nucleosides before they can exit.

What happens to the membrane added to the lysosomes by the vesicles continuously arriving from the endosome? Some of it may be recycled to the plasma membrane, but probably much of it is degraded in the lysosome. The membrane movements needed are rather similar: in one vesicles could bud outward into the cytoplasm: in the other inward into the lysosome. It has been suggested[6] that the extreme curvature of these postulated intralysosomal vesicles might render their membrane components more susceptible to attack by the lysosomal enzymes than in their normal configuration.

CONCLUSION

Until quite recently the lysosome seemed the only realistic target for drug-macromolecule conjugates. The lysosome is still an important target, with particular potential for drug delivery, but it is no longer the only one. As surveyed above, recent research has revealed a rather sophisticated pattern of vesicular traffic-flow. An understanding of how this intracellular traffic is controlled must be the next major objective: the answers may have profound implications for targeted drug delivery.

ACKNOWLEDGEMENT

I thank Dr. R.L. Brent for his hospitality at the Stein Research Center, Jefferson Medical College, Philadelphia, where this paper was written.

REFERENCES

1. A. Hasilik and K. von Figura, Processing of lysosomal enzymes in fibro-
       blasts, in: "Lysosomes in Biology and Pathology", Vol. 7, J.T. Dingle,
       R.T. Dean and W.S. Sly, eds, Elsevier, Amsterdam, (1984).
2. K.E. Creek and W.S. Sly, The role of the phosphomannosyl receptor in
       the transport of acid hydrolases to lysosomes, in: "Lysosomes in
       Biology and Pathology", Vol. 7, J.T. Dingle, R.T. Dean and W.S. Sly,
       eds., Elsevier, Amsterdam, (1984).
3. C. De Duve, The lysosome concept, in: "Ciba Foundation Symposium on
       Lysosomes", Churchill, London, (1963).
4. S.C. Silverstein, R.M. Steinman and Z.A. Cohn, Endocytosis, Ann. Rev.
       Biochem. 46:669 (1977).
5. M.K. Pratten, R. Duncan and J.B. Lloyd, Adsorptive and passive pino-
       cytic uptake, in: "Coated Vesicles", C.D. Ockleford and A. Whyte,
       eds., Cambridge University Press, Cambridge, (1980).
6. J.B. Lloyd, Insights into mechanisms of intracellular protein turnover
       from studies on pinocytosis, in: "Ciba Foundation Symposium on
       Protein Degradation in Health and Disease", Excerpta Medica,
       Amsterdam, (1980).
7. R.M. Steinman, I.S. Mellman, W.A. Muller and Z.A. Cohn, Endocytosis
       and the recycling of plasma membrane, J. Cell Biol. 96:1 (1983).
8. M.K. Pratten and J.B. Lloyd, The effect of particle size on rate of

endocytic capture by rat peritoneal macrophages, Eur. J. Cell. Biol. Suppl 1:34 (1983).

9. C.R. Hopkins, The importance of the endosome in intracellular traffic, Nature 305:684 (1984).

10. M.C. Willingham and I. Pastan, Receptosomes, endosomes, CURL: different terms for the same organelle system, Trends Biochem. Sci. 10:190 (1985).

11. C. Watts, Rapid endocytosis of the transferrin receptor in the absence of bound transferrin, J. Cell Biol. 100:633 (1985).

12. J.B. Lloyd and K.E. Williams, Non-specific adsorptive pinocytosis, Biochem. Soc. Trans. 12:527 (1984).

13. J.B. Lloyd, M.K. Pratten, R. Duncan, T. Kooistra and S.A. Cartlidge, Substrate selection and processing in endocytosis, Biochem. Soc. Trans. 12:977 (1984).

14. I. Mellman, Membrane recycling during endocytosis, in: "Lysosomes in Biology and Pathology", Vol. 7, J.T. Dingle, R.T. Dean and W.S. Sly, eds., Elsevier, Amsterdam, (1984).

15. R. Duncan and M.K. Pratten, Membrane economics in endocytic systems, J. Theoret. Biol. 66:727 (1977).

16. J.B. Lloyd and K.E. Williams, Lysosomal digestion of endocytosed proteins: opportunities and problems for the cell, in: "Protein Turnover and Lysosome Function," H.L. Segal and D.J. Doyle, eds., Academic Press, New York, (1978).

17. L.H. Rome, Curling receptors, Trends Biochem. Sci. 10:245 (1985).

18. C. de Duve, The lysosome revisited, Eur. J. Biochem. 137:391 (1983).

19. A.J. Barrett, Proteolytic and other metabolic pathways in lysosomes, Biochem. Soc. Trans. 12:899 (1984).

20. J. Tager, L.V.M. Jonsson, J.M.F.G. Aerts, R.P.J. Oude-Elferink, A.W. Schram, A.H. Erickson and J.A. Barranger, Metabolic consequences of genetic defects in lysosomes, Biochem. Soc. Trans. 12:902 (1984).

21. J.P. Reeves, The mechanism of lysosomal acidification, in: "Lysosomes in Biology and Medicine", Vol. 7, J.T. Dingle, R.T. Dean and W.S. Sly, eds., Elsevier, Amsterdam, (1984).

22. M. Merion, P. Schlesinger, R.M. Brooks, J.M. Moehring, T.J. Moehring and W.S. Sly, Defective acidification of endosomes in Chinese hamster ovary cell mutants "cross-resistant" to toxins and viruses, Proc. Natl. Acad. Sci. USA 80:5315 (1983).

23. D.J. Reijngoud and J.M. Tager, The permeability properties of the lysosomal membrane, Biochim. Biophys. Acta 472:419 (1977).

24. J.B. Lloyd, Penetration of small molecules across the lysosome membrane: the 'classical view'. Biochem. Soc. Trans. 12:906 (1984).

25. H.N. Christensen, Organic ion transport during seven decades. The amino acids, Biochim. Biophys. Acta 779:255 (1984).

26. J.A. Schneider, A.J. Jonas, M.L. Smith and A.A. Greene, Biochem. Soc. Trans. 12:908 (1984).

27. R.L. Pisoni, J.G. Thoene and H.N. Christensen, Detection and charact-erization of carrier-mediated cationic amino acid transport in lysosomes of normal and cystinotic human fibroblasts, J. Biol. Chem. 260:4791 (1985).

28. R.T. Dean, W. Jessup and C.R. Roberts, Effects of exogenous amines on mammalian cells, with particular reference to membrane flow, Biochem. J. 217:27 (1984).

29. D.K. Miller, E. Griffiths, J. Lenard and R.A. Firestone, Cell killing by lysosomotropic detergents, J. Cell Biol. 97:1841 (1983).

PEPTIDES AS TARGETS AND CARRIERS

P.S. Ringrose and M.J. Humphrey

Pfizer Central Research
Sandwich, Kent, U.K.

INTRODUCTION

The discovery of diverse neurohormonal and other biologically active peptides has increased interest in exploiting peptide mimetics as therapeutic drugs.[1-4] The high potency, local synthesis/release and rapid inactivation of natural peptides can cause particular problems in designing synthetic drug molecules that are both selective and have reasonable duration of action. Selective drug targeting is usually achieved by specific agonist/antagonist activity for a particular cell surface receptor or by specific interaction with an enzyme active site on a peptidase, which in turn is involved in either activating or deactivating a biologically active peptide. Synthetic peptides also have utility in facilitating the selective uptake of other drug molecules into target tissues by exploiting permease specificities and in some cases the specific tissue localization of peptidases. These mechanisms will be discussed with examples of carrier peptides and peptides targeted on specific receptors. Particular comment will, however, also be made of the pharmacokinetic problems encountered with peptide drugs the route of drug administration, intestinal absorption, metabolic stability, target tissue distribution and renal vs. biliary excretion.

CARRIER PEPTIDES

The use of peptides in masking and carrier functions as a means of delivering potent neuropeptides to their target receptors and as a way of smuggling cytotoxic molecules into micro-organisms has been well exploited in nature.[5-7] Virtually all neuropeptides are produced initially as prepropeptides and require an N-terminal signal sequence as part of the secretory process.[8] The liberated propeptide is further processed by proteases to give the active neuro- or hormonal peptide. Some propeptides even act as a source for three or more distinct neuropeptides (e.g. pro-opiomelanocortin[9]). Targeting of natural neuropeptides therefore involves the use of (i) location or tissue-selective peptidases to release the active moiety where it is needed, (ii) recognition by specific cell surface receptors, which in the case of polypeptide hormones are subsequently internalized via a coated pit endocytotic process[10] and (iii) rapid destruction by other peptidases in order that the elicited response is not unnecessarily prolonged or rendered nonselective by wide systemic exposure (Fig. 1).

The use of small peptides in antibiotic molecules to achieve selective uptake by the target microorganism and in some cases intracellular release of a cytotoxic warhead moiety is also known.[2,5] The application of these targeting principles has so far been attempted in three main areas of drug

TARGETING OF NATURAL PEPTIDES

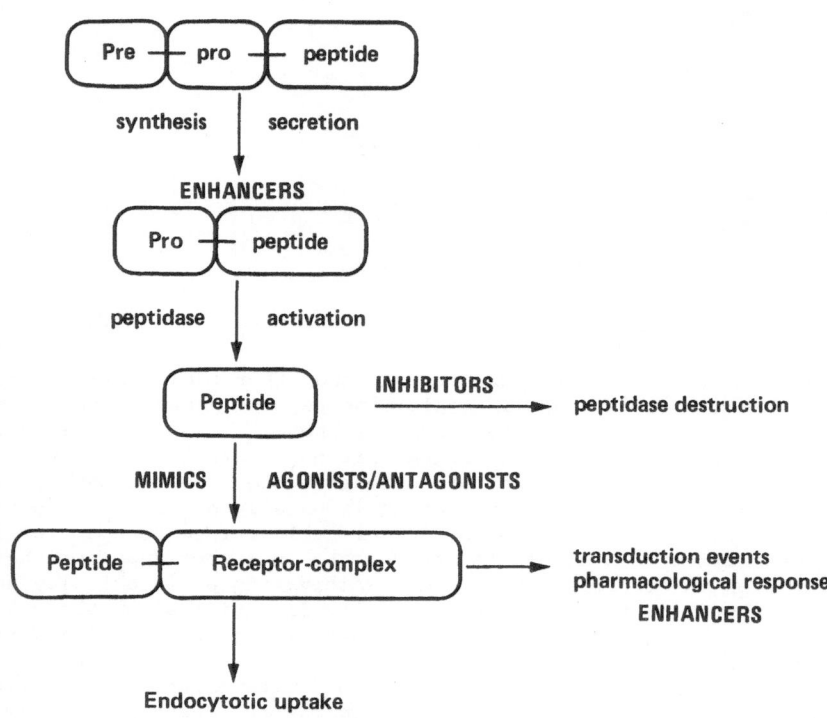

Fig. 1.   Scheme showing points of targeted drug intervention in the natural process of peptide precursor processing and receptor interaction.

design; antimicrobial warhead delivery systems, peptides capable of releasing cytotoxic agents in tumors and renally-selective dopamine peptides[1] (Fig. 2).

Antimicrobial Peptides

Of the many peptide antibiotics known to act as "warhead delivery" systems,[2,5] few have been designed to act in this way and an even smaller number have subsequently been shown to have useful activity in vivo. Alafosfalin is an example of a synthetic peptide mimetic which has clinical activity.[11-16] Mechanistic studies with this phosphonate analog of L-alanyl-L-alanine have shown that the N-terminal L-Ala moiety serves purely as a carrier for the 1-aminoethylphosphonate (AlaP) warhead moiety.[14,16] Only when disguised as a peptide is it possible for the warhead to be transported into the bacterial cell, where it is rapidly released by aminopeptidase cleavage. The underivatized AlaP warhead is essentially inactive antibacterially but when delivered via peptide transport, it accumulates intracellularly and subsequently inhibits Ala-racemase and consequently peptidoglycan biosynthesis.[14] Although alafosfalin is hydrolysed to a

TARGETING OF PEPTIDE DRUGS

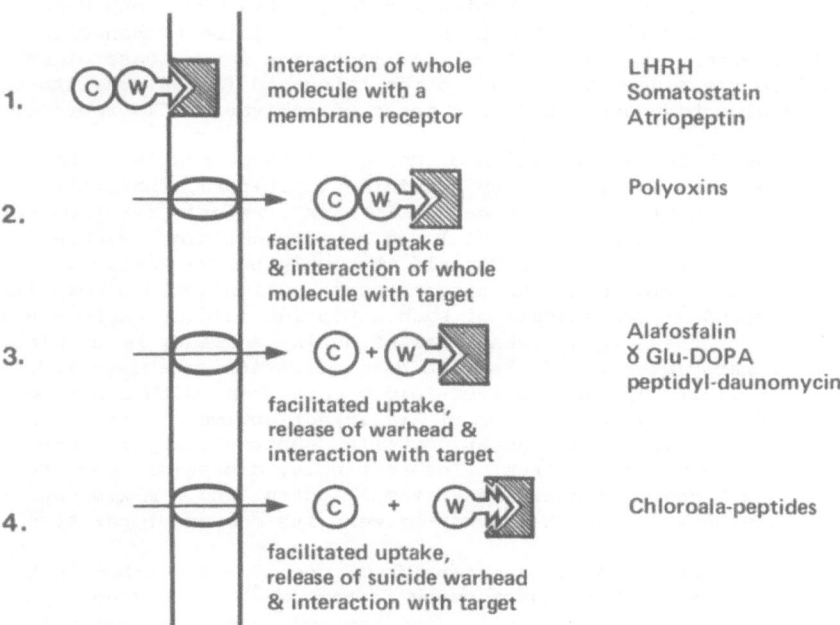

Fig. 2. Scheme showing 4 mechanisms for targeting peptide drugs (1) to surface receptors, (2) by facilitating uptake and interaction of intact molecule with intra-cellular target site, (3) by facilitating uptake and intracellular release of a warhead moiety and (4) by facilitating uptake and release of a suicide warhead which is only activated after interacting with its target. A fifth possibility is when the warhead is released extracellularly by a peptidase in the local-ity of the target cell.

limited extent in man, sufficient intact dipeptide reaches the site of infection so as to exert its potent antibacterial effect.[17] (see later).

Others have also attempted to exploit this mechanism of "illicit or portage transport"[18,19] by linking short peptides to D-aminoxyalanine,[20-22] the glutamine-antagonist phosphinothricin,[23] nalidixic acid,[24] D-cycloser-ine,[25] 5-fluorocytosine,[26-28] polyoxins[29-31] and isonicotinic acid hydra-zide.[32] Implicit in this approach is the need to conform to the overall requirements of the bacterial or fungal peptide permease(s) together with a selective susceptibility to intracellular microbial peptidases. Thus although structurally diverse peptide side chains are usually tolerated, specific overall peptide size limitations, L-stereochemistry and a free amino-terminus are normally required for effective transport and intracellular hydrolysis.[33,34] The recently reported transport of peptides containing D-aminoxyalanine is therefore of some note since normally overall L-L stereoselectivity is required.[21,22] This peptide is also cleaved by intracellular bacterial peptidases whilst remaining relatively resistant to mammalian peptide hydrolysis.[20] An alternative means of intracellular warhead release from a transported peptide has recently been described.[35] Thiol-containing warheads may be taken into bacterial cells using cysteine containing di or tripeptides. When the warhead (2-mercaptopyridine or 4-N-(2-mercaptoethyl) aminopyridine-2,6-dicarboxylic acid) is attached via a disulfide linkage to a peptide cysteine residue, activity is enhanced 100-fold. The mechanism relies on the high rate of disulfide exchange in bacterial thiol-rich cytoplasm. A related mechanism for release of warheads linked to the side chain of a peptide aminoacid (alpha-substituted glycines) has also been described as a means of delivering fluorouracil.[27,36]

Perhaps the ultimate expression of molecular targeting is to have peptide-mediated warhead delivery of a suicide inhibitor. The active species of warhead would thus have a "double mask", relying first on selective uptake into its target cell followed by intracellular release of the suicide inhibitor. The warhead would then only become "activated or primed" after interaction with its target enzyme, which would become irreversibly inactivated.[37] An example of such a peptide-suicide warhead has been described for the suicide substrates of D-Ala racemase, beta-chloroalanine and propargylglycine.[38] Peptide derivatization increases their antibacterial activity as much as 4000-fold. Consistent with SAR studies with analogs of alafosfalin, N-terminal Ala is not optimal as the carrier function and N-terminal methionine or norvaline substantially increases antibacterial potency and spectrum. Interestingly, dipeptides composed of warheads at both N and C termini are extremely potent and a mixed double-warhead molecule shows apparent synergy between its component parts.[38]

Clearly for a targeted warhead-peptide of the type described to be effective as an antimicrobial agent, many factors must be considered and the appropriate design features built into the molecule. The peptide must not only conform to the specific structural requirements of permease-mediated transport through the cytoplasmic membrane, it must ideally also penetrate the Gram-negative outer membrane. Once inside the cell, the warhead moiety must be rapidly liberated by peptidase, reduction or other means. The warhead should also not penetrate the cell membrane unaided so as to allow intracellular accumulation to take place. Its eventual target enzyme should be specific to the infecting microorganism and for in vivo activity the peptide-mimetic must be sufficiently stable to animal tissue hydrolases to survive intact before reaching the bacteria. Fortunately there are peptides (e.g. alafosfalin) that, whilst being efficiently transported across the intestinal mucosa and into bacteria cells, are poorly transported by other mammalian cells. Likewise stability to mammalian peptidases is sufficiently greater than against microbial peptidases for significant antimicrobial activity to be observed. This, however, does not seem to be the case for the peptide antibiotics bacilysin[39] or the polyoxins.[31]

Antitumor Peptides

The potential of peptides as nontoxic masking agents or prodrugs of cytotoxic antitumor drugs has been extensively researched as a means of localizing drug action to the immediate area of a tumor.[40-45] Tumor cells produce many hydrolytic enzymes, some of which are in part selective for certain tumor types e.g. gamma-glutamyl transpeptidase, plasmin, plasminogen activator protease, cathepsin B.[42,46,47] Peptide derivatization of many cytotoxic agents generally renders the drug innocuous until it is in the vicinity of the tumor or even inside the tumor cell, where the carrier is cleaved away to reveal the toxic "warhead". Peptide derivatives of daunorubicin have been most closely studied, e.g. Leu-Leu-daunorubicin.[41,44] In addition, gamma-glutamyl-aminophenol, D-Val-Leu-Lys-glutamine antagonists, peptidyl-phosphonoacetyl aspartic acid, peptidyl-avicin and peptidyl-phenylenediamine mustard have also been shown to exert some selectivity although an improved therapeutic index is often difficult to show conclusively.[40,42,45] Peptide derivatization not only masks nonselective toxicity but also in certain cases increases lipophilicity and lowers pKa such that uptake into the tumor is facilitated.[44] Attempts to exploit "piggyback endocytosis" by linking peptides (e.g. carbobenzoxyglycyl-Phe) to lysosomotropic amines have also been tried with some success, as a means of further increasing the selective action of lysosomotropic agents against lysosome-rich tumor cells.[43]

Renally Selective Dopamine-Peptides

In order to target the vasodilatory and natriuretic action of dopamine to the kidney and thus avoid the central and emetic actions of dopamine,[48,49] various peptidyl and aminoacyl derivatives of dopamine and of L-DOPA have been examined.[50,51] Because gamma-glutamyl transpeptidase is primarily restricted to the proximal convoluted renal tubule (also present in certain tumors),[47] gamma-glutamyl derivatives of DOPA and dopamine were synthesized. Gamma-glutamyl-DOPA was particularly effective at selectively delivering dopamine to the kidney due to the high levels of transpeptidase and DOPA-decarboxylase in this tissue.[50] The concentration of dopamine in the kidney was five times higher after glutamyl-DOPA than after an equivalent dose of DOPA. Renal blood flow was increased without systemic effects. However the poor oral absorption of this compound has prevented its full clinical evaluation. Gamma-glutamyl derivatives of other drugs (e.g. sulfamethoxazole) have also been tried as kidney selective prodrugs with some success.[52]

As can be seen from the above examples, carrier peptides rely largely on their ability to release an active drug moiety or its precursor at or inside the target cell or tissue due to the selective action of peptidases. The targeting of peptide carriers because of either specific surface receptors or selective transport mechanisms is however an ideal which is seldom realised. The following section on the other hand deals with intrinsically active peptides which target their action because of a specific "homing in" on either cell surface receptors or on enzyme active sites.

PEPTIDES TARGETED TO SPECIFIC RECEPTORS AND ENZYME ACTIVE SITES

Two types of intrinsic molecular targeting will be described for peptide mimetics; first those that seek out specific cell surface receptors such as analogs of opiate peptides, thyrotropin releasing hormone (TRH), luteinising hormone releasing hormone (LH-RH), somatostatin and atriopeptin; second those that owe their high degree of selectivity and hence low toxicity to specific high affinity binding to the active site of a particular target enzyme, e.g. inhibitors of angiotensin converting enzyme and bacterial DD-carboxypeptidases.

Analogs of Neurohormonal Peptides

Despite the apparent molecular complexity of most small molecular weight neuro- and hormonal peptides, the region of amino acids involved in interaction with the receptor is usually no more than a tetrapeptide[3,53-55] which can be at the C terminus as in the case of angiotensin, atriopeptin, CCK and substance P, at the N terminus as in the case of enkephalins, endorphins, and VIP, and in the middle of a loop as with somastatin. The rest of the peptide molecule is necessary either for maintaining a particular conformationally restricted configuration of the agonist sequence or sometimes for reasons of drug delivery or metabolic protection. There are now at least 30 neuropeptides identified, each exerting specific actions at one or more central and/or peripheral receptors (Fig. 3). Although some "neuropeptides" were first isolated in the brain (enkephalins, neurotensin, substance P) and others initially recognized as either intestinal hormones (cholecystokinin CCK, vasoactive intestinal polypeptide VIP) or blood-derived peptides (angiotensin, bradykinin, atriopeptin), it is now believed that many of these neuropeptides act both in the CNS and at peripheral target sites.[53,56] The neuropeptides can act either directly as neurotransmitters on peptidergic neurones[57] alone or more likely in concert with other transmitters or even as modulators of nonpeptide aminergic transmitters e.g. enkephalins and noradrenalin, CCK and dopamine, substance P and serotonin, VIP and acetylcholine.[53] Other biologically more stable peptides act systemically as hormones exerting subtle interactive regulatory roles. Some neurohormonal peptides are able to act at all these levels and at a multiplicity of potential receptor sites. Clearly in terms of drug design it is important to exploit by molecular modification a particular action (central or peripheral) at a specific receptor, whilst also incorporating the appropriate physiochemical properties consistent with satisfactory duration of action and route of administration.

Although much work has been published on the exploitation of peptide drug molecules based on the opiate group of neuropeptides (enkephalins, endorphins, dynorphins, etc) and their selectivity for delta, mu, kappa and sigma receptor sites re analgesia, mood modulation and GI-antisecretory/antimotility activities,[58-63] synthetic analogs of the hypothalamic releasing hormones (TRH, somatostatin and LH-RH) have been pursued more extensively in clinical studies as examples of peptide drugs targeted to specific receptors.[64]

Thyrotropin releasing hormone (TRH) was the first of the three hypothalamic neuropeptides to be identified and is also the smallest, being a tripeptide (pyroGlu-His-Pro-NH2).[65] Its structure has increased stability towards peptidases in the gut relative to most natural peptides having a protected cyclized N-terminus and a C-terminal proline amide, although in brain, TRH is rapidly metabolised. A major metabolite is His-Pro-diketopiperazine which also may have a role as a regulatory peptide.[66] Owing to its small size and stability in the gut, it is one of the few neuropeptides to be intrinsically orally absorbed and analogs, e.g. RX 77368, have attempted to increase CNS selectivity by improving biological stability[67] (see later). TRH acts on central receptors to provoke release of thyroid stimulating hormone (TSH) and prolactin and is believed to act in concert with serotonin and substance P.[68] Interest in TRH analogs has extended to their ability to act as mood elevators in man and growth promotors in animals.

Luteinizing hormone releasing hormone (LH-RH) is a decapeptide which acts on receptors of the gonadotrophic cells in the adenohypophysis modulating the release of luteinizing hormone (LH) and follicle stimulating hormone (FSH).[69] Use of low dose agonist analogs has attracted much interest as contraceptive agents since they have the advantage of being devoid of steroidal side effects, have no adverse metabolic action and are rapidly metabolised after carrying out their function.[70] The structure of LH-RH decapeptide is believed to involve a beta turn configuration and analogs incorporating gamma-lactams, D-amino acids and N-alkyl amino acids have been synthesized in order to conformationally restrain the structure which is believed to occur at the receptor site.[71] Alternative means of delivery of LH-RH analogs have been investigated for treatment of prostatic tumors using injectable microcapsules and biodegradable polymers as peptide de-

NEUROHORMONAL PEPTIDES

Peptide	Amino acid Number	Therapy area for Agonist/Antagonist	Cotransmitter
TRH	3	Growth enhancer Antidepressant	HT
ENK Enkephalin	5	Analgesic, mood-modulation, obesity	NA, A, D
CCK Cholecystokinin	8	Anoretic, GI-antispasmodic	D
AII Angiotensin II	8	Antihypertensive Heart failure	
BK Bradykinin	9	Antihypertensive Inflammation	
AVP Arg-Vasopressin	9	Memory-enhancer	
OTC Oxytocin	9	(Memory-enhancer)	
LH-RH Luteinizing hormone-releasing hormone	10	Contraceptive Antitumor	
SP Substance P	11	Analgesic	HT
NT Neurotensin	13	Mood modulation	NA
SS Somatostatin	14	Antitumor, GI-agent Anti-diabetic	NA
AP Atriopeptin III	24	Hypertension, Heart failure	
VIP Vasoactive intestinal Polypeptide	28	Pulmonary disease, diarrhoea, stroke	ACh
END Beta endorphin	31	Mood modulation Analgesia	
ACTH Corticotropin	39	Mood modulation	

[A=Adrenalin; NA=Noradrenalin; D=Dopamine; HT=Serotonin; ACh=Acetyl-choline]

Fig. 3. Neurotransmitter and hormonal peptides and the potential therapeutic use of their agonists/ antagonists.

ivery of LH-RH analogs have been investigated for treatment of prostatic tumors using injectable microcapsules and biodegradable polymers as peptide delivery systems.[72,73] LH-RH analogs (e.g. ICI 118630, zoladex) lower testosterone levels more effectively than stilbesterol but intermittant bolus administration on a chronic basis may raise rather than lower testosterone levels. The aim of these slow release depot formulations is to give low levels of peptide which in early clinical studies would seem to be effective.[72] Indeed early studies showed that use of long acting analogs by prolonged infusion were ineffective as fertility agents and short duration episodic bursts more closely mimics natural LH-RH.[69] The ability of LH-RH agonists to enhance or decrease LH, FSH and testosterone levels therefore seems to depend less on the intrinsic nature of receptor interaction than on pulsed bursts of receptor stimulation (enhances fertility) versus low level continuous stimulation (antifertility and antitumor).

Somatostatin is the third hypothalamic releasing hormone neuropeptide. It is a looped tetradecapeptide and occurs at high levels in the cerebral cortex along with the "intestinal" peptides CCK and VIP.[74] Somatostatin modulates growth hormone, somatomedin, glucagon and insulin release.[69] It even modulates angiotensin II stimulated release of aldosterone.[75] The original use of analogs (SMS 201-995) to treat acromegaly[76] has now progressed to use in juvenile diabetes, gastro-intestinal endocrine tumors, analgesia and upper GI-bleeding.[77-81] Synthetic work has concentrated on producing truncated analogs of long duration (see later). The active moiety of this disulfide-bridged tetradecapeptide is a tetrapeptide (Phe-Trp-Lys-Thr) that is located in the middle of a loop and is forced by the rest of the molecule into a type II' beta turn.[82,83] Downsized analogs have concentrated on retaining the conformationally restricted bioactive tetrapeptide whilst replacing the rest of the molecule with either a Cys-(7-amino-heptanoyl)-Cys tripeptide bridge or with a Phe-Pro dipeptide to give an active cyclic hexapeptide.[83] Most clinical work has been reported with the octa-peptide analog SMS 201-995 which contains D-Phe and D-Trp residues.

Atrial natriuretic factor (ANF) and the atriopeptins[84] are a very recent area of peptide research. These hormonal peptides are already attracting much attention in terms of seeking ways of potentiating their action. ANF is unlike other hormonal peptides in that it is secreted by the atrial cells of the heart. This occurs in response to physical activation of baroreceptors which trigger release of various atriopeptins,[85,86] the smallest of which is a 24 amino acid looped peptide atriopeptin III which can be degraded at the C terminus to the smaller active atriopeptin II and largely inactive atriopeptin I.[84] All the atriopeptins appear to be derived from a common polypeptide precursor molecule[87] and there is a remarkable conservation of amino acid primary structure, with rat differing from human ANF by isoleucine for methionine in the loop region. In this way the heart targets messages to specific receptors in the kidneys and vasculature by means of atriopeptins III and II, which are potent natriuretic and vaso-dilatory agents.[88,89] Interaction with the receptor(s) triggers guanyl cyclase and consequently a pharmacological response.[90] Atriopeptins III and II are also very potent inhibitors of induced aldosterone secretion.[84] The loss of the C-terminal dipeptide (Phe-Arg) from atriopeptin II to I, results in loss of most activity and indicates a very specific interaction between the target receptor(s) and this region of the molecule. The role of the 17 amino acid loop is uncertain but does not appear to significantly influence bioactivity.[87] Synthetic mimetics are already under clinical evaluation but the potential of this area to fundamental treatment of hypertension and heart failure has yet to be determined.[91]

Peptides Targeted at Peptidase Active Sites

An alternative means of "targeting" peptides for a specific action involves the use of specific inhibitors of peptidases.[92,93] Examples of this mechanism are the angiotensin converting enzyme inhibitors which selectively inhibit the Zn-dependent carboxydipeptidase involved in converting the essentially inactive decapeptide angiotensin I to the potent vasoconstrictor octapeptide angiotensin II by cleavage of the C-terminal His-Leu dipeptide.[92] By analogy with carboxypeptidase A and the ACE dipeptidyl carboxypeptidase, thiol containing mimics of alanyl-proline were designed to interact with the Zn ion at the enzyme active site.[94] The result of this work led to the development of captopril which is now being used successfully in the treatment of hypertension and congestive heart failure.[95] Concern about thiol-related side effects led to the design of N-carboxyphenylpropyl transition state analogs and the discovery of enalapril.[96] Design of extremely selective and potent (Ki's of $10^{-10} - 10^{-11}$M) inhibitors has led to complete inhibition of AI to AII conversion with a consequent antihypertensive effect without any mechanism-related side effects of significance.[97,98] The extent to which inhibition of ACE peptidase inactivation of the vaso-dilator nonapeptide bradykinin contributes to the antihypertensive effect is uncertain however.

An alternative way of interfering with the renin-angiotensin axis, is to inhibit the aspartyl protease renin, which converts the precursor protein angiotensinogen to angiotensin I.[99,100] Renin is a highly selective enzyme and appears to have only one substrate. Targeted inhibitors of this enzyme have been designed by (a) replacing the Leu-Val sequence, that is cleaved by human renin, by "non-cleavable" peptide bond analogs in decapeptide mimetics[101] and (b) inserting the potent transition state inhibitor statine into hepta-peptide analogs of the renin substrate.[102,103] The problem of retaining potency, selectivity and duration of action in these large pro-tease inhibitors, whilst attempting to decrease overall size for increased oral absorption and decreased biliary excretion (see later) remains un-resolved.

Although ACE and renin inhibitors are highly selective and effective in terms of interaction with their respective enzyme active sites, other "designed" inhibitors have been less successful due to the involvement of a multiplicity of different enzymes acting on the substrate of interest. This has been the case with attempts to inhibit "enkephalinase".[104] Enke-phalins are now known also to be degraded by aminopeptidase, endopeptidase and ACE as well as the carboxypeptidase, enkephalinase.[58]

Perhaps the best example of selective targeting of peptidase inhibit-ors are the beta-lactam group of antibiotics.[2,105] The beta-lactams spec-ifically inhibit a group of D-alanyl-D-alanine carboxy- and transpeptidases, ubiquitous in bacteria but absent in all higher life forms. This group of antibiotics can therefore be regarded as being targeted against invading bacteria, since although being present in many tissue sites after admin-istration to man, they are only effective as cytotoxic drug molecules against bacteria at the site of infection. This is even more surprising when one realises that beta-lactam antibiotics are irreversible enzyme inhibitors, i.e. site directed acylating agents, being 'activated' after conformational changes in the molecule as result of initial noncovalent interaction with the target enzyme.[105-107] Earlier proposals that the beta-lactam nucleus may resemble a transition state analog of the natural sub-strate are now thought unlikely.[108,109]

Another intriguing highly site-targeted peptide molecule is vancomycin which acts as a minireceptor for the acyl D-alanyl-D-alanine ligand and thus blocks DD-carboxypeptidase action and incorporation of precursor muramyl pentapeptide into bacterial cell wall peptidoglycan polymer.[110] This tricyclic heptapeptide is unfortunately too large for intestinal ab-sorption and penetration of gram-negative outer membrane pores. It is surprising that the specificity of targeting of both these types of mole-cules has not yet been exploited by directing related peptide mimetics against targets in other therapy areas, e.g. alternative molecular express-ions of vancomycin could be seen as mini "monoclonal antibodies" directed at specific sequences of relevant neurohormonal peptides and thus blocking their receptor interaction.

BIOAVAILABILITY OF PEPTIDE DRUGS

Despite the great potential for the discovery of new peptide mimetics with novel biology, there are potential problems associated with the use of

these molecules as drugs. Naturally-occurring peptides are rapidly degraded by tissue peptidases and, therefore, their biological half-lives
are normally too short to be relevant to therapeutic use. In addition,
their bioavailability following oral administration is low due to poor
absorption and instability in the gastro-intestinal tract. Nevertheless
there are therapeutically useful peptide drugs, which have successfully
overcome the problems of bioavailability. Some of the relevant design
features are outlined below.

Absorption

An ingested peptide enters a hostile environment where it may be
destroyed by chemical and enzymatic action. Most polypeptides are hydrolysed in the gut lumen by brush border enzymes. Some approaches used to
design more metabolically stable peptides are discussed later. Intestinal
mucosal uptake of peptides is mostly limited to di- and tripeptides[111] and
involves facilitated diffusion by an active transport system, distinct from
that used by amino acids.[112,113] Preference is shown for peptides with
bulky side chains, component L-stereoisomer amino acid residues, and unsubstituted N and C termini.

Thyrotropin releasing hormone (TRH) and its tripeptide analogs[67] are
absorbed to the extent of a few percent of dose by the oral route. However,
absorption of TRH is limited to the upper regions of the intestine and is
saturable.[114] Studies in vitro have demonstrated that the process is Na^+-
dependent and is inhibited by the dipeptide glycylglycine. In contrast,
the close structural TRH analog, DN-1417, is absorbed in all parts of the
intestine and is not saturable or inhibited by glycylglycine.[115] Thus DN-
1417 seems to be absorbed by a simple diffusion process (1-10% of dose) and
illustrates how slight modification in chemical structure may change the
absorption characteristics of peptides.

Oral absorption of alafosfalin in the upper intestine is 100% but
bioavailability of intact drug is only 20-40%.[17] Interestingly, bioavailability in man is considerably greater than that in animals. Phosphono-
dipeptides are actively transported across the intestinal membrane and are
in part coupled with metabolic hydrolysis to their component amino acids.
Peptide-mediated tubular reabsorption also appears to occur in the kidney
thus contributing to prolonged plasma levels of drug.

Amino-penicillins are also dipeptide mimetics which are effectively
absorbed (up to 80%) from the small intestine, although they are all poorly
lipophilic and the zwitterion is the dominant ionic species in the small
intestine.[116-118] Investigations of the mechanism of amino-beta-lactam
antibiotic absorption suggest that absorption is saturable and involves
carrier-mediated transport.[119-120] Most of the amino-beta-lactam antibiotics, ampicillin excepted, inhibit intestinal absorption of each other,
and thus, there appears to be a common carrier system for these drugs.

Carrier-mediated peptide drug transport systems in the intestine is
still, however, poorly characterized and differences between species, effects of age, disease and nutrition need to be considered. However with a
knowledge of carrier-mediated specificities there is potential to produce
by chemical or pharmaceutical design, better oral forms of peptide-related
drugs.

Absorption of peptides other than by carrier-mediated transport may
occur by (a) diffusion through aqueous pores or inter-cellular spaces,
(b) absorption through membrane lipids, or (c) uptake into epithelial cells
by a process of endo- or pinocytosis. Some penicillins, for example, are
sufficiently lipophilic to be absorbed passively through the lipid membrane.

However, ampicillin pro-drugs derivatized on the carboxy function to increase lipophilicity, significantly increase the extent of absorption.[118]
Amoxycillin, on the other hand, is an hydroxyl analog of ampicillin in which
oral absorption is increased by a factor of at least two fold.[119] For the
carboxy-beta-lactam antibiotics like carbenicillin, instability in gastric
contents and low absorption by passive diffusion is overcome by esterific-
ation of the alpha-carboxylic acid function,[121] the phenol ester showing
some 30 times the bioavailability of the parent drug. Mutual pro-drugs have
also been found to be effective, as in the case of sultamicillin, which
facilitates the absorption of both its component parts of the beta-lactamase
inhibitor, sulbactam and the antibiotic, ampicillin.[122]

Some peptide inhibitors of angiotensin converting enzyme have also encountered absorption problems. Enalapril, the first of the more potent non-
thiol compounds, is a mono-ester pro-drug of the parent diacid (MK-422)
which is the active enzyme inhibitor. Pharmacological studies showed that
much higher doses of MK-422 were required orally than intravenously to
obtain an inhibition of the pressor response to exogenously administered
angiotensin I, suggesting poor absorption.[123] Oral bioavailability of the
diacid in dogs was 11% compared to 64% for the prodrug. A similar prodrug
approach has been used with ramipril (Hoe-498).[125]

For larger peptides absorption after oral administration may be very
low, for example, the pentapeptide, pepstatin A and its glycine analog are
only absorbed to the extent of about 1%.[125] Similarly for downsized somatostatin analogs, which are cyclic hexa- and octa-peptides, low oral absorption has prompted research into alternative routes of administration.[126,127]
However, there are exceptions as exemplified by the immunosuppressive drug
cyclosporin, a cyclic undecapeptide, which has bioavailability in the range
20% to 50%.[128] The mechanism of absorption of this large hydrophobic peptide is not clearly understood, but may involve a carrier-mediated process.[129] Absorption via the lymphatics, however, has been found to be only
about 2% of total absorbed drug.

Overall, the absorbtion of intact peptides can vary greatly. Clearly,
if the peptide is vulnerable to peptidase attack, oral bioavailability will
be severely compromised. Thus design of metabolically stable peptides is
the first step in achieving effective oral activity.

Alternative Routes of Administration

Besides the intestinal tract, there are other mucous membranes which
can be used as potential sites of drug delivery. The various membrane
sites have different permeabilities to water and peptide molecules because
of the different junctions and pores between the cells. The sites which
have shown the most potential for peptide absorption are the vagina, lung
and nasal mucosa. The absorption of a potent LH-RH analog has been shown
to be about 20% by the vaginal route.[130] Variations in absorption may,
however, occur during the estrous cycle. Intratracheal administration of
an aerosol formulation of insulin has been shown to give about 40% bioavailability.[131] The extensive absorption of a number of peptide drugs,
including vasopressin, LH-RH, sulbenicillin and cephalosporin have been
demonstrated following intranasal administration.[132-134] The absorption of
insulin was about 30%.[135] Although such routes for peptides do not give
complete bioavailability as for parenteral routes of administration, they
do offer greater convenience and potentially much greater absorption than
for the oral route.

Distribution

The kinetic principles which apply to drug distribution to cells and organs are also generally applicable to peptides. Exceptions are due to active transport into or out of tissues, which may be saturable, or uptake into cells by an absorptive endocytosis mechanism.[10] The ability of peptides to cross cell membranes is suggested by the rapidity with which some peptides undergo metabolism and elimination.[136,137] However, for peptides crossing into cells by passive diffusion, the physicochemical properties relevant to gastro-intestinal absorption will also influence rate of tissue membrane diffusion. Thus for the pro-drug, enalapril, the hepatic extraction is 16 times that of the parent diacid.[138] Of course, there are specific mechanisms for the transport of peptides into invading microbial cells and these have been used in the design of a variety of chemotherapeutic agents.[2]

For polar peptides slow rates of diffusion across membranes can be expected, particularly for the blood-brain barrier, and indeed tissue distribution studies using whole-body autoradiography for the tripeptide TRH analog RX77368,[139] the hexapeptide pepstantinyl glycine analog[125] and the octapeptide somatostatin analog[126] confirm low penetration into the CNS. Thus, the stimulant properties of TRH analogs are 100,000-fold greater following intracerebral ventricular administration than for the intravenous route indicating limited penetration of the blood-brain barrier.[139] One approach to increasing the availability of peptides to the CNS is to increase the rate of passive diffusion by using a more lipophilic pro-drug as the carrier molecule. Thus, for example the t-butyl ester of glycine is a pro-drug of glycine capable of enhancing central glycinergic activity.[140] A similar approach has been applied in the design of GABA pro-drugs.[141]

Tissue specific delivery of peptides, as described earlier, is possible by identification of selective membrane transport systems, specific localization of tissue binding sites, or local metabolising activity. An example of the application of the latter being the exploitation of the kidney, which contains high concentrations of gamma-glutamyl transpeptidase.[47,50,52]

Metabolism and Excretion

A major design goal in the synthesis of biologically active peptides is to enhance metabolic stability, oral bioavailability and duration of action. Approaches which have been used to stabilise peptides to hydrolytic attack include: (a) steric constraints by use of unnatural, D-isomer or N-methyl aminoacids. Thus the incorporation of D-Ala, MeTyr, MePhe in Met-enkephalin, together with sulfoxidation of the Met residue increases activity to a level equivalent to morphine;[142] (b) covalent constraints, i.e. the use of cyclic amino acids, bridged dipeptides or cyclic peptides;[143] (c) retro-enantiomer analogs, i.e. the direction of the peptide backbone is reversed and the chirality of each amino acid is inverted.[144] This approach is limited if the peptide links are important for receptor interaction; (d) modification of the N-terminus by alkylation or acylation of the carboxy terminus by reduction or amide formation. The cyclised N-terminal pyroglutamic acid residue and the C-terminal amide of TRH hinder exopeptidases,[114] while the dimethyl groups on the proline in the analog RX 77368 prevent initial amidase attack.[137]

Chemical modification of peptides to achieve stability against proteolysis is, however, likely to reduce absorption by carrier-mediated mechanisms but overall tends to increase oral activity. Examples of successfully designed orally-active derivatives of natural peptides include analogs of LH-RH,[130,145] Met-enkephalin[142] and ACTH.[146]

Peptide drugs which are stable to peptidase activity may still undergo metabolism by other detoxication systems in the body. For example, the cyclic undecapeptide cyclosporin is metabolised to at least 9 metabolites and suggest the involvement of the cytochrome-P450 monooxygenase system.[147]

Metabolism of carbapenems by a renal dehydropeptidase can be blocked by coadministration of the designed inhibitor cilastatin.[148]

Achievement of metabolic stability with a peptide structure does not however, guarantee a long biological half-life. Rapid elimination from the body can occur by renal clearance (active and/or passive), as for the penicillins, and by hepatobiliary secretion. Biliary excretion of small peptides has not, however, been thoroughly investigated. The early ACE inhibitors are extensively excreted in the urine, although it would seem that the more recent lipophilic ACE inhibitors with ring-expanded Pro residues, for example remipril, show increasing preference for the biliary route.[124] An example of a larger peptide being excreted unchanged in both urine and bile is pepstatinyl-glycine, a hexapeptide.[125] Thus for acidic peptides, biliary excretion occurs probably by the same anionic transport process described for many organic acids, this role of elimination favoring high molecular weight compounds.[149]

TRH analogs, are tripeptides with a basic histidine moiety and all seem to be cleared predominantly unchanged in the urine. However, there are octapeptide inhibitors of renin containing basic residues which lack oral activity due to 'first pass' clearance by hepatic biliary excretion.[102] Similarly for octa- and hexapeptide somatostatin analogs, rapid clearance of the intact peptides by biliary excretion (60% dose in one hour) is reported to be a problem in the design of effective therapeutic agents.[126,127] Interestingly, the parent peptide is not excreted in bile perhaps because of rapid degradation.[136] The fact that these basic compounds are secreted in bile against a high concentration gradient points to a specific carrier-mediated active transport process, such as that known to exist for organic cations. It would appear that the properties of the cyclic peptides described, which facilitate active excretion in bile, have high lipid solubility combined with some degree of basicity and residual polarity.

CONCLUSIONS

The therapeutic application of small peptide mimetics generally results from their ability to act as either carriers, neuropeptide agonists/antagonists or peptidase inhibitors. The use of carrier peptides for targeting drugs to specific tissues or cells, is more likely to be achieved by exploiting local specificity differences of peptidases rather than selective transport systems or surface receptors. Realization of "targeted" peptide carrier prodrugs in clinical terms remains somewhat elusive and would seem to be largely restricted to delivery of specific cytotoxic agents to invading microorganisms or tumor cells. Peptide portage of suicide substrates may be the ultimate molecular expression of a targeted drug, but this has yet to be realized in a therapeutically useful molecule.

Targeting of peptide drugs can also occur by virtue of high specificity of a neuropeptide super-agonist or antagonist for a particular membrane receptor or a peptidase inhibitor for the active site of a very selective peptidase. Examples of the former are being actively pursued in terms of synthetic mimetics of CCK, enkephalins, LH-RH, substance P, somatostatin, VIP, etc.[150] Experimental clinical studies are still at an early stage but on a wide therapeutic front. Initial use is, however, likely to be in the antitumor area probably using slow-release depot formulations. Peptide mimetics targeted against specific peptidases are the most successful examples of drugs in this area, as exemplified by ACE inhibitors and beta-lactam antibiotics. (Fig. 4)

Problems which have been encountered in the use of peptides as drugs are low oral absorption, poor metabolic stability and rapid clearance by

Tissue Target	Ideal Kinetic Property	Drug and/or Formulation
Tongue — sweetness receptors	Local action	Aspartame
Intestine	Local action	Delta-optiates CCK-analogs
Immune system (T-cells)	Oral/Systemic activity	Cyclosporin/emulsion
Renin-Angiotensin system	Oral/Systemic activity	ACE pro-drugs
Prostatic cancer	Controlled release from i.m. injection	LHRH polymer
CNS	Sustained blood levels and passage across BB. Barrier	TRH analogues
Hypothalmus	Sustained systemic activity	s.c. administration of somatostatin analogs
Kidney	Tissue selective delivery & peptidase release	γ-glutamyl DOPA
Systemic infection	Oral activity Normal kinetics in hepatic/renal failure	β-lactam pro-drugs Cefoperazone/ moxalactam
Urinary tract	Oral activity and renal clearance	carbenicillin pro-drugs alafosfalin

Fig. 4. Summary of different delivery forms of peptide drug molecules and their corresponding target tissues.

the renal and hepatobiliary routes. For some larger super-agonists or antagonists of neuropeptides, which suffer from less than ideal pharmaco-kinetic properties, alternative routes of absorption and slow-release matrix-delivery formulations can offer a viable approach. The future of this area should, however, see better rational design of nonpeptide hetero-cyclic mimetics of natural peptides based on a clearer understanding of receptor interaction and molecular conformation.

REFERENCES

1. P.S. Ringrose, Small peptides as carriers and targets in human therapy, Biochem. Soc. Trans. 11:804 (1983).
2. P.S. Ringrose, Warhead-delivery and suicide substrates as concepts in antimicrobial drug design, Soc. Gen. Microbiol. Sumposium 35, F. O'Grady and D. Greenwood, eds., Cambridge University Press, Cambridge, (1985).
3. V.J. Hruby, Design of peptide superagonists and antagonists, conform-ational and dynamic considerations, Amer. Chem. Soc. 251:9 (1984).
4. V.J. Hruby, J.L. Krstenansky and W.L. Cody, Recent progress in the rational design of peptide hormones and neurotransmitters, Ann. Rep. Med. Chem. 19:303 (1984).
5. P.S. Ringrose, Peptides as antimicrobial agents, in: "Microorganisms

and Nitrogen Sources". J.W. Payne, ed., John Wiley & Sons, Chichester (1980).

6. I. Chopra and P. Ball, Transport of antibiotics into bacteria, Adv. Microb. Physiol. 23:183 (1982).

7. P.S. Ringrose, Expolitation of the bacterial envelope: rational design of antibacterial agents, Med. Microbiol. 3:179 (1983).

8. D. Steiner, D. Quinn, S. Chan, J. Marsh and H. Tager, Processing mechanisms in the biosynthesis of proteins, Ann. NY. Acad. Sci. 343:1 (1980)

9. F.E. Bloom, The endorphins: A growing family of pharmacologically pertinant peptides, Ann.Rev. Pharmacol. Toxicol. 23:151 (1983).

10. R.B. Dickson, Endocytosis of polypeptides and their receptors, Trends in Pharm. Sci. 6:164 (1985).

11. J.G. Allen, F.R. Atherton, M.J. Hall, C.H. Hassall, S.W. Holmes, R.W. Lambert, L.J. Nisbet and P.S. Ringrose, Phosphonopeptides, a new class of synthetic antibacterial agents, Nature 272:56 (1978).

12. J.G. Allen, F.R. Atherton, M.J. Hall, C.H. Hassall, S.W. Holmes, R.W. Lambert, L.J. Nisbet and P.S. Ringrose, Phosphonopeptides as antibacterial agents: alaphosphin and related phosphonopeptides, Antimicrob. Agents Chem. 15:684 (1979).

13. F.R. Atherton, M.J. Hall, C.H. Hassall, R.W. Lambert and P.S. Ringrose, Phosphonopeptides as antibacterial agents: rationale, chemistry, and structure-activity relationships. Antimicrob. Agents Chem. 15:677 (1979).

14. F.R. Atherton, M.J. Hall, C.H. Hassall, R.W. Lambert, W.J. Lloyd and P.S. Ringrose, Phosphonopeptides as antibacterial agents: mechanism of action of alaphosphin, Antimicrob. Agents Chem. 15:696 (1979).

15. F.R. Atherton, M.J. Hall, C.H. Hassall, S.W. Holmes, R.W. Lambert, W.J. Lloyd and P.S. Ringrose, Phosphonopeptide antibacterial agents related to alafosfalin: design, synthesis and structure-activity relationships, Antimicrob. Agents Chem. 18:897 (1980).

16. F.R. Atherton, M.J. Hall, C.H. Hassall, R.W. Lambert, W.J. Lloyd, A.V. Lord, P.S. Ringrose and D. Westmacott, Phosphonopeptides as substrates for peptide transport systems and peptidases of E.coli, Antimicrob. Agents Chem. 24:522 (1983).

17. J.G. Allen, L. Havas, E. Leicht, I. Lenox-Smith and L.J. Nisbet, Phosphonopeptides as antibacterial agents: metabolism and pharmacokinetics of alafosfalin in animals and humans, Antimicrob. Agents Chem. 16:306 (1979).

18. B.N. Ames, F.L. Ames, J.D. Young, D. Tsuchiya and J. Lecocq, Illicit transport: the oligopeptide permease, Proc. Natl. Acad. Sci. USA 70:456 (1973).

19. C. Gilvarg, Portage transport, in: "The Future of Antibiotherapy and Antibiotic Research", Ninet, L., Bost, P.E., Bouanchaud, D.H. and Florent, J., eds., Academic Press, London (1981).

20. J.W. Payne, J.S. Morley, P. Armitage and G.M. Payne, Transport and hydrolysis of antibacterial peptide analogues in E.coli: Backbone modified aminoxy peptides, J. Gen. Microbiol. 130:2253 (1984).

21. J.S. Morley, J.W. Payne and T.D. Hennessey. Antibacterial activity and uptake into E.coli of backbone-modified analogues of small peptides, J. Gen. Microbiol. 129:3701 (1983).

22. J.S. Morley, T.D. Hennessey and J.W. Payne, Backbone-modified analogues of small peptides: transport and antibacterial activity. Biochem. Soc. Trans. 11:798 (1983).

23. H. Diddens, M. Dorgerloh and H. Zahner, Metabolic products of microorganisms. On the transport of small peptide antibiotics in bacteria, J. Antibiotics 32:87 (1979).

24. T. Kametani, K. Kigasawa, M. Hiiragi, K. Wakisaka, S. Haga, H. Sugi, K. Tanigawa, Y. Suzuki, K. Fukara, O. Irino, O. Saita and S. Yamabe, Studies on the synthesis of chemotherapeutics, Heterocycles 16:1205 (1981).

25. R.A. Payne and C.H. Stammer, Cycloserine peptides, J. Org. Chem. 33:2421 (1968).

26. A.S. Steinfeld, F. Naider and J.M. Becker, Anticandidal activity of 5-fluorocytosine-peptide conjugates, J. Med. Chem. 22:1104 (1979).
27. W.D. Kingsbury, J.C. Boehm, R.J. Mehta, S.F. Grappel and C. Gilvarg, A novel peptide delivery system involving peptidase activated pro-drugs as antimicrobial agents. Synthesis and biological activity of peptidyl derivatives of 5-fluorouracil, J. Med. Chem. 27:1447 (1984).
28. J.S. Ti, A.S. Steinfeld and F. Naider, Anticandidal activity of pyrimidine-peptide conjugates, J. Med. Chem. 23:913 (1980).
29. W.D. Lichliter, F. Naider and J.M. Becker, Basis for the design of anticandidal agents from studies of peptide utilization in Candida albicans, Antimicrob. Agents Chem. 10:483 (1976).
30. P.J. McCarthy, P.F. Troke and K. Gull, Mechanism of action of nikkomycin and the peptide transport system of Candida albicans, J. Gen. Microbiol. 131:775 (1985).
31. F. Naider, P. Shenbagamurthi, A.S. Steinfeld, H.A. Smith, C. Boney and J.M. Becker, Synthesis and biological activity of tripeptidyl polyoxins as antifungal agents, Antimicrob. Agents Chem. 24:787 (1983).
32. H. Bruckner, G. Jung, R.G. Werner and K.R. Appel, Synthesis and biological activities of the tri-L-alanine derivative of isonicotinic acid hydrazide, Arzneimittel-Forschung 33:1630 (1983).
33. J.W. Payne, Transport and utilization of peptides by bacteria, in: "Microorganisms and Nitrogen Sources". J.W. Payne, ed., J. Wiley & Sons, Chichester, (1980).
34. C.L. Hermsdorf and S. Simmonds, Role of peptidases in utilization and transport of peptides by bacteria, in: "Microorganisms and Nitrogen Sources". Payne, J.W. ed., J. Wiley & Sons, Chichester, (1980).
35. J.C. Boehm, W.D. Kingsbury, D. Perry and C. Gilvarg, The use of cysteinyl peptides to effect portage transport of sulfhydryl-containing compounds in E.coli, J. Biol. Chem. 258:14850 (1983).
36. W.D. Kingsbury, J. Boehm, D. Perry, and C. Gilvarg, Portage of various compounds into bacteria by attachment to glycine residues in peptides, Proc. Natl. Acad. Sci., U.S.A. 81:4573 (1984).
37. C.T.Walsh, Suicide substrates: mechanism-based enzyme inactivators: Recent Developments, Ann. Rev. Biochem. 53:493 (1984).
38. K-S Cheung, S.A. Wasserman, E. Dudek, S.A. Lerner and M. Johnston, Chloroalanyl and propargylglycyl dipeptides. Suicide substrate containing antibacterials, J. Med. Chem. 26:1733 (1983).
39. D. Perry and E.P. Abraham, Transport and metabolism of bacilysin and other peptides by suspensions of Staphylococcus aureus, J. Gen. Microbiol. 115:213 (1979).
40. P.L. Carl, P.K. Chakravarty, J.A. Katzenellenbogen and M.J. Weber, Protease-activated 'prodrugs' for cancer therapy, Proc. Natl. Acad. Sci. U.S.A. 77:2224 (1980).
41. P.K. Chakravarty, P.L. Carl, M.J. Weber and J.A. Katzenellenbogen, Plasmin-activated prodrugs for cancer chemotherapy. 2. Synthesis and biological activity of peptidyl derivatives of doxorubicin, J. Med. Chem. 26:638 (1983).
42. P.K. Chakravarty, P.L. Carl, M.J. Wever and J.A. Katsenellenbogen, Plasmin-activated prodrugs for cancer chemotherapy. 1. Synthesis and biological activity of peptidylacivicin and peptidylphenylenediamine mustard, J. Med. Chem. 26:633 (1983).
43. R.L. Firestone, J.M. Pisano, P.J. Bailey, A. Sturm, R.J. Bonney, P. Wightman, R. Devlin, C.S. Lin, D.L. Keller and P.C. Tway, Lysosomotropic agents. 4. Carbobenzoxyglycylphenylalanyl, a new protease-sensitive masking group for introduction in cells. J. Med. Chem. 25:539 (1982).
44. M. Masquelier, R. Baurain and A. Trouet, Amino acid and dipeptide derivatives of daunorubicin. 1. Synthesis, physiocochemical properties, and lysosomal digestion. J. Med. Chem. 23:1166 (1980).

45. M.M. Ponpipom, R.L. Bugianesi, J.C. Robbins, T.W. Doebber and T.Y. Shen, Cell-specific ligands for selective drug delivery to tissues and organs, J. Med. Chem. 24:1388 (1981).
46. R.H. Goldfarb, Proteases in tumor invasion and metastasis, in: "Tumor Invasion and Metastasis", Liotta, L.A. and Hart, I.R., eds., Martinus Nijhoff, The Hague (1982).
47. S.D.J. Magnan, F.N. Shirota and H.T. Nagasawa, Drug latentiation by gamma-glutamyl transpeptidase, J. Med. Chem. 25:1018 (1982).
48. T.M. Dolak and L.I. Goldberg, Renal blood flow and dopaminergic agonists, Ann. Rep. Med. Chem. 11:103 (1981).
49. M.R. Lee, Dopamine and the Kidney, Clin. Sci. 62:439 (1982).
50. S. Wilk, H. Mizoguchi and M. Orlowski, Gamma-glutamyl Dopa: A kidney-specific dopamine precursor, J. Pharm. Exp. Ther. 206:227 (1978).
51. D. Worth, J. Brown, J. Cooke, J. Harvey and M.R. Lee, The effect of intravenous gamma-glutamyl L-DOPA on renal function in normal volunteers, Clin. Sci. 66:13 (1984).
52. M. Orlowski, H. Mizoguchi and S. Wilk, N-Acyl-gamma-glutamyl derivatives of sulfamethoxazole as models of kidney-selective prodrugs, J. Pharm. Exp. Ther. 212:167 (1980).
53. L.L. Iversen, Nonopioid neuropeptides in mammalian CNS, Ann. Rev. Pharmacol. Toxicol. 23:1 (1983).
54. I. Sami and M.D. Said, Vasoactive peptides: state of the art review, Hypertension, 5 (Suppl.1):1 (1983).
55. J.S. Morley, Modulation of the action of regulatory peptides by structural modification, Trends in Pharm. Sci. 1:463 (1980).
56. G.M. Makhlouf, Enteric neuropeptides: Role in neuromuscular activity of the gut, Trends in Pharm. Sci. 6:214 (1985).
57. T. Hokfelt, O. Johansson, A. Ljungdahl, J.M. Lundberg and M. Schultzberg, Peptidergic neurones, Nature 284:515 (1980).
58. S. Undenfriend and D.L. Kilpatrick, Biochemistry of the enkephalin- and enkephalin-containing peptides. Arch. Biochem. Biophys. 2211: 309 (1983).
59. J. DiMaio and P.W. Schiller, A cyclic enkephalin analog with high in-vitro opiate activity, Proc. Natl. Acad. Sci. U.S.A. 77:7162 (1980).
60. J. DiMaio, T.M-D. Nguyen, C. Lemieux and P.W. Schiller, Synthesis and pharmacological characterization in-vitro of cyclic enkephalin analogues: Effect of conformational constraints on opiate receptor selectivity, J. Med. Chem. 25:1432 (1982).
61. J.L. Krstenansky, R.L. Baranowski and B.L. Currie, A new approach to conformationally restricted peptide analogs: Rigid beta-bends. 1. Enkephalin as an example, Biochem. Biophys. Res. Comm. 109: 1368 (1982).
62. A. Camerman, D. Mastropaolo, I. Karle, J. Karle and N. Camerman, Crystal Structure of Leucine-Enkephalin, Nature 306:447 (1983).
63. G.D. Smith and J.F. Griffin, Conformation of [Leu5]-enkephalin from X-ray diffraction: Features important for recognition at opiate receptor, Science 199:1214 (1978).
64. A.F. Spatola, Peptides of the hypothalamus, Ann. Rep. Med. Chem. 19:199 (1981).
65. A.J. Prange and C.B. Nemeroff, Peptides in the central nervous system: focus on thyrotropin releasing hormone and neurotensin, Ann. Rep. Med. Chem. 17:31 (1982).
66. A. Peterkofsky, F. Battaini, Y. Koch, Y. Takahara and P. Dannies, Histidyl-proline diketopiperazine: Its biological role as a regulatory peptide, Mol. Cell. Biochem. 42:45 (1982).
67. M. Hichens, A comparison of thyrotropin - releasing hormone with analogues: influence of disposition upon pharmacology, Drug Met. Rev. 14:77 (1983).
68. N.A. Sharif, Diverse roles of thyrotropin-releasing hormone in brain, pituitary and spinal function, Trends in Pharm. Sci. 6:119 (1985).

69. J.C. Buckingham, Hypothalamic releasing hormones, _Trends in Pharm. Sci._ 2:335 (1981).

70. H.M. Fraser, A new class of contraceptives, _Nature_ 296:391 (1982).

71. R.M. Freidinger, D.F. Veber, D.S. Perlow, Bioactive conformation of luteinizing hormone-releasing hormone: evidence from a conformationally constrained analog, _Science_ 210:656 (1980).

72. W.K. Burn, D. Machin, W.E. Waters, Biodegradable polymer luteinising hormone releasing hormone analogue for prostatic cancer: Use of a new peptide delivery system, _Br. Med. J._ 289:1580 (1984).

73. T.W. Redding, A.V. Schally, T.R. Tice and W.E. Meyers, Long-acting delivery systems for peptides: Inhibition of rat prostate tumors by controlled release of [D-Trp^6] luteinizing hormone-releasing hormone from injectable microcapsules, _Proc. Natl. Acad. Sci. U.S.A._ 81:5845 (1984).

74. S.H. Snyder, Brain peptides as neurotransmitters, _Science_ 209:976 (1980).

75. G. Aguilera, J.P. Harwood and K.J. Catt, Somatostatin modulates effects of aniotensin II in adrenal glomerulosa zone, _Nature_ 292:262 (1981).

76. L.J. Chang, L.M. Sandler, M.E. Kraenzlin, J.M. Burrin, G.F. Joplin and S.R. Bloom, Long term treatment of acromegaly with a long acting analogue of somatostatin, _Br. Med. J._ 290:284 (1985).

77. J.E. Gerich, Somatostatin modulation of glucagon secretion and its importance in human glucose homeostasis, _Metabolism_ 27:1283 (1978).

78. P.M. Maton, T.M. O'Dorisio, B.A. Howe, K.E. McArthur, J.M. Howard, J.A. Cherner, T.B. Malarkey, M.J. Collen, J.D. Gardner and R.T. Jensen, Effect of a long-acting somatostatin analogue (SMS 210-995) in a patient with pancreatic cholera, _N. Engl. J. Med._ 312:17 (1985).

79. K. von Werder, M. Losa, O.A. Muller, L. Schweiberer, R. Fahlbusch and E. del Pozo, Treatment of metastasising GRF-producing tumour with a long-acting somatostatin analogue, Lancet, 2:282 (1984).

80. I. Whitehouse, C. Beglinger, M. Fried and K. Gyr, The effect of an octapeptide somatostatin analog and somatostatin on pentagastrin-stimulated gastric acid secretion in man, _Hepatogastroenterology_ 31:227 (1984).

81. I. Magnusson and T. Ihre, Does somatostatin help in upper gastrointestinal bleeding?, _Lancet_ 1:337 (1985).

82. D.F. Veber, F.W. Holly, W.J. Paleveda, R.F. Nutt, S.J. Bergstrand, M. Torchiana, M.S. Glitzer, Saperstein and R. Hirschmann, Conformationally restricted bicyclic analogs of somatostatin, _Proc. Natl. Acad. Sci U.S.A._ 75:2636 (1978).

83. D.F. Veber, R.M. Freidinger, D.S. Perlow, W.J. Paleveda, F.W. Holly, R.G. Strachan, R.F. Nutt, B.H. Arison, C. Homnick, W.C. Randall, M.S. Glitzer, R. Saperstein and R. Hirschmann, A potent cyclic hexapeptide analogue of somatostatin, _Nature_ 292:55 (1981).

84. R. Palluk, W. Gaida and W. Hoefke, Atrial natriuretic factor, _Life Sci._ 36:1415 (1985).

85. M.G. Currie, D. Sukin, D.M. Geller, B.R. Cole and P. Needleman, Atriopeptin release from the isolated perfused rabbit heart, _Biochem. Biophys. Res. Comm._ 124:711 (1984).

86. J. Gutkowska, K. Horky, G. Thibault, P. Januszewics, M. Cantin and J. Genest, Atrial natriuretic factor is a circulating hormone, _Biochem. Biophys. Res. Comm._ 125:315 (1984).

87. K. Kangawa, A. Fukuda and H. Matsuo, Structural identification of beta and gamma-human atrial natriuretic polypeptides, _Nature_ 313:397 (1985).

88. A.A. Seymour, E.H. Blaine, E.K. Mazack, S.G. Smith, I.I. Satabilito, A.B. Haley, M.A. Napier, M.A. Whinnery and R.F. Nutt, Renal and systemic effects of synthetic atrial natriuretic factor, _Life Sci._ 36:13 (1985).

89. M.A. Napier, R.L. Vandlen, G. Albers-Schonberg, R.F. Nutt, S. Brady, T. Lyle, R. Winquist, E.P. Faison, L.A. Heinel and E.H. Blaine, Specific membrane receptors for atrial natriuretic factor in renal and vascular tissues, Proc. Natl. Acad. Sci. U.S.A. 81:5946 (1984).

90. J. Tremblay, R. Gerzer, P. Vinay, S.C. Pang, R. Beliveau and P. Hamet, The increase of cGMP by atrial natriuretic factor correlates with the distribution of particular guanylate cyclase, FEBS Lett. 181:17 (1985).

91. A.M. Richards, H. Ikram, T.G. Yandle, M.G. Nicholls, M.W.I. Webster and E.A. Espiner, Renal, haemodynamic, and hormonal effects of human alpha atrial natriuretic peptide in healthy volunteers, Lancet 1:545 (1985).

92. M.A. Ondetti and D.W. Cushman, Design of protease inhibitors. Biopolymers 20:2001 (1981).

93. U. Brodbeck, ed., "Enzyme Inhibitors", Verlag Chemie, Basel (1980).

94. M.A. Ondetti, M.E. Condon, J. Reid, E.F. Sabo, H.S. Cheung and D.W. Cushman, Design of potent and specific inhibitors of carboxypeptidases A and B, Biochemistry 18:1427 (1979).

95. G. Mackaness, The Future of angiotensin-converting enzyme inhibitors, J. Cardiovasc. Pharmacol. 7:S30 (1985).

96. A.A. Patchett, E. Harris, E.W. Tristram, M.J. Wyvratt, M.T. Wu, D. Taub, E.R. Peterson, T.J. Ikeler, J. ten Browke, L.G. Payne, D.L. Ondeyka, E.D. Thorsett, W.J. Greenlee, N.S. Lohr, R.D. Hoffsommer, H. Joshua, W.V. Ruyle, J.W. Rothrock, S.D. Aster, A.L. Maycock, F.M. Robinson, R. Hirschmann, C.S. Sweet, E.H. Ulm, D.M. Gross, T.C. Vassil and C.A. Stone, A new class of antiotensin-converting enzyme inhibitors, Nature 288:280 (1980).

97. C.S. Sweet, A.A. Patchett, E.H. Ulm and D.M. Gross, Structure-activity studies with angiotensin converting enzyme inhibitors related to enalapril and MK-521, Roy. Soc. Chem. 50:36 (1984).

98. H.R. Brunner, J. Nussberger and B. Waeber, The present molecules of converting enzyme inhibitors, J. Cardiovasc. Pharmacol. 7:S2 (1985).

99. K.G. Hofbauer and J.M. Wood, Inhibition of renin: Recent immunological and pharmacological advances, Trends in Pharm. Sci. 6:173 (1985).

100. D.H. Rich, F.G. Salituro, M.W. Holladay and P.C. Schmidt, Design and discovery of aspartyl protease inhibitors, mechanistic and clinical implications, Am. Chem. Soc. (Symp. Series) 251:211 (1984).

101. M. Szelke, B. Leckie, A. Hallett, D. Jones, J. Sueiras, B. Atrash and A.F. Lever, Potent new inhibitors of human renin, Nature, 299:555 (1982).

102. J. Boger, N.S. Lohr, E.H. Ulm, M. Poe, E.H. Blaine, G. Fanelli, T. Lin, L.S. Payne, T.W. Schorn, B.I. Lamont, T.C. Vassil, I.I. Stabilito, D.F. Veber, D.H. Rich and A.S. Bopari, Novel renin inhibitors containing the amino acid statine. Nature 303:81 (1983).

103. D.H. Rich, E.T.O. Sun and E. Ulm, Synthesis of analogues of the carboxyprotease inhibitor pepstatin. Effect of structure on inhibition of pepsin and renin, J. Med. Chem. 23:27 (1980).

104. M-C. Fournie-Zaluski, P. Chaillet, E. Soroca-Lucas, H. Marcais-Collado, J. Costentin and B. P. Roques, New carboxyalkyl inhibitors of brain enkephalinase: Synthesis, biological activity, and analgesic properties, J. Med. Chem. 26:6 (1983).

105. J.M. Frere, C. Duez, J. Dusart, J. Coyette, M. Leyh-Bouille, J.M. Ghuysen, O. Dideberg and J. Knox, Mode of action of beta-lactam antibiotics at the molecular level, in: "Enzyme Inhibitors as Drugs", Sandler, M., ed., Macmillan, London (1980).

106. J. Lamotte-Brasseur, G. Dive and J-M. Ghuysen, On the structural analogy between D-alanyl-D-alanine terminated peptides and beta-lactam antibiotics, Eur. J. Med. Chem. 19:319 (1984).

107. D.J. Waxman, R.R. Yocum and J.L. Strominger, Penicillins and cephalosporins are active site-directed acylating agents: evidence in support of the substrate analog hypothesis, Trans. Roy. Soc. London, Series B. 289:257 (1980).

108. D.B. Boyd, Transition state structures of a dipeptide related to the mode of action of beta-lactam antibiotics, <u>Proc. Natl. Acad. Sci. U.S.A.</u> 74:5239 (1977).

109. P. Charlier, O. Dideberg, J-C. Jamoulle, J-M. Frere, J-M. Ghuysen, G. Dive and J. Lammotte-Brasseur, Active-site-directed inactivators of the Zn^{2+}-containing D-alanyl-D-alanine-cleaving carboxypeptidase of S.albus G, <u>J. Biochem.</u> 219:763 (1984).

110. M.P. Williamson, D.H. Williams and S.J. Hammond, Interactions of vancomycin and ristocetin with peptides as a model for protein binding, <u>Tetrahedron</u> 40:569 (1984).

111. S. Adibi and E. Morse, The number of glycine residues which limits intact absorption of glycine oligopeptides in human jejunum, <u>J. Clin. Invest.</u> 60:1008 (1977).

112. D.B. Silk, Peptide transport, <u>Clin. Sci.</u> 60:607 (1981).

113. D.M. Matthews, Intestinal absorption of peptides, <u>Physiol. Rev.</u> 55: 538 (1975).

114. S. Yokohama, T. Yoshioka, K. Yamashita and N. Kitamore, Intestinal absorption mechanisms of thyrotrophin-releasing hormone, <u>J. Pharm. Dyn.</u> 7:445 (1984).

115. T. Kimura, Transmucosal absorption of small peptide drugs, <u>Pharm. Int.</u> 5:75 (1984).

116. K.H. Jones, P.F. Langley and L.J. Lees, Bioavailability and metabolism of talampicillin, <u>Chemotherapy</u> 24:217 (1978).

117. D.A. Spyker, R.J. Rugleski, R.L. Vann and W. O'Brien, Pharmacokinetics of amoxicillin dose dependence after intravenous, oral and intramuscular administration, <u>Antimicrob. Agents Chem.</u> 11:132 (1977).

118. C.G. Hertz, Serum and urinary concentrations of cyclacillin in humans, <u>Antimicrob. Agents Chem.</u> 4:361 (1973).

119. T. Kimura, H. Endo, M. Yoshikawa, S. Muranishi and H. Sezaki, Carrier-mediated transport systems for aminopenicillins in rat small intestine, <u>J. Pharm. Dyn.</u> 1:262 (1978).

120. E. Nakashima, A. Tsuji, S. Kagatani and T. Yamana, Intestinal absorption mechanism of amino-beta-lactam antibiotics. III. Kinetics of carrier-mediated transport across the rat small intestine in situ, <u>J. Pharm. Dyn.</u> 7:452 (1984).

121. J.P. Clayton, M. Cole, S.W. Elson, K.D. Hardy, L.W. Mizen and R. Sutherland, Preparation, hydrolysis and oral absorption of alpha-carboxy esters of carbenicillin, <u>J. Med. Chem.</u> 18:172 (1985).

122. S. Hartley and R. Wise, A three way crossover study to compare pharmacokinetics and acceptability of sultamicillin at two dose levels with that of ampicillin, <u>J. Antimicrob. Chem.</u> 10:49 (1982).

123. D.J. Tacco, A. deLuna, A.E. Duncan, T.C. Vassil and E.H. Ulm, The physiological disposition and metabolism of enalapril maleate in laboratory animals, <u>Drug Met. Disp.</u> 10:15 (1982).

124. H.G. Eckert, M.J. Badian, D. Gantz, H.M. Kellner and M. Volz, Pharmacokinetics and biotransformation of Hoe-498 in rat, dog and man, <u>Arzneim. Forschung.</u> 34:1435 (1984).

125. D.A. Grant, T.F. Ford and R.J. McCulloch, Distribution of pepstatine and statine following oral and intravenous administration in rats. Tissue localisation by whole body autoradiography, <u>Biochem. Pharm.</u> 31:2302 (1982).

126. J.P. Baker, B.H. Kemmenoe, C. McMartin and G.E. Peters, Pharmacokinetics, distribution and elimination of a synthetic octapeptide analogue of somatostatin in the rat, <u>Reg. Peptides.</u> 9:213 (1984).

127. J. Bell, G.E. Peters, C. McMartin, N.W. Thomas and C.G. Wilson, Estimation of gut absorption of peptides by biliary sampling, <u>J. Pharm. Pharmacol.</u> 36:88P (1984).

128. A.J. Wood, G. Maurer, W. Niederberger and T. Beveridge, Cyclosporine: pharmacokinetics metabolism, and drug interactions, <u>Transplant</u> Proc. 15:2409 (1983).

129. C.T. Veda, M. Lemaire, G. Gsell and K. Nussbaumer, Intestinal lymph-atic absorption of cyclosporin following oral administration in an olive oil solution in rats, Biopharm. Drug Disp. 4:113 (1983).

130. H. Okada, I. Yamazaki, Y. Ogawa, S. Hirai, H. Okada, T. Yashiki and H. Mima, Vaginal absorption of a potent luteinizing hormone-releas-ing hormone analog (Leuprolide) in Rats I: Absorption by various routes and absorption enhancement, J. Pharm. Sci. 71:1367 (1982).

131. H. Yoshida, K.O. Kumura, R. Hori, T. Anmo and H. Yamaguchi, Absorpt-ion of insulin delivered to rabbit trachea using aerosol dosage form, J. Pharm. Sci. 68:670 (1979).

132. S. Hirai, T. Yashiki, T. Matsuzawa and H. Mima, Absorption of drugs from the nasal mucosa of rat, Int. J. Pharm. 7:317 (1981).

133. C. Bergquist, S.J. Nillius and L. Wide, Intranasal gonadotropin-releasing hormone agonist as a contraceptive agent, Lancet 2:215 (1979).

134. M.K. Ward and T.R. Fraser, DDAVP in treatment of vasopressin-sensitive diabetes insipidus, Brit. Med. J. 3:86 (1974).

135. S. Hirai, T. Yashiki and H. Mima, Effects of surfactants on the nasal absorption of insulin in rats, Int. J. Pharm. 9:165 (1981).

136. G.E. Peters, Distribution and metabolism of exogenous somatostatin in the rat, Reg. Peptides 3:361 (1982).

137. D. Brewster, M.J. Humphrey and M.V. Wareing, Metabolism and pharmaco-kinetics of TRH and an analogue with enhanced neuropharmacological potency, Neuropeptides 1:153 (1981).

138. K.S. Pang, W.F. Cherry, J.A. Terrell and E.H. Ulm, Disposition of enalapril and its diacid metabolite enalaprilat in a perfused rat liver preparation. Presence of a diffusional barrier for enal-aprilat into hepatocytes, Drug Met. Disp. 12:309 (1984).

139. G. Metcalf, P.W. Dettmar, A. Lynn, D. Brewster and M.E. Havler, Thyro-trophin-releasing hormone (TRH) analogues show enhanced CNS select-ivity because of increased biological stability, Reg. Peptides, 2:277 (1981).

140. S. Sarhan, M. Kolb and N. Seiler, The amolification of the anticon-vulsant effect of vinyl GABA (4-aminohexenoic acid) by esters of glycine, Arzneim. Forsch. 34:687 (1984).

141. P.S. Callery, L.A. Geelhaar, M.S. Balachandran Nayar, M. Stogniew and K. Gurudarh Rao, Pyrrolines as pro-drugs of gamma-aminobutyric acid analogues, J. Neurochem. 38:1064 (1982).

142. D. Roemer and J. Pless, Structure activity relationship of orally active enkephalin analogues as analgesics, Life Sci. 24:621 (1979).

143. R.M. Freidinger and P.F. Veber, Design of novel cyclic hexapeptide somatostatin analogs from a model of the bioactive conformation, Am. Chem. Soc. (Symp. Series) 251:169 (1984).

144. R. Freidinger and D. Veber, Design of novel cyclic hexapeptide somato-statin analogs from a model of the bioactive conformation, Am. Chem. Soc. (Symp. Series) 251:169 (1984).

145. R.H. Rippel, E.S. Johnson, W.F. White, M. Fujino, T. Fukuda and S. Kobayashi, Ovulation and gonadotrophin-releasing activity of [D-Leu[6],desGlyNH$_2$[10],Pro-ethylamide[9]]-GNRH(38715), Proc. Soc. Exp. Biol. Med. 148:1193 (1975).

146. K. Wiedhaup, The stability of small peptides in the gastrointestinal tract, in: 'Topics in Pharmaceutical Sciences', Breimer, D. and Speiser, P., eds., Elsevier/North Holland Biomedical Press, Amsterdam, (1981).

147. G. Maurer, H.R. Loosli, E. Schreier and B. Keller, Disposition of cyclosporine in several animal species and man. Structural elucid-ation of its metabolites, Drug Met. Disp. 12:120 (1984).

148. H. Kroppe, J.G. Sundelof, R. Hajdu and F.M. Kahan, Metabolism of thienamycin and related carbapenem antibiotics by the renal dipep-tidase, dehydropeptidase-I, Antimicrob. Agents Chem. 22:62 (1982).

149. C.D. Klaassen and J.B. Watkins, Mechanism of bile formation hepatic

uptake and biliary excretion, Pharmacol. Rev. 36:1 (1984).

150. V.J. Hruby, Design of peptide hormone and neurotransmitter analogues, Trends in Pharm. Sci., 6:259 (1985).

HEPATIC FUNCTIONS IN HEALTH AND DISEASE: IMPLICATIONS IN DRUG CARRIER USE

Neil McIntyre

Academic Department of Medicine, Royal Free Hospital

London NW3 2QG, U.K.

GENERAL ANATOMICAL AND PHYSIOLOGICAL CONSIDERATIONS

The liver is an unusual organ in that it receives blood not only from the aorta but an even larger amount via the portal vein, the main vein draining the intestine, spleen and pancreas. The arterial blood and portal venous blood mingle within the liver and return to the heart via the hepatic veins and inferior vena cava. The liver receives in total about a quarter of the output of the heart, but only 20-25% of it is arterial blood.

The liver secretes lipid-rich bile which passes via the biliary system into the second part of the duodenum, where it emulsifies fats, assists their intra-luminal digestion and solubilizes the products of fat soluble compounds. The bile is also a medium for the excretion of many endogenous and exogenous compounds.

The liver has a lymphatic system.[1] About 80% of hepatic lymph leaves the liver via hilar lymphatics; it is filtered by hilar lymph nodes before passing to the cisterna chyli and thence into the thoracic duct. The remaining 20% goes through lymph vessels following the hepatic venous drainage and subsequently travels in retrosternal lymphatics which run, along the course of the internal mammary artery, to the root of the neck where they enter the great veins.

Drugs or other materials injected intravenously in liposomes or other carriers must traverse the capillaries of the lungs to reach the arterial blood. On the first circulation about 25% reaches the liver but most of this first passes through either the spleen (which receives about 6% of the cardiac output), pancreas or intestine. In the small intestinal wall it might interact with absorbed lipids (in chylomicrons and VLDL); conceivably material in the intestinal wall could also return to the right side of the heart via lymph nodes and lymphatic vessels. Within the liver the carrier will be exposed to hepatic cells which may take it up; some may then be rapidly excreted in the bile.

The carrier may pass straight through the hepatic circulation without being taken up, or could be returned to the heart via hepatic lymphatics. Carrier which traverses the liver unchanged will of course return to the liver and have a further chance of hepatic uptake on subsequent recirculations. But for the time that liposomes and other carriers remain in the circulation they will be subject to alteration by the various substances, particles and cells which make up the blood and to the effects of transit through other tissues of the body where they may also be altered or removed. The liver may play a role in intravascular changes of these carriers as it secretes lipoproteins into the circulation, the lipids of which would be expected to interact with lipid carriers such as liposomes; it contains a hepatic lipase which has phospholipase activity; and it secretes an enzyme, lecithin cholesterol acyltransferase or LCAT, which plays an important role in intravascular lipoprotein metabolism, and which affects the cholesterol and lecithin concentrations of lipoproteins and of cell membranes.[2] By changing the lipid composition of cell membranes the liver may therefore indirectly affect the handling of drug carriers by peripheral tissues.

While these general anatomical and physiological factors clearly play an important role in the fate of particles such as liposomes, the main interest has been in the role of liver cells themselves in hepatic uptake of exogenous material. The liver is made up of a variety of cell types. Parenchymal cells make up about 80% of the total volume of the liver parenchyma and for the purpose of simplification we can consider these as the cellular framework of the liver.[3] They are arranged in branching cords or plates, one cell thick (about 18-20 μm), which run from a portal tract (which contains the terminal branches of the hepatic artery and portal veins, and small bile ducts and lymphatic vessels which drain in the opposite direction, i.e. towards the hilum) towards the central veins, which are the small veins which join up into larger venous radicles emptying into the hepatic veins. Between the cords or plates are the hepatic capillaries or sinusoids which connect the terminals of the hepatic arterioles and the portal vesicles with the central vein. The sinusoids occupy 10-13% of the parenchymal volume. Their diameter, which is difficult to measure accurately, is between 5 and 12 μm. The sinusoids are lined by endothelial cells but these are separated from the parenchymal cells by a gap, the space of Disse. The space of Disse occupies about 5% of the parenchymal volume. The sinusoids and the space of Disse are in direct contact because the thin wall of the endothelial cells is perforated by numerous fenestrations which have a diameter of approximately 1-2 μm in man.[4]

The parenchymal cells have 3 types of cell surface: 1. The surface in contact with the space of Disse (the sinusoidal surface) has numerous microvilli projecting from it which vary in length and have a diameter of about 70-120 nm. If the surface area of the microvilli is ignored then the sinusosidal surface of parenchymal cells occupies about 37-42% of the cell surface (but if it is taken into account then about 70% of the cell membrane is in contact with the space of Disse). There is sinusoidal surface on two sides of the cells. About 50-54% of the surface is in contact with other parenchymal cells, and about 6-13% (not including microvilli) is on the surface of the bile canaliculi. These are small canals running between parenchymal cells; they receive newly formed bile and run along the cords or plates of parenchymal cells to empty into the bile spaces and ductules of the portal tract.[4]

There are cell types other than parenchymal and endothelial cells which make up the liver substance. Kupffer cells are solitary macrophages which lie on or within the endothelial lining; they are numerous in the periportal areas where they bulge into the sinusoidal lumen and may interfere with blood flow. They are part of the mononuclear phagocytic system: they are derived from bone marrow, are highly phagocytic and can ingest very large particles such as red blood cells; they have membrane receptors for the Fc part of immunoglobulin molecules and for the C3b component of complement; and characteristically express HLA-DR antigens on their surface.[5]

Fat storing cells (Ito cells) are found in the space of Disse, normally

in recesses between parenchymal cells.[6] They have a characteristic morphology, appear to be involved in the storage of vitamin A, and may also be
involved in collagen production. They are evenly distributed throughout
the liver lobules, occupying about 2% of the parenchymal volume and appear
fixed in position.

The last of the sinusoidal cells is the 'pit' cell which appears to
be embedded within the endothelial lining of the sinusoid. These cells
are scarce in sinusoids, and they may not be present in human liver. They
have a characteristic polarity with most organelles lying on one side of
the cells and they contain granules, the appearance of which suggest an
endocrine function. Their function is unknown.

The blood supply of the sinusoids is not a simple matter of entry of
blood at the portal tract and straight flow through the sinusoid into the
central vein. There is direct inflow of blood from the terminals of the
hepatic arteries and portal veins, but some arterial branches go first to
a peribiliary plexus supplying the small bile ducts in the portal tract;
blood from these plexuses then drains into the sinusoidal network.[4] This
suggests the possibility of a feedback control in the secretory mechanisms
of the hepatocyte. The sinusoids are not all straight communications with
the central vein. There are also branching sinusoids and interconnecting
sinusoids. The rate of blood flow varies in sinusoids and there is evidence that flow may sometimes reverse in individual sinusoids. There may
be control of flow in sinusoids by changes in the size of endothelial cells,
and particularly of Kupffer cells which are situated in peripheral areas
and can show marked changes in size.

REMOVAL OF MATERIAL BY LIVER CELLS

With these anatomical details in mind we can now consider the way in
which the liver functions in the removal of material from the blood stream.
Goresky has shown, using labelled red cells and labelled markers of various
sizes, that the sinusoidal epithelium is freely permeable to all dissolved
markers and that it is possible to calculate a space for their distribution
within the liver.[7] Molecules are not equally free to permeate the space
of Disse, probably because the Disse space contains a mesh of very large
molecules, such as proteoglycans, which has its own interstices, and which
permit[6] entry of water and small molecules but tends to exclude larger
particles. This free permeability of the endothelial cells, which is due
to their numerous fenestrations, allows hepatic parenchymal cells free
contact with dissolved material and with many of the smaller particles contained in the blood stream; it also allows such material to enter the blood
stream easily from liver cells.

The narrowness of the sinusoids probably enhances access of plasma
contents to the Disse space. As the sinusoid is only about 7 μm in diameter
the passage of single red and white cells along it will result in fluctuation in pressure in the plasma between the cells, with forced sieving of
the plasma into and out of the Disse space and good mixing of plasma and
interstitial fluid.[3]

Uptake of material by parenchymal cells can take place in a variety
of ways. Substances such as water and urea will enter or leave fairly
freely and net movement depends on the presence of osmotic or concentration
gradients. Some small molecules, particularly charged particles, cannot
diffuse easily through the lipid bilayer or through small channels in the
membrane; they are either excluded from the cells or are transported by
specialized mechanisms at the surface.[8] It has been widely assumed, correctly in many instances, that these substances are picked up from the

water phase of plasma but recent studies suggest that the uptake of small molecules which are strongly bound to albumin, such as fatty acids and bilirubin, may be facilitated by binding of the albumin-ligand complex to the cell surface.[9] A conformational change in the albumin may then release the ligand in close proximity to the membrane thus promoting its uptake by the appropriate transport mechanism. As yet the putative albumin receptor has not been identified.

Other uptake mechanisms involve endocytosis. Currently there is great interest in the process of receptor mediated endocytosis.[10] This involves the binding of extracellular ligands such as apoprotein B, differic transferrin, or IgA, to specific receptors on the cell surface. These then cluster in pits on the cell surface which are surrounded by a layer of the protein clathrin. These pits are internalised by a process involving membrane fusion and transported to other vesicular structures within the cell such as lysosomes. The ligands may suffer a variety of fates. Apo B, for example, is degraded. Diferric transferrin simply loses its iron and the apotransferrin is usually returned to the plasma by a process of exocytosis.[11] IgA may traverse the cell and be excreted directly via the canaliculus. Recent studies by La Russo and his colleagues suggest that there may be direct biliary secretion of many of the ligands entering the cell by endocytosis.[12]

It is clear that the invagination of clathrin coated vesicles would inevitably result in the internalisation of a certain amount of extracellular water together with the substances dissolved in it. This involves a relatively passive process without the concentration of the incoming material which results from binding to receptors. Not all pits on the surface of the cells have clathrin coats and it is possible that pinocytotic vesicles may enter hepatocytes without receptors for specific ligands being present on the surface. Fluid phase endocytosis, whether or not it involves vesicles containing receptors, is a major process in cellular function. It has been estimated that macrophages ingest 25% of their own volume each hour, or about 3% of their plasma membrane each minute.[8] The figure is probably much smaller for the hepatic parenchymal cell. This mechanism, which occurs in cells throughout the body, is probably a major pathway for the removal of circulating albumin, for which no receptor-mediated removal, or major site of removal, has yet been demonstrated.[13]

These are not the only methods by which material may be taken up by parenchymal cells. Lipids on the surface of lipoproteins exchange with material in the plasma membrane of various types of cell and if a concentration gradient exists between the two then net movement of lipids such as cholesterol and lecithin will occur.[2] This may result in a change in the cholesterol:phospholipid ratio of the membrane or in the relative proportion of the different phospholipids. Movement of phospholipid in this way may also be a method of inter-organ transfer of fatty acids such as arachidonic acid.

Finally, it is possible that drug carriers and other materials having a lipid bilayer surface, such as liposomes or LP-X (to which I shall return later) may be incorporated into cells by fusion of their surface bilayer with that of the cell. A number of studies suggest that fusion occurs but there are many problems in interpretation of such studies.[14] Certainly the fluidity and chemical composition of the bilayer would be an important determinant of whether or not fusion would occur.

The above discussion on cellular uptake mechanisms has concentrated on uptake mechanisms by parenchymal cells, but uptake by these cells is also affected by the presence of other cells in the liver, and by the position within the lobule of individual parenchymal cells.[15,16] It is now clear

that there is heterogeneity of hepatocyte function depending on the sinusoidal position of parenchymal cells. Cells close to the portal tract will,
by comparison with those close to the central vein, receive blood relatively
rich in oxygen and other substances removed by parenchymal cells; 'central
cells', however, see higher concentrations of substances which have been
released into sinusoidal blood by the periportal cells. The environment
of the cells may dictate their functional adaptation. This certainly seems
to be the case for the metabolic processes of glycolysis and ketogenesis
(which normally predominate in pericentral cells) and for oxygen uptake
and gluconeogenesis (which usually predominate in periportal cells); if
the blood flow through an isolated liver is artificially reversed by retro-
grade perfusion then metabolic activities shift to the opposite end of the
sinusoid. This is not true of glucoronidation and mono-oxygenation which
predominate in the pericentral zone regardless of the direction of flow of
arterialized blood. There is little information whether heterogeneity of
parenchymal cell function affects handling of liposomes or other drug
carriers.

There is also marked heterogeneity in the uptake of particles by the
different liver cell types. Kupffer cells and endothelial cells, (like
parenchymal cells), can remove small particles from the blood stream, and
it is important that they are in closer contact with the circulating blood
than parenchymal cells and have the opportunity to remove material before it
traverses the fenestrations in the endothelial cell wall. Indeed, particles
larger than the endothelial fenestrations would presumably remain in the
sinusoids and fail to make contact with parenchymal cells. There appears
to be functional heterogeneity within the population of Kupffer cells as
these in the periportal region show more active endocytosis and have large
lysosomes which are particularly rich in lysosomal enzymes.[17]

The main difference between Kupffer cells and other liver cells is that
they can phagocytose large particles, a process which is enhanced by the
addition to the particle surface of recognition factors called 'opsonins';
these substances may be immunospecific (e.g. IgG or IgM antibodies) or
non-immunospecific (e.g. fibronectin). Phagocytosis is essentially a
function of macrophages but for this process to occur the cells and the
large particles must come into contact. When sheep red blood cells are
injected intravenously into mice, about 90% enter Kupffer cells in the
liver; it has been inferred from this that about 90% of the body's macrophages are in the liver. But this is not true. In tissues other than the
liver (including the lungs) macrophages usually lie outside the capillaries
and have far less contact with circulating particles than Kupffer cells.[5]

In one study various substances were injected intravenously into
rats.[18] Only endotoxin appeared to be exclusively taken up by Kupffer
cells. PVP, colloidal albumin, antimony sulphur colloid and heparin were
also taken up by endothelial and parenchymal cells, and except for PVP,
which is taken up simply by fluid phase endocytosis, they appeared to be
removed mainly by absorptive endocytosis.

Many circulating proteins are glycoproteins and the nature of the
terminal sugar of the carbohydrate chains is important in determining the
rate of removal of the glycoproteins from plasma. When sialic acid occupies
the terminal positions the glycoproteins tend to have a relatively long
half life.[19] If the sialic acid residues have been removed the carbohydrate
chains terminate in other sugars such as galactose, N-acetyl glucosamine or
mannose. Parenchymal cells have a specific receptor for galactose-terminated glycoproteins. Endothelial cells appear to have a receptor for mannose
and N-acetylglucosamine. In either case injected proteins or desialylated
endogenous proteins are removed more rapidly and are presumably catabolised
in the cells which remove them.[20] One of the ways in which particles could

theoretically be targeted to different cell types within the liver is by exposing galactose or mannose residue[6] on the surface (which would lead to parenchymal or endothelial cell uptake respectively) or by using particles which would have to be removed by phagocytoses, a property of the Kupffer cells.

IMPLICATIONS IN THE USE OF DRUG CARRIERS

Do the above mechanisms have any practical implications for the hepatologist or others managing patients with liver disease? Our knowledge in this field is scanty but there are several exciting areas. In mice and rats injected with frog virus 3 (FV3) there is initial damage to Kupffer cells. Subsequently hapatocytes are damaged, possibly through failure of Kupffer cells to neutralise toxins presented to the liver, particularly endotoxin. Particulate matter may attack the hepatocytes directly because the protective layer of the endothelium is breached with large gaps appearing in the sinusoidal lining. As a consequence, parenchymal cells themselves acquire the ability to endocytose colloidal carbon and latex particles.[21]

In a number of inherited disorders there is an accumulation of various compounds in Kupffer cells, endothelial cells and/or parenchymal cells because there is a deficiency of the lysosomal enzyme which normally degrades these compounds.[22] These diseases are of particular interest as theoretically the missing enzyme could be incorporated into carriers such as liposomes and targeted to the appropriate cells. This approach has already been tried using injections of glucorebrosidase in patients with Gaucher's disease with encouraging results. It is of interest that there is also parenchymal cell damage in Gaucher's disease (primarily a disease of Kupffer cells); it has been suggested that this results from failure of Kupffer cell protection, as postulated for experimental infection with FV3.

As noted above, many circulating proteins are glycosylated and desialylation is thought to be important in their ordered removal from plasma, because exposure of galactose, N-acetylglucosamine or mannose residues would enhance uptake by parenchymal or endothelial cells. In patients with various kinds of liver disease asialoglycoproteins (such as desialyated orosomucoid and α_1-antitrypsin) have been shown to be present in excess in the plasma. There might be secretion of abnormally glycosylated proteins and there is evidence for this in some liver diseases.[20,23] Alternatively, the uptake of the asialoglycoproteins may be impaired because of liver damage per se, because of specific damage to relevant receptors, or because there is accumulation of substances which inhibit the receptor-mediated uptake of the asialyglycoproteins.

In cirrhotic patients there is a profound disturbance of the lobular architecture and an interference with the passage of blood through the liver. An important consequence of this is that much of the portal venous blood is shunted past the liver to the heart in collateral vessels. There is also evidence of shunting within the liver itself so that arterial and portal venous blood reaches the hepatic venous radicals without going through the sinusoids. Functional heterogeneity of hepatocytes persists in experimental cirrhosis, oxygen uptake and gluconeogenesis being greater at the portal end of the sinusoids while glycolysis and ketogenesis predominate at the hepatic venous end[15], but we still have much to learn about the microcirculatory implications of cirrhosis, especially in man. There appears to be no endothelial cell fenestrations in cirrhotic sinusoids,[24] but this needs confirmation and exploration of its relevance for the clearance of particles such as chylomicron remnants. Another factor which may affect parenchymal cell uptake of material in cirrhosis is collagenization of the space of Disse which has been observed in biopsies from cirrhotic patients.

I have concentrated mainly on anatomical and physiological factors affecting the uptake of material by hepatic cells in health and disease. But in patients with severe liver disease another factor may be important in the handling of liposomes injected intravenously. In liver disease there are changes in the structure and composition of plasma lipoproteins.[25] Several factors may be involved in the genesis of these changes but one of the most important functions is a fall in plasma LCAT activity. This affects both the core and the surface of lipoprotein particles. There is an increase in plasma free cholesterol and lecithin concentrations (a predictable consequence of diminished LCAT activity); these are surface components of both lipoproteins and cell membranes which exchange such molecules freely. As a consequence the surface membrane of red cells and platelets, and probably of many other cells show an increased cholesterol: phospholipid ratio and an increase in the proportion of lecithin relative to other phospholipids.[26] Patients with obstruction to the biliary tract regurgitate a large amount of biliary lecithin into plasma; this accumulates in the plasma, together with free cholesterol and apoproteins, as bilayer structures called lipoprotein X, or LP-X. Much of this material appears to take the form of vesicles which are essentially unilamellar liposomes. The high cholesterol content of LP-X (approximately equimolar with lecithin) appears to be due to leaching out of cholesterol from tissues because of an initial excess of lecithin in the regurgitated bile; and a similar incorporation of cholesterol must take place in phospholipid liposomes which are injected intravenously. As cholesterol stabilises phospholipid bilayers, makes them less leaky and prolongs their half life in plasma, this process clearly has important implications for liposome use in vivo. It has also been claimed that cholesterol rich liposomes are more resistant to intracellular digestion by lysosomes.[27]

Studies have been done on the fate of LP-X in vivo.[28] "LP-X" containing radiodinated albumin was prepared from bile lipids and injected into rats. It disappeared rapidly from plasma, particularly in the initial period, and on a per gram of tissue basis much more appeared in the spleen than in the liver. Within the liver, non-parenchymal cells appeared to remove much more LP-X than parenchymal cells. Unfortunately, these studies were performed with an "artificial" LP-X injected into normal rats. While similar data were apparently obtained with LP-X isolated from the serum of cholestatic rats,[29] and by injected "LP-X" into cholestatic rats one must be cautious about concluding that these results represent the fate of plasma LP-X in patients with biliary obstruction whose lipoproteins and tissues may have 'equilibrated' with LP-X. The picture appears similar to that obtained with injection of liposomes into normal animals.

There are a couple of other interesting aspects about LP-X. It is thought to inhibit the uptake of chylomicron remnants by the liver but the mechanism for this effect is not understood.[28] It is also believed that LP-X can fuse with erythrocyte membranes thus expanding the red cell bilayer.[30]

The presence of LP-X in plasma is not essential for expansion of the red cell bilayer as cholesterol and lecithin will transfer from the surface of lipoproteins enriched in these substances. It is often assumed that uptake of cholesterol by nucleated cells occurs only by uptake of LDL. But LDL receptors are down regulated in cholesterol-rich cells. We have found that cholesterol is transferred to fibroblasts from patients with familial homozygous hypercholesterolaemia, who have no receptor for LDL when they are incubated with cholesterol rich LDL.[31] The added cholesterol is esterified, thus maintaining a gradient for the transfer of cholesterol from the LDL, and we believe that this accounts for the development of cholest-eryl-rich cutaneous xanthomata in some patients with biliary obstruction.

The red cell membrane from patients with severe liver disease has a high cholesterol:phospholipid ratio and this results in a decrease in mem-brane fluidity.[32] There is a large and growing literature on the effect of fluidity change on various cellular functions. We found ouabain-insens-itive Na efflux to be decreased in red cells from patients with liver disease.[33] Others showed that it was frusemide-sensitive Na efflux that was impaired, and that anion exchange was similarly affected;[34,35] both effects may be due to the effect of reduced membrane fluidity on the func-tion of the Band 3 transmembrane protein. We believe that a change in the composition of cell membranes may mediate many functional abnormalities in patients with liver disease.[26]

Red cells from patients with liver disease tend to fuse more easily in the presence of exogenous fusogens.[36] This raises some very important questions. If membrane changes are widespread in body cells in patients with liver disease does this affect the membrane fusion events which are so important in the processes of endocytosis and phagocytosis; are there corresponding changes in the composition of intracellular membranes which would affect fusion events within the cell? If liposomes are to be used in patients with liver disease, we must ask whether they would behave in the same way as if they were injected into patients without the lipoprotein changes of liver disease. When lecithin containing liposomes are injected into rats there is a marked transfer of lecithin to HDL and this seems to be an important factor causing early intravascular release of material contained within the liposomes.[37] Does a similar effect occur in man - and if so, does it occur to the same extent in patients with liver disease in whom the lipoproteins are already enriched in lecithin. Are liposomes removed by the liver affected by the structural changes which occur in the liver in cirrhosis, or by the enlargement of the spleen which is an import-ant consequence of obstruction in portal blood flow? These and many other questions remain to be answered.

I will finish by considering the problem of hepatocellular carcinoma, a relatively common and important complication of cirrhosis. Because of the high incidence of cirrhosis and hepatitis B in many parts of the world it is one of the commonest human cancers. The prognosis of this condition is very poor and there is no really effective therapy. Targeting of anti-tumour drugs to hepatocellular carcinoma is an attractive proposition and the search is on to find appropriate methods of treating such tumours. It is clear that a therapeutic approach is feasible. Human hepatocellular carcinoma can be made to grow in nude mice. If these animals are given diptheria toxin, which attacks human but not mouse cells, the tumour virtu-ally disappears.[38] Other compounds such as ricin will kill tumour cells. The problem is to make them enter the tumour without damaging normal cells. At present the most promising line of approach is to use monoclonal anti-bodies directed against surface components which are specific for the tumour cells. At the Royal Free we have found an antigen (Ll), which is present on hepatocellular carcinoma cells but not on normal liver cells, but it is also present on the luminal surface of pancreatic and breast acinar cells and the surface of small intestinal epithelial cells.

Preliminary studies suggest that monoclonal antibodies against Kl do not provide the precise localization that would be necessary for targeting such powerful agents as ricin. One factor limiting progress in this area is our poor understanding of the ultrastructure of liver tumours, of the contribution of endothelial and Kupffer cells to their structure, and of the way in which uptake mechanisms are affected in the different cell types. I hope that work in these areas will lead to progress in our attempts to treat this serious condition.

REFERENCES

1. J.A. Barrowman, Physiology of the gastro-intestinal lymphatic system, Cambridge University Press, Cambridge (1978).
2. J.S. Owen, N. McIntyre, and M.P.T. Gillett, Lipoproteins, cell membranes and cellular functions, Trends Biochem. Sci. 9:238 (1984).
3. E. Wisse, and A.M. De Leeuw, Structural elements determining transport and exchange processes in the liver, in: "Microspheres and Drug Therapy, Pharmaceutical, Immunological and Medical Aspects", S.S. Davis, L. Illum, J.G. McVie and E. Tomlinson, eds., Elsevier, Amsterdam (1984).
4. P. Motta, M. Muto, and F. Tsuneo, The Liver: an atlas of scanning electron microscopy, Igaku-Shoin, Tokyo (1978).
5. J.W.B. Bradfield, The reticulo-endothelial system and blood clearance, in: "Microspheres and Drug Therapy. Pharmaceutical, Immunological and Medical Aspects", S.S. Davis, L. Illum, J.G. McVie and E. Tomlinson, eds., Elsevier, Amsterdam (1984).
6. E. Wisse and D.L. Knook, The investigation of sinusoidal cells: a new approach to the study of liver function, in: "Progress in Liver Diseases" Vol. 6, H. Popper and F. Schaffner, eds., Grune and Stratton, New York (1979).
7. C.A. Goresky, The processes of cellular uptake and exchange in the liver, Fed. Proc. 41:3033 (1982).
8. B. Alberts, D. Bray, J. Lewis, M. Raff, K. Roberts and J.D. Watson, Molecular Biology of the Cell, Garland, New York (1983).
9. R. Weisiger, J. Gollan, and R. Ockner, Receptor for albumin on the liver cell surface may mediate uptake of fatty acids and other albumin-bound substances, Science 211:1048 (1981).
10. M.S. Brown, R.G.W. Anderson, and J.L. Goldstein, Recycling receptors: the round trip of migrant membrane proteins, Cell 32:663 (1983).
11. P. Aisen, Transferrin metabolism and the liver, Seminars in Liver Disease 4:193 (1984).
12. N.F. La Russo, Proteins in bile: how they get there and what they do, Am. J. Physiol. 247:G199 (1984).
13. J. Munniksma, M. Noteborn, T. Kooistra, S. Stienstra, J.M.W. Bouma, M. Gruber, A. Brouwer, D. Praaning-van-Dalen, and D.L. Knook, Fluid endocytosis by rat liver and spleen, Biochem. J. 192:613 (1980).
14. G. Poste, The interaction of lipid vesicles (liposomes) with cultured cells and their use as carriers for drugs and macromolecules, in: "Liposomes in Biological Systems", G. Gregoriadis and A.C. Allison, eds., John Wiley, Chichester (1980).
15. K. Jungermann and N. Katz, Functional hepatocyte heterogeneity, Hepatology 2:385 (1982).
16. R.G. Thurman and F.C. Kaufmann, Sublobular compartmentation of pharmacologic events (SCOPE): Metabolic fluxes in periportal and pericentral regions of the liver lobule, Hepatology 5:144 (1985).
17. E.C. Sleyster and D.L. Knook, Relation between localization and function of rat liver Kupffer cells, Lab. Invest. 47:484 (1982).
18. D.P. Praaning-van Dalen, A. Brouwer and D.L. Knook, Clearance capacity of rat liver Kupffer, endothelial and parenchymal cells, Gastroenterology 81:1036 (1981)
19. G. Ashwell and A.G. Morell, The role of surface carbohydrates in the hepatic recognition and transport of circulating glycoproteins, Adv. Enzymol. 41:99 (1974).
20. H. Popper, W. Reutter, F. Gudat, and E. Kottgen (eds.), "Structural carbohydrates in the liver" MTP Press, Boston (1983).
21. A. Kirn, J.P. Gut, J-L. Gendrault, Interaction of viruses with sinusoidal cells, in: "Progress in Liver Diseases" Vol. 7., H. Popper and F. Schaffner, eds., Grune and Stratton, New York (1982).

22. R.O. Brady, S.P. James, and J.A. Barranger, The liver in lipid storage disease: biochemical basis of pathogenesis and clinical features, in: "Progress in Liver Diseases", Vol. 7. H. Popper and F. Schaffner, eds., Grune and Stratton, New York (1982).

23. E. Kottgen, R. Buschel, and C. Bauer, Glycoproteins in liver disease, in: "Liver in Metabolic Diseases", L. Bianchi, W. Gerok, L. Landmann, K. Sickinger and G.A. Stalder, eds, MTP Press, Boston (1983).

24. F. Schaffner and H. Popper, Capillarization of hepatic sinusoids in man, Gastroenterology 44:239 (1963).

25. R.C. Day, D.S. Harry, and N. McIntyre, Plasma lipoproteins and the liver, in: "Liver and Biliary Disease", R. Wright, K.G.G.M. Alberti, S. Karran, and G.H. Millward-Sadler, eds., W.B. Saunders, London (1979).

26. D.S. Harry, J.S. Owen, and N. McIntyre, Lipids, Lipoproteins and Cell Membranes, in: "Progress in Liver Disease", Vol. 7, H. Popper and F. Schaffner, eds., Grune and Stratton, New York (1982).

27. I.R. McDougall, Liposomes as diagnostic tools, in: "Liposomes in Biological systems", G. Gregoriadis and A.C. Allison, eds., John Wiley, Chichester (1980).

28. A.K. Walli and D. Seidel, Role of lipoprotein X in the pathogenesis of cholestatic hypercholesterolaemia, J. Clin. Invest. 74:867 (1984)

29. D. Seidel, H.V. Buff, U. Fauser, and U. Bleyl, On the metabolism of lipoprotein X, Clin. Chim. Acta 66:195 (1976).

30. A.J. Verkleij, I.L.D. Nauta, J.M. Werre, J.G. Mandersloot, B. Reinders, P.H.J.Th. Ververgaert, and J. De Gier, The fusion of abnormal plasma lipoprotein-X and the erythrocyte membrane in patients which cholestasis studied by electron microscopy, Biochim Biophys. Acta 436:366 (1976).

31. J.S. Owen and M.P.T. Gillet, Plasma lipids, lipoproteins and cell membranes, Biochem. Soc. Trans. 11:336 (1983).

32. J.S. Owen, K.R. Bruckdorfer, R.C. Day, and N. McIntyre, Decreased erythrocyte membrane fluidity and altered lipid composition in human liver disease, J. Lipid Res. 23:124 (1982).

33. J.S. Owen and N. McIntyre, Erythrocyte lipid composition and sodium transport in human liver disease, Biochim. Biophys. Acta 510:168 (1978).

34. P. Jackson and D.B. Morgan, The relation between membrane cholesterol and phospholipid and sodium efflux in erythrocytes from health subjects and patients with chronic cholestasis, Clin. Sci. 62:101 (1982).

35. P. Jackson and D.B. Morgan, The relation between the membrane cholesterol content and anion exchange in the erythrocytes of patients with cholestasis, Biochim. Biophys. Acta 693:99 (1982).

36. M.J. Hope, K.R. Bruckdorfer, J.S. Owen and J.A. Lucy, Chemically induced cell fusion in vitro of erythrocytes from patients with liver diseases, Biochem Soc. Trans. 5:1144 (1977).

37. G. Scherphof, F. Roerdink, D. Hoekstra, J. Zborowski, and E. Wisse, Stability of liposomes in presence of blood consitituents: consequences for uptake of liposomal lipid and entrapped compounds by rat liver cells, in: "Liposomes in Biological Systems", G. Gregoriadis and A.C. Allison, eds., John Wiley, Chichester (1980).

38. M.J. Bassendine, U. Shipton, N. Wright, H.C. Thomas, and S. Sherlock, Cytokinetic studies of human primary liver cell cancer in athymic mice: effect of protein toxins, Abstract 24, 15th Meeting of European Association for the Study of the Liver (1980).

39. K.H. Wiedmann, L.K. Trejdosiewicz, and H.C. Thomas, Human hepatocellular carcinoma: cross-reactive and idiotypic antigens associated with malignant transformation of epithelial cells. Submitted for publication (1985).

TARGETING WITH SYNTHETIC POLYMERS: A REALISTIC GOAL

J.B. Lloyd

Biochemistry Research Laboratory, Department of Biological
Sciences, University of Keele, Staffordshire, England

INTRODUCTION

This short article has two concerns. The first is to counter a
currently prevalent scepticism about the possibility of targeting drugs
with macromolecules. The second and subsidiary concern is to present an
advocacy for the synthetic (over against the natural) macromolecule for this
purpose.

SOLUBLE MACROMOLECULES ARE NOT HANDLED LIKE PARTICLES

After a decade of high expectations it is now clear that liposomes
and other microparticles have limited potential for targeting in the whole
animal. The reason is the existence of the reticuloendothelial system, a
network of cells whose raison d'etre appears to be the removal of particles,
such as bacteria, from the bloodstream. The cells of this system perform
their task with enthusiasm, and efficiently engulf potentially therapeutic
particles before they can reach their intended target. Furthermore the
parenchymal cells of most tissues outside the reticuloendothelial system
have little phagocytic capacity and are also rather inaccessible to particles in the bloodstream because the capillary endothelium provides an
effective barrier.

A period of disenchantment with liposomes followed the years of hyper-
bole, and this was intensified by the recognition that in many instances of
liposomes failing to fulfil their devotees' hopes, this outcome could have
been predicted.[1] Despondency is now in turn giving way to a cautious appreciation that (a) manipulation of the size and surface characteristics of
liposomes and other microparticles can alter their distribution within the
reticuloendothelial system, and that targeting to subsets of this system,
e.g. in the lung or bone marrow, has some therapeutic potential; (b) de-
laying uptake by the reticuloendothelial system permits targeting to lymphocyte populations and other intravascular sites.

Soluble macromolecules are handled by the body quite differently from
particles, and the factors that limit the targeting potential of the latter
do not apply. First, soluble macromolecules encounter fewer barriers to
their movement around the body; they can pass into many organs by transport
(diacytosis) across capillary endothelium or (in the liver) by passage

97

through the fenestrations connecting the sinusoidal lumen to the space of Disse. Secondly, and even more important, uptake of soluble macromolecules is by pinocytosis not phagocytosis. The differences between these two processes are discussed more fully elsewhere in this volume,[2] but the most significant is that the reticuloendothelial system has no monopoly of pinocytosis. Whereas any foreign particle introduced into the bloodstream will be quickly removed by the reticuloendothelial system, it is possible to introduce a soluble macromolecule and find it still circulating many hours later. Pinocytosis, although it occurs in most cell types, is a rather inefficient process, unless the substrate binds to the plasma membrane of the pinocytosing cell. Such adsorption depends on the nature of the substrate and on the presence of "receptors" for that substrate on the cell surface.

Adsorptive pinocytosis displays specificity of two types. There is substrate specificity and cell-type specificity. Substrate specificity may be broad or narrow. Proteins are captured more efficiently if they possess exposed hydrophobic domains;[3] this is an example of non-specific adsorptive pinocytosis. An example of much more specific adsorptive (receptor-mediated) pinocytosis is the enhanced uptake (observed in many cell types) of glycoproteins if they possess an exposed mannose-6-phosphate moiety.[4] It is however cell-type specificity that has potential for targeting. The most celebrated example is the uptake by mammalian hepatocytes of glycoproteins that bear a terminal galactose residue.[5] The receptor appears to be present in significant density only on the liver parenchymal cell, and it leads to the uptake by this cell-type of natural or synthetic glycoproteins bearing the relevant structural feature. Trouet and his colleagues[6] have utilized this hepatocyte receptor to target daunorubicin and other chemotherapeutic agents to hepatocytes and hepatocyte-derived neoplasms; their drug carrier is a "neoglycoprotein", lactosylated serum albumin, to which the drug is attached by a lysosomally-degradable tetrapeptide. An essentially similar approach, but using a synthetic macromolecule, polyhydroxypropylmethacrylamide, as carrier, was reported by this laboratory in the preceding volume in this series.[7]

Few other clear-cut examples of cell-type specific adsorptive pinocytosis can be offered at the present time. However it is unlikely that the hepatocyte-galactose system is a unique phenomenon.

Targeting is particularly desirable in cancer chemotherapy, where therapeutic indices are notoriously low and the conditions to be treated are common and life-threatening. Success would seem to demand the existence on tumour cells of receptors that are absent or in significantly lower density on the corresponding normal cell, and the search for glycoprotein and glycolipid[8] tumour markers and for other tumour-specific antigens is currently very active. The increasing awareness that the individual cells within a tumour often display an unexpected heterogeneity must however inject a note of caution into the quest for targeting regimes based on highly specific tumour markers.

Quite apart from the active targeting possible with soluble macromolecules, there are major advantages ascribable to "negative" targeting. Trouet and colleagues have shown that their daunorubicin-albumin conjugate is much less cardiotoxic in man than the free drug (P. Pirson, personal communication). This welcome finding is presumably due to the poor pinocytic capacity of the heart muscle as well as to the specific uptake by liver. In unpublished work we too have found a decrease in daunorubicin cardiotoxicity in mice when the drug was administered as a conjugate with polyhydroxypropylmethacryamide.

To date, attempts to target using soluble macromolecules have been based on an assumed delivery to the lysosomes following pinocytic uptake. The ability of the lysosomal enzymes to cleave drug from carrier and, preferably, to degrade the carrier have been necessary components of conjugate design. While the lysosome remains the most accesible target, recent advances in cell biology (see ref. 1) make it clear that some pinocytosed ligands cross the endosome membrane into the cytoplasm and thence to the nucleus or ribosomes. The mechanisms underlying endosomal ligand sorting are not yet understood but, when they are, it may be possible to harness them to effect drug delivery.

SOLUBLE SYNTHETIC POLYMERS AS DRUG CARRIERS

At Cape Sounion natural beauty and man-made beauty compete for our admiration, the seascape and the contours of the coastline being brilliantly complemented by the Temple of Poseidon on the cliff top. Similarly, as we contemplate the use of soluble macromolecules for drug targeting, natural and synthetic polymers present themselves as candidates. In spite of the advocacy of Ringsdorf[9] and Kopecek,[10] it seems that the considerable advantages of synthetic macromolecules are not widely appreciated. I should like to redress this balance.

Chemical Considerations

Here synthetic polymers seem to have a clear advantage. In comparison with natural macromolecules such as proteins, they are easier and cheaper to produce in quantity and high purity. Furthermore they can be tailor-made to pre-determined specifications, so that the characteristics of the carrier (e.g. molecular size, charge, hydrophobicity, capacity for drug attachment) can be optimized to a degree quite impossible if we use nature's own macromolecules. Synthetic polymers are more robust, with consequently greater stability during manipulation and storage and less laborious procedures for chemical attachment of drug and targeting moieties. Only in one respect do proteins have a slight advantage over synthetic macromolecules: they are monodisperse, whereas the latter (as also polysaccharides) are, to a greater or lesser degree, polydisperse.

Biological Considerations

Here the advantages are more evenly balanced, but synthetic polymers seem to have a slight edge. Work[7] in this laboratory and from that of Trouet[6] has demonstrated that it is possible to prepare conjugates based on either albumin or polyhydroxypropylmethacrylamide with the following properties: receptor-mediated pinocytosis utilizing the hepatocyte receptor for galactose moieties (see above); intralysosomal release of drug moieties attached to the polymer by an appropriately designed oligopeptide. Acute or chronic toxicity need pose no problem with synthetic polymers: the most promising candidates such as polyvinylpyrrolidone and polyhydroxypropyl-methacrylamide were developed originally as plasma expanders and extensively tested in vivo. Immunogenicity is a potential hazard with all macromolecular carriers, being minimal with homopolymers but increasing with heterogeneity in the backbone or due to pendent groups. Proteins with their highly heterogeneous composition are intrinsically much more immunogenic than most synthetic polymers (or homopolysaccharides).

A disadvantage of most synthetic polymers is their non-biodegradability. A molecule such as polyhydroxypropylmethacrylamide, having released its charge of drug, will remain in the lysosomes of the target cell, probably for the life of the cell. By contrast a protein, and many polysaccharides, will be degraded. Even when a non-degradable polymer is released on the death of the cell, it will still remain in the body because, to be useful in targeting, its molecular weight must be above the renal threshold. The importance of this problem posed by the accumulation of carrier in the body depends on the dosage regime contemplated. It would be a major drawback for any chronic administration, but more acceptable for the acute therapy of a life-threatening disease. Nevertheless the search for synthetic polymers that are lysosomally degradable to fragments able to escape from the lysosome (and the cell) is a worthwhile objective. Meanwhile a partial solution has been reported:[11,12] small chains of polyhydroxypropylmethacrylamide joined by lysosomally degradable cross-links. In intact form this carrier is too large to enter the glomerular filtrate, but lysosomal processing in the target cell will both release the drug and also degrade the carrier to excretable fragments.

CONCLUSIONS

Soluble macromolecules have considerable potential for drug delivery. Negative targeting, avoiding undesirable side-effects of drugs, has already been demonstrated, and positive targeting is a realistic aim. Synthetic macromolecules have many advantages over natural macromolecules as potential drug carriers.

ACKNOWLEDGEMENTS

The author's own research on drug targeting is supported by the Cancer Research Campaign. I also thank Dr. R.L. Brent for his hospitality at the Stein Research Center, Jefferson Medical College, Philadelphia, where this paper was written.

REFERENCES

1. G. Poste, Drug targeting in cancer therapy, in: "Receptor-Mediated Targeting of Drugs", G. Gregoriadis, G. Poste, J. Senior and A. Trouet, eds., Plenum Press, New York and London, (1984).
2. J.B. Lloyd, Endocytosis and lysosomes: recent progress in intracellular traffic, in: "Targeting of Drugs with Synthetic Systems", G. Gregoriadis, G. Poste, J. Senior and A. Trouet, eds., Plenum Press, New York and London, (1986).
3. J.B. Lloyd and K.E. Williams, Non-specific adsorptive pinocytosis, Biochem. Soc. Trans., 12:527 (1984).
4. K.E. Creek and W.S. Sly, The role of the phosphomannosyl receptor in the transport of acid hydrolases to lysosomes, in: "Lysosomes in Biology and Pathology", Vol. 7, J.T. Dingle, R.T. Dean and W.S. Sly, eds., Elsevier, Amsterdam, (1984).
5. G. Ashwell and J. Harford, Carbohydrate-specific receptors of the liver, Annu. Rev. Biochem., 51:531 (1982).
6. Y.-J. Schneider, J. Abarca, E. Aboud-Pirak, R. Baurain, F. Ceulemans, D. Deprez-de Campaneere, B. Lesur, M. Masquelier, C. Otte-Slachmuylder, D. Rolin-van Swieten and A. Trouet, Drug targeting in human cancer chemotherapy, in: "Receptor-Mediated Targeting of Drugs", G. Gregoriadis, G. Poste, J. Senior and A. Trouet, eds., Plenum Press, New York and London, (1984).
7. J.B. Lloyd, R. Duncan, J. Kopecek and P. Rejmanova, Targeting and lysosomal handling of polymethacrylamide-oligopeptide conjugates, in: "Receptor-Mediated Targeting of Drugs", G. Gregoriadis, G. Poste, J. Senior and A. Trouet, eds., Plenum Press, New York and London, (1984).
8. S. Hakomori, Tumor-associated glycolipid markers: possible targets for drug and immunotoxin delivery, in: "Targeting of Drugs with

Synthetic Systems", G. Gregoriadis, G. Poste, J. Senior and A. Trouet, eds., Plenum Press, New York and London, (1986).

9. H. Ringsdorf, Structure and properties of pharmacologically active polymers, J. Polymer Sci. Symposium, 51:135 (1975).

10. J. Kopecek, Biodegradation of polymers for biomedical use, in: "IUPAC Macromolecules", H. Benoit and P. Rempp, eds., Pergamon Press, Oxford and New York, (1982).

11. J. Kopecek, I. Clifkova, P. Rejmanova, J. Strohalm, B. Obereigner and K. Ulbrich, Polymers containing enzymatically degradable bonds, 4. Preliminary experiments in vivo, Makromol. Chem., 182:2941, (1981).

12. S.A. Cartlidge, R. Duncan, J.B. Lloyd, P. Rejmanova and J. Kopecek, Soluble, crosslinked N-(2-hydroxypropyl)methacrylamide copolymers as potential drug carriers I. Pinocytosis by rat visceral yolk sacs and rat intestine cultured in vitro. Effect of molecular weight on uptake and intracellular degradation, J. Controlled Release, in press.

DRUG-POLY(LYSINE) CONJUGATES: THEIR POTENTIAL FOR CHEMOTHERAPY AND FOR THE STUDY OF ENDOCYTOSIS

Hugues J.-P. Ryser and Wei-Chiang Shen

Department of Pathology, Boston University School of Medicine, 80 E. Concord Street, Boston, MA 02118, U.S.A.

ENDOCYTOSIS OF POLY(LYSINE)

Poly(L-Lysine) (poly(Lys)), a strongly cationic macromolecule, is efficiently transported into cells by endocytosis (Ryser et al., 1982). This uptake is preceeded by a strong adsorption to the cell surface, which is due to a non-specific interaction of the polymer's positive charges with the negative charges present at the surface of most mammalian cells. The membrane transport of poly(Lys), therefore, can be defined as non-specific adsorptive endocytosis, as opposed to fluid-phase endocytosis or to receptor-mediated endocytosis (Fig. 1). In fluid-phase endocytosis macromolecules do not bind to any sites at the cell surface, and their transport occurs only through the internalization of a small quantum of medium engulfed by the constitutive process of membrane vesiculation. In receptor-mediated endocytosis a macromolecule binds to a specific site usually located in clathrin coated area of the membrane. Such specialized area gives rise to clathrin coated pits and baskets, which pinch off to form clathrin coated vesicles. The number of binding sites for a specific ligand is usually limited (1×10^4 to 1×10^6/cell) which accounts for the saturable nature of the transport process (Fig. 1). Specificity is established by demonstrating competition with unlabeled ligand and lack of competition with chemically related molecules. The specific sites are commonly called membrane receptors, even when they are not yet associated with specific cellular functions. The kinetics of these three forms of endocytosis are predictably quite different, as illustrated in Fig. 1. When uptake is plotted against concentration, it assumes a linear non-saturable pattern in the case of fluid-phase and of adsorptive non-specific endocytosis. Fluid-phase endocytosis is the least efficient of the three kinds of transport. Efficiency, however, is a relative concept that varies with concentrations of the considered macromolecules in the medium. At low concentration the receptor-mediated process is by far the most efficient. At higher concentration, however, receptor-mediated endocytosis levels off, but non-specific adsorptive endocytosis keeps increasing and may end up transporting more macromolecules than are transported with the help of receptors (Fig. 1). Even transport by fluid-phase endocytosis at very high concentrations may in theory exceed that of receptor-mediated endocytosis. Macromolecules can be modified in manners that will change their endocytosis either quantitatively or qualitatively. We have discribed an instance where a modification of poly(Lys), namely its complexing with heparin, results in an uptake that resembles receptor-mediated endocytosis (Morad et. al, 1984). We have also

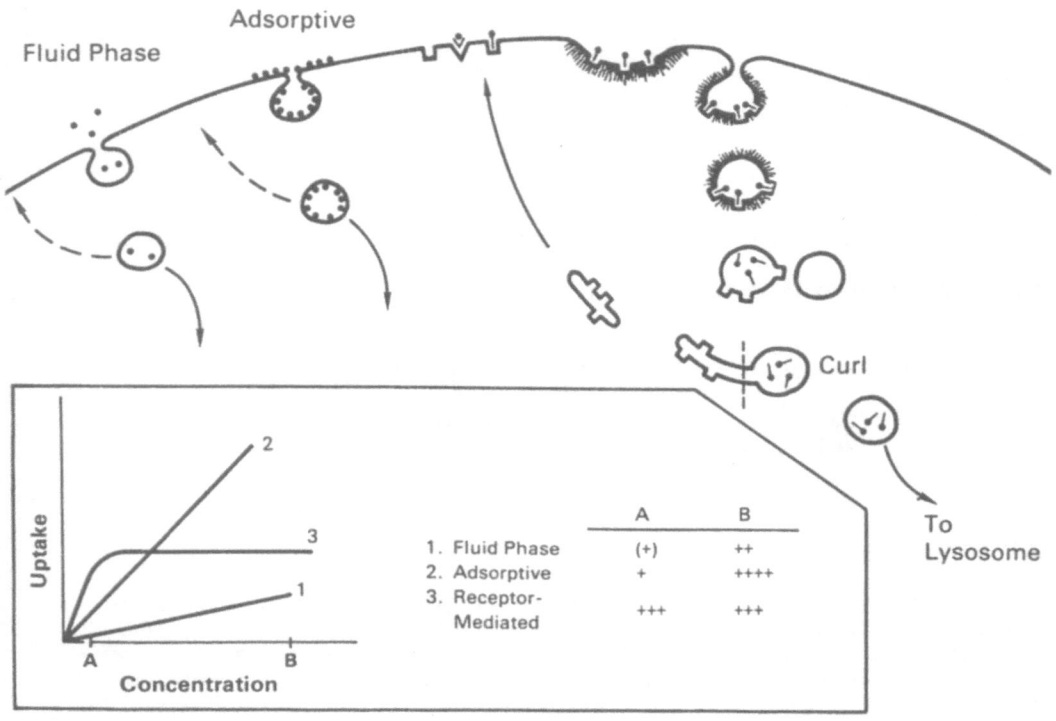

Fig. 1. Diagram contrasting the three main forms of endocytosis (see text).

described cases where macromolecules that are ordinarily taken up by fluid-
phase endocytosis (human serum albumin (HSA) and horseradish peroxidase
(HRP)) are modified by the addition of a fragment of poly(Lys) and are
then transported by adsorptive endocytosis (Shen and Ryser, 1978). Fig. 2
shows that in the case of HRP, the conjugation of a poly(Lys) of Mr 13,000
to an enzyme that is several times larger results in a 1,000-fold increase
in the rate of enzyme uptake (Ryser et al., 1982). Poly(Lys) acts here as
a carrier for another macromolecule. Poly(Lys) not surprisingly can also
serve as carrier for small molecules like drugs as will be examined in some
details below.

     In all three forms of endocytosis (with only few exceptions) the ves-
icle-bound macromolecules reach lysosomes, where they are degraded by lysosomal hydrolases, as shown in Fig. 3.

     Primary endocytotic vesicles are rapidly acidified to pH 5.0 - 5.5,
(Tycko and Maxwell, 1982). At this pH many ligands will dissociate from
their specific receptors (Helenius et al., 1983). Vesicles also fuse with
each other, undergo major changes in shape and form tubular extensions
(Fig. 3). It is assumed that these complex structures represent compartments (CURL) where some receptor-ligand complexes are dissociated (Geuze
et al., 1983). They fragment into structures that go two separate routes,
namely tubular structures that carry unoccupied receptors back to the cell
surface (receptor-recycling) and a compartment that carries dissociated
ligands as well as the bulk of other internalized macromolecules to lysosomes. Following fusion with either primary or secondary lysosomes, these

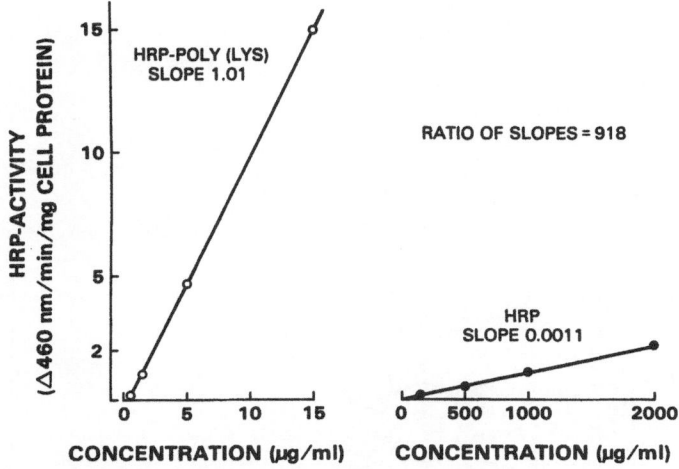

Fig. 2. Uptake of horseradish peroxidase (HRP) and HRP-
poly(Lys) as a function of HRP-concentration. Exposure was 60 min at 37°C. Slopes were calculated
by regression analysis. From Ryser et al., 1982.

vesicles become digestive vacuoles in which foreign macromolecules can be
enzymatically degraded. The breakdown products of the internalized macromolecules finally diffuse through the lysosomal membrane. Acidification
of endosomes is also credited with initiating the escape of certain viruses
and of the active moiety of diphtheria toxin (Helenius et al., 1983) which
thus pass undegraded from endosomes into the cytoplasm (Fig. 3). If the
macromolecule is not dissociated from its binding site at acid pH it may
itself return to the cell surface, as illustrated by the case of apotransferrin. The membrane binding of poly(Lys) is not decreased at acid pH
(Morad et al., 1984), hence a fraction of internalized poly(Lys) can be
expected to return to the cell surface in the same way, as shown in Fig. 3.
Another fraction of ingested poly(Lys) reaches lysosomes where it is degraded. Drugs conjugated to poly(Lys) are thus freed inside lysosomes
following poly(Lys) degradation and, by diffusion, reach other parts of the
cell.

USE OF POLY(LYSINE) AND OTHER POLYCATIONS AS CARRIERS FOR METHOTREXATE

The anti-cancer drug methotrexate (MTX) was conjugated through one of
its carboxyl groups to epsilon-amino groups of poly(Lys) at a ratio of one
drug molecule per 29 amino groups (Ryser and Shen, 1978). Surprisingly,
the cellular uptake of ^{3}H-MTX by cultured cells was greater when the drug
was given as a conjugate than when given as free drug. The difference was
particularly marked in cell lines defective in MTX transport, such as CHO
PRO[-]3 MTX ^{R}11 5-3 (Ryser and Shen, 1978) or the tumor line M5076 (Fig. 4).
Although such mutants are resistant to therapeutic concentrations of MTX,
they were killed by MTX-poly(Lys) (Figs. 4 and 5). The transport defect
in these cases is circumvented by the endocytotic pathway that leads to

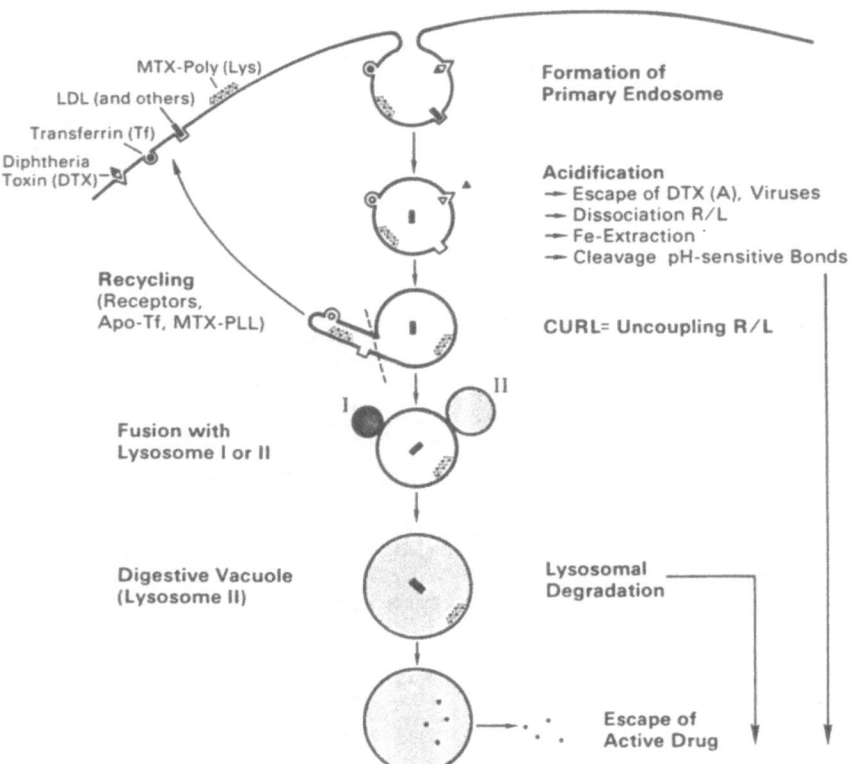

Fig. 3. Different fate of macromolecules internalized
by receptor-mediated endocytosis. ▲ = A-
fragment of diptheria toxin; O= apotransferrin;
▮ = low density lipoprotein (LDL); CURL = compartment of dissociation of ligand (L) from receptor (R); ⊔ = unoccupied LDL-receptor; I, II
= primary and secondary lysosomes.

the intracellular release of drug following lysosomal degradation of the
poly(Lys) carrier. Cells proficient in MTX transport were about equally
sensitive to free and conjugated MTX (Ryser and Shen, 1978). Gel filtration of the lysate of cells exposed to ^{3}H-MTX-poly(Lys) revealed a small
molecular ^{3}H-MTX fraction eluting slightly ahead of a MTX-marker (Shen and
Ryser, 1979), indicating that MTX freed from poly(Lys) inside cells carried
one or a few lysine residues. This adduct inhibited dihydrofolate reductase (DHFR) in vitro. When reisolated from exposed cells and tested in vitro
at a concentration of 1.25×10^{-8}M MTX, it had about 60% of the inhibitory
activity of free MTX, while the intact conjugate was not inhibitory at all,
but became partially active following trypsin treatment in vitro (Table 1).
The ID_{50}'s of free and conjugated MTX for DHFR differed by a factor of 500
(Shen and Ryser, 1979). These findings demonstrate that the cytotoxicity
of MTX-poly(Lys) must be caused by the small molecular adduct tested in
Table 1.

Cytotoxic effects comparable to those of MTX-poly(Lys) were obtained
when MTX was attached to a natural cationic protein, lysine-rich histone
(Shen and Ryser, 1981d). At comparable MTX-concentrations, MTX-histone
and MTX-poly(Lys) were taken up to the same extent by CHO cells and had
identical cytotoxic effects (Fig. 5), indicating that MTX was adequately

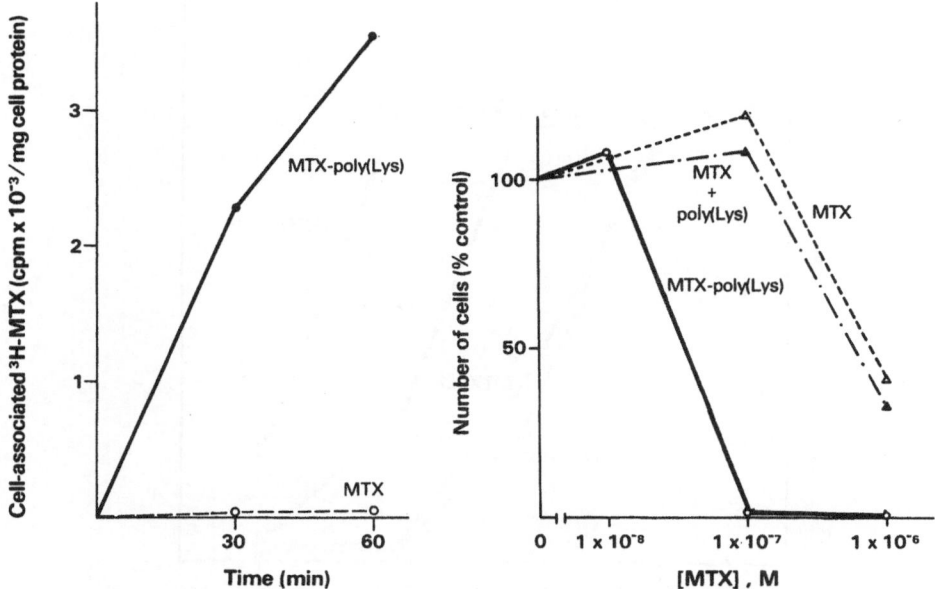

Fig. 4. Cellular uptake and cytocidal effect of MTX and
MTX-poly(Lys) in M5076 tumor cells. Low uptake
of free MTX by M5076 cells (left panel) is related to their 20-fold resistance to free MTX
compared to MTX-poly(Lys) (right panel). Procedures were as described for CHO cells in Ryser
and Shen, 1978.

Table 1. Inhibition of DHFR[a] by MTX-poly(Lys) Digests

1.25×10^{-8}M MTX as	% Inhibition
MTX-poly(Lys)[a]	0
MTX-poly(Lys) (trysin)[a]	18.7
MTX-poly(Lys) (cellular)[b]	33.0
MTX	56.0

[a]Determination of dihydrofolate reductase (DHFR) activity, Mr and trypsin
treatment of poly(Lys) were as described in Shen and Ryser, 1979.

[b]Cellular digest was recovered from lysate of cells exposed to 1×10^{-6}M
MTX as ³H-MTX-poly(Lys) for 24 h, as described in Shen and Ryser, 1979.

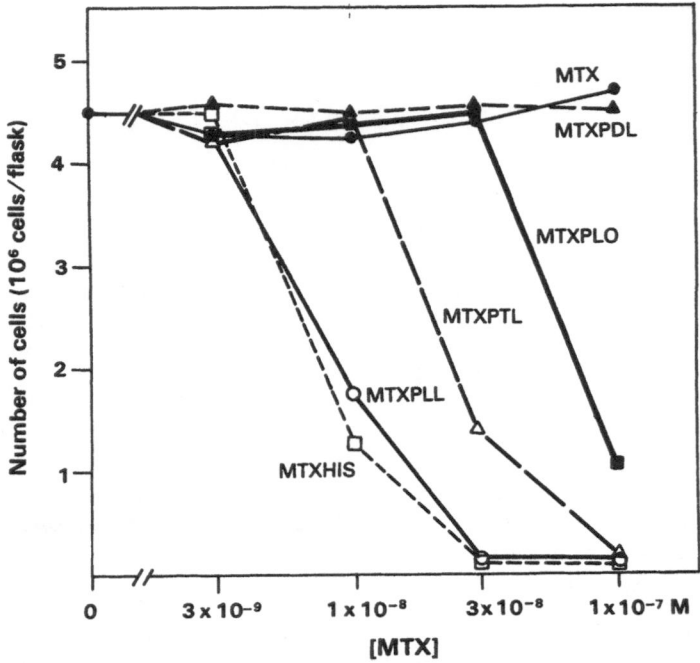

Fig. 5. Comparative cytocidal effect or 5 basic polypeptidic
conjugates on MTX on MTX-resistant CHO cells. MTX,
free drug; MTXPLL and MTXPDL, MTX-conjugated L- and
D-isomers of poly(Lys); MTXPLO, MTX-poly(L-ornithine);
MTXPTL, MTX poly(L-tyrosine:L-lysine) (1:1); MTXHIS,
MTX conjugated to lysine-rich histone. The conjugation ratios were as described for MTX-poly(Lys) by
Shen and Ryser, 1979. The cellular uptakes of all
five conjugates were comparable, and their different
effect reflect different rates of intracellular breakdown.

released from internalized MTX-histone. By contrast, MTX conjugated to
poly(D-lys) was pharmacologically totally inactive as shown by the upper
line of Fig. 5. The D-isomer, a peptide made of an unnatural amino acid,
is not susceptible to lysosomal enzymes and consequently does not release
MTX in active form inside cells. This absence of release was demonstrated
by gel filtration of the lysate of cells exposed to ^{3}H-MTX-poly(D-Lys)
(Shen and Ryser, 1979). Fig. 5 also shows that poly(L-ornithine) and a
co-polymer of lysine:tyrosine (1:1) assumed intermediate positions with regard to their carrier properties (Shen and Ryser, 1981d). In all cases,
the cells could be protected from the cytotoxic effect of MTX-conjugates
by large doses of leucovorin, a MTX-antagonist (Shen and Ryser, 1981a).
(Leucovorin, a tetrahydrofolate, is an analog of the product of the enzymatic reaction blocked by MTX). This demonstrates that the cytocidal effect
of MTX-conjugates is due to the interaction of intracellularly released
MTX with its cytoplasmic target, DHFR. As shown by Fig. 5, of all poly-
cationic carriers tested, poly(Lys) and lysine-rich histone were most and
equally active, while poly(D-lys) was inactive.

While poly(D-lys) is a useless carrier for MTX when the drug is conjugated directly to its epsilon-amino group, it turned out to be an ideal
tool to study linkages between drugs and carriers.

Table 2. Effect of Triglycine Spacer on Cytocidal Action of MTX-poly (D-lys)[a]

[MTX] (M)	Number of cells (in millions) after 4 days in presence of			
	MTX	MTX–GGG	MTX-poly(D-Lys)	MTX–GGG-poly (D-lys)
0	5.85	--	--	--
1×10^{-7}	5.79	--	5.96	0.93
1×10^{-7}	5.48	5.68	--	0.21
3×10^{-7}	5.84	--	6.17	0.31
1×10^{-6}	1.63	--	6.18	0.30

[a]The Mr of poly(D-lys) and procedures were as described in Shen and Ryser, 1979, and Shen et al., 1985.

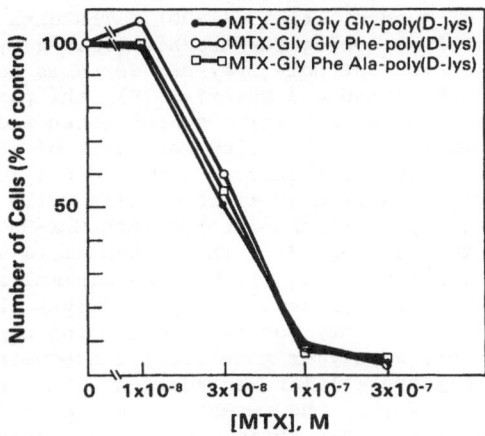

Fig. 6. Cytocidal effect of three conjugates in which MTX
was linked to poly(D-lys) through three different
tripeptides. Experiment was carried out on MTX-
resistant CHO cells. MTX as well as MTX-poly(D-
lys) were ineffective at the tested concentrations.
Procedure as described in Shen and Ryser, 1979.
The tripeptide conjugates were prepared as described
for the triglycine conjugate in Shen et al., 1985.

Taking advantage of the fact that poly(D-lys) is efficiently transported, but is not itself degraded inside cells, several types of spacers, inserted between drug and carrier, were studied for their susceptibility to intracellular cleavage. Beyond their practical implication for drug delivery, spacers have the potential of identifying intracellular environments and biochemical reactions encountered by an internalized macromolecule in the course of its intracellular pathway. Three different types of spacers were considered, with specific sensitivities to either (a) proteases, (b) acid pH, (c) reduction (disulfide bond).

Protease-Sensitive Spacers

The simplest protease sensitive linkage conceivable is the tripeptide triglycine. It was inserted between MTX and poly(D-lysine) in a two step synthesis, using MTX-Gly-Gly-Gly as intermediate. As shown by Table 2 the intermediate as well as free MTX had no effect on the MTX-resistant CHO. The MTX-Gly-Gly-Gly-Poly(D-lys), however, was very effective, (Shen and Ryser, 1981d, Shen and Ryser, 1982, Shen et al., 1985). Does the composition of a tripeptide influence its effectiveness as a protease sensitive spacer? Two other tripeptides, namely Gly-Gly-Phe and Gly-Phe-A-a were used in similar fashion, with exactly the same results (Fig. 6). Oligopeptides have been used as spacers between daunorubicin and a protein carrier by Trouet et al., (1982) who found that in their system a tetrapeptide with the sequence Ala-Leu-Ala-Leu was optimal. This requirement, we believe, is peculiar to anthracycline-protein linkages which use the sugar-amino group of the drug as linking point. It is known that this amino group is essential for drug action and must be functionally restored upon cleavage of a peptide spacer. The level of restoration may vary with the nature of the amino acid(s) left on the drug upon cleavage. Other models of drug conjugates may require peptide spacers with other sequences, as suggested by the work of Duncan, et al. (1980). The data of Fig. 6 indicate that the MTX-poly(D-lys) conjugates are not limited by such requirements. Since we have shown that poly(Lys) can serve as carrier for proteins such as HSA and HRP, (Shen and Ryser, 1978), the possibility was tested to use HSA as a protease sensitive spacer (Shen and Ryser, 1982). MTX was first conjugated to HSA at a molecular ratio of about 18:1 to yield a MTX-HSA intermediate which was then conjugated to poly(D-Lys). The MTX-HSA-poly(D-Lys) was very effective in killing cells while both the direct linkage of MTX to poly(D-Lys) and the intermediate MTX-HSA were ineffective at the concentrations tested (Fig. 7). The intermediate was inactive because of insufficient uptake (Fig. 7), but the minimal amount of MTX-HSA internalized was efficiently degraded. MTX-poly(D-lys) behaved in opposite fashion: its uptake was excellent, but its degradation was nil. Thus, to qualify as a good carrier, a polymer must combine adequate uptake and adequate susceptibility to intracellular degradation. Compared to the direct conjugation of MTX to poly(Lys), the two-step conjugation in which the drug is first attached to a protein has three substantial advantages. First, the protein may accept a large number of drug molecules and act as a "drug-reservoir". Second, an increase in the drug-carrier ratio can occur without any change in the cationic character of the poly(Lys) vector. Third, the drug-substituted protein may be cleaved from poly(Lys) and subsequently follow a different route than MTX-poly(Lys); it may for instance be less likely to recycle to the cell surface. It is conceivable, furthermore that HSA-poly(Lys) may have a lesser non-specific charge-related cytotoxicity than poly(Lys).

The properties of the tripeptide and HSA spacers can be compared to those of the two direct MTX-poly(Lys) conjugates in a cartoon using the metaphor of the Trojan horse (Fig. 8). In cells defective in MTX-transport,

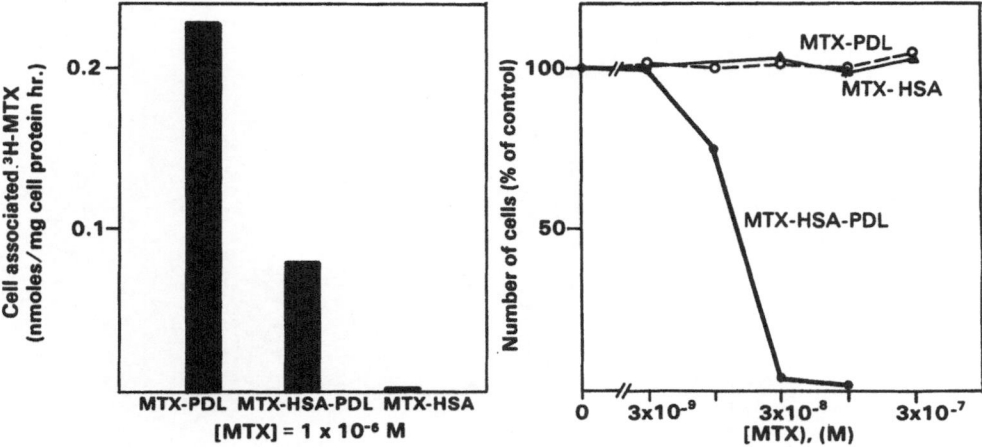

Fig. 7. Cellular uptake and cytocidal effect of MTX-poly
(D-lys), before and after insertion of human serum
albumin as a drug-carrying spacer. Uptake (left
panel) and growth inhibitory effect (right panel)
were measured as described in Shen and Ryser (1979,
1982). PDL, poly(D-lys); MTX-HSA, intermediate in
which MTX is conjugated to human serum albumin.
Experiment was carried out on MTX-resistant CHO cells.

the conjugation of MTX to poly(Lys) initiates a new mode of entry into the
"City of Troy". In the case of poly(Lys), the molecules of MTX are freed
and capable of acting as effective killers. In the case of poly(D-lys),
they remain tied to "the horse" and unable to act. When attached to poly
(D-lys) through a cleavable spacer, they are freed and effective. In yet
another case, they are tied to in a chariot pulled by the horse but are
released from the chariot and able to do their job. The latter ploy would
surely have been too obvious to fool the Trojan, but it is good enough to
fool the MTX-resistant cell. In all cases, drug is released from the internalized carrier through the proteolytic action of lysosomal enzymes.

Acid-Sensitive Spacers

Another remarkable property of lysosomes, besides their high content in
hydrolytic enzymes, is their acidic pH of 4.5 to 5.5. We explored the possibility of exploiting the acidic milieu of lysosomes to release drugs attached to poly(D-lysine) through an acid sensitive linkage (Shen and Ryser,
1981b).

Tricarboxylic acids in which two of the carboxyls are in cis-configur-
ation, as in cis-aconitic acid, can be conjugated, at one end, to the amino
group of a drug like daunomycin (DM) and at the other end, to a carrier,
leaving the middle carboxyl group unoccupied. In such a conjugate, the
cis-aconityl-DM bond is acid labile. As carrier, we used either an Affigel
701 bead (AFB) that can easily be centrifuged, or poly(D-Lys). A DM-ca-AFB
preparation was exposed to buffers of decreasing pH, and the free DM was
determined in the supernatant. Incubation at acid pH caused rapid drug
release while at neutral pH the conjugate was stable. As a control, we used
a conjugate in which cis-aconitic acid was replaced by the corresponding
cis-dicarboxylic acid, namely maleic acid (ma), to form DM-ma-PDL. Exposing cells to DM-ca-PDL caused a strong dose-related cell killing, while the
acid-stable DM-ma-PDL had no effect (Fig. 9) (Shen and Ryser, 1981b). Since

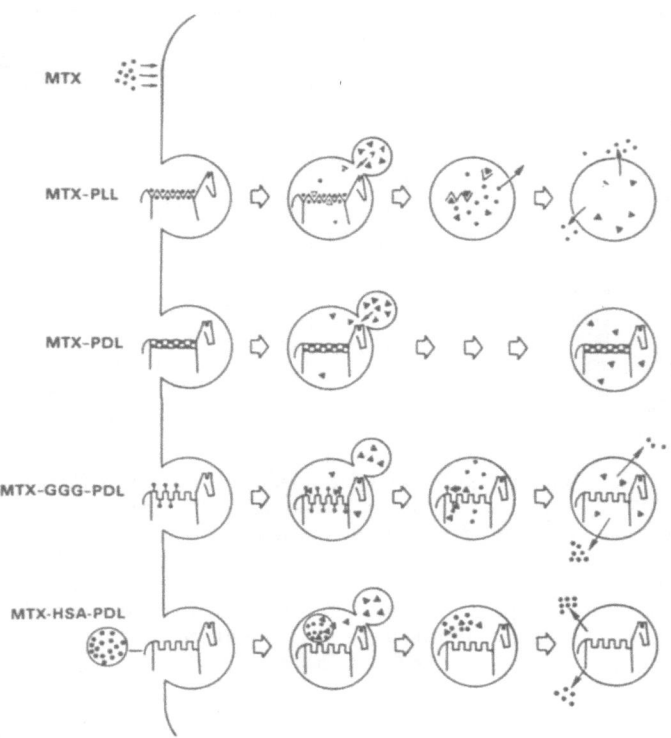

MTX

MTX-PLL

MTX-PDL

MTX-GGG-PDL

MTX-HSA-PDL

Fig. 8.   Trojan horse metaphor. Transport of free drug (●)
          is defective, but defect is circumvented by endo-
          cytosis of drug-carrier (horse). To be effective,
          however, drug must be released from carrier inside
          cells. The four modifications of the Trojan horse
          illustrate the different intracellular fates of MTX
          when conjugated to poly(L-lys), poly(D-lys), trigly-
          cine-poly(D-lys), and human serum albumin - poly (D-
          lys).

PDL is not degraded, the drug must be cleaved at the level of the spacer. Since
only the acid-sensitive conjugate is active, it can be concluded that the
conjugate is carried to acidic intracellular environment. That cleavage
occurred inside cells and not in the medium is shown by the fact that, at
a concentration of $3 \times 10^{-7}M$ daunamycin, the addition of DM-ca-AFB to the
medium of growing cells caused only a 20% decrease in the number of the cells
after a 3-day exposure at pH 7.0, while at the same concentration, DM-ca-
poly(D-lys) caused a 98% inhibition. There are at least 2 distinguishable
acidic intracellular compartments in which cleavage could occur in mammalian cells, namely prelysosomal endosomes and lysosomes. It is not yet
established whether DM-ca-poly(D-lys) is cleaved in the first, the second,
or both.

Disulfide-Spacers

     It is known that the intracellular degradation of proteins requires
the reduction of disulfide bond, although the site at which the reduction
takes place and the exact nature of the reactions are not known. It appeared to us that a conjugate in which MTX was linked to poly(D-lys) through
a disulfide bridge would be of considerable interest for two reasons,
namely to explore the potential of disulfide linkages for intracellular

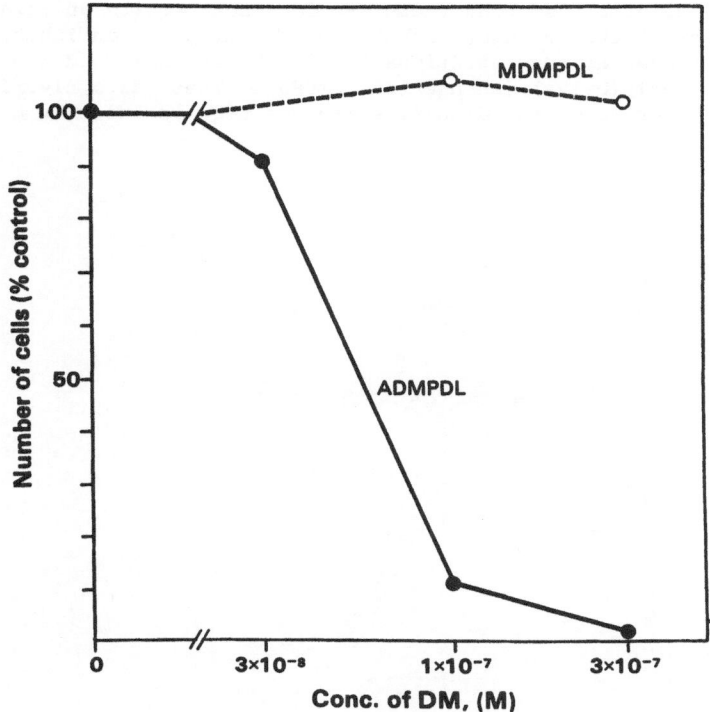

Fig. 9. Cytocidal effect of two poly(D-lys) conjugates,
in which daunomycin (DM) is conjugated to carrier
through an acid-sensitive, or an acid-insensitive
spacer. ADMPDL, DM-cisaconityl-poly(D-lys) (acid-
sensitive); MDMPDL, DM-maleyl-poly(D-lys) (acid-
insensitive). From Shen and Ryser, 1981b.

drug release, and to study the sites and mechanisms of disulfide reduction (Shen et al., 1985). MTX was reacted with cystamine in presence of
1-ethyl-3-dimethyl aminopropyl carbodiimide and reduced to MTX-thioethylamide, a derivative that is not quite, but nearly, as active as MTX in
inhibiting dihydrofolate reductase. Poly(D-lysine) was reacted with SPDP,
to yield a poly(D-lys) derivative containing a disulfide bond with a good
thiopyridine leaving group. The two derivatives were reacted so as to replace the thiopyridine leaving group with MTX-thioethylamide and form a
new disulfide bond with poly(D-lys) thiopropionamide, to yield the sequence
MTX-NH-CH$_2$-CH$_2$-CO-NH-poly(D-lys). This conjugate caused a dose-dependent
inhibition of cell growth in MTX-transport defective cells as seen in Fig.
10 (Shen et al. 1985). This Figure, in addition, compares the effect of
the MTX-disulfide-conjugate with that of MTX-triglycine-poly(D-lys) and
shows that both conjugates have comparable activity. This indicates that
comparable MTX-activity is released through cleavage of the disulfide bond
or cleavage of the triglycine bond under the conditions of this experiment.
Free MTX and MTX-poly(D-lys) had no effect. When the disulfide conjugate
was pretreated with 2-mercaptoethanol, its effect was abolished. Comparable
treatment of the triglycine conjugate caused no change. Cell treatment with
NH$_4$Cl markedly decreased the effect of the triglycine conjugate, but left
the effect of the disulfide conjugate unchanged. At the concentrations used,
NH$_4$Cl markedly inhibits the effect of MTX-triglycine-poly(D-lys) by raising
the lysosomal pH beyond the value required for optimal proteolytic action.

The cleavage of the disulfide bond therefore does not require an acidic pH. Nor was the effect of the conjugate inhibited by leupeptin, an inhibitor of thiol proteases, given at concentrations that markedly inhibited the effect of MTX-triglycine-poly(D-lys) (Shen et al., 1985). These data clearly indicate that the cleavage of the disulfide linkage requires neither a pro-

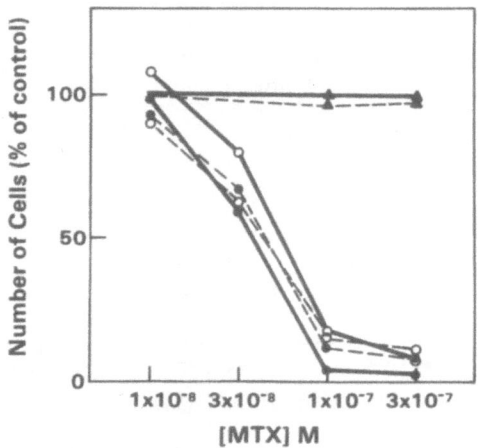

Fig. 10. Cytocidal effect of increasing concentrations of MTX given as three different poly(D-lys) conjugates. Effect of MTX-poly(D-lys) (▲), MTX-S-S-poly(D-lys) (O) and MTX-triglycine-poly(D-lys) (●) on MTX-transport proficient (- - -) or defective (——) cells. From Shen et al., 1985.

teolytic nor an acidic environment. They do not exclude the possibility that cleavage occurs in lysosomes, but the exact nature of the intracellular site(s) at which it occurs is not yet defined.

ENDOCYTOSIS AND DRUG-CARRIER PROPERTIES OF POLY(LYS): HEPARIN COMPLEX

Poly(Lys) is strongly toxic when injected intravenously to animals and its potential as a drug carrier is, therefore, limited. We found, not surprisingly, that complexing poly(Lys) with excess heparin abolishes the toxicity. These experiments led to an unexpected finding, namely that the resulting complex, although devoid of net positive charge, was still taken up by endocytosis and was transported according to different kinetics resembling those of receptor-mediated endocytosis (Morad et al., 1984).

We have pursued this observation and have investigated both the membrane transport of this new macromolecular entity and its potential as a drug-carrier. The uptake of ^{3}H-MTX-poly(Lys) by CHO cells decreased linearly in presence of increasing concentration of unlabeled heparin, an observation

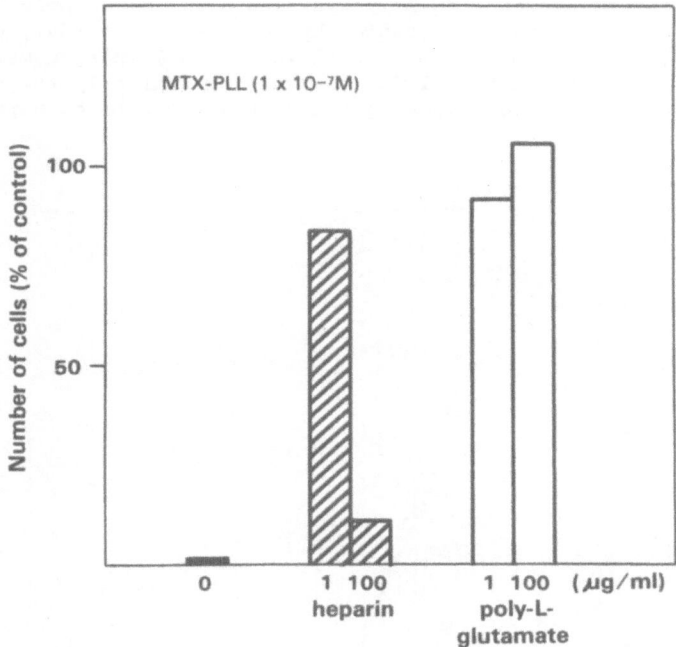

Fig. 11. Cytocidal effect of MTX-poly(Lys) in presence of
          heparin, or poly(glutamic acid). The pharmacologic
          effect of 1×10^{-7}M MTX-poly(Lys) (left column) is
          inhibited by 1 µg/ml heparin, but reappears in 100
          µg/ml heparin. Poly(glutamic acid) is inhibitory
          at both concentrations. Procedures were as described
          in Shen and Ryser, 1981c.

that is easily explained by the polyanion-polycation interaction leading to
a decreasing level of free poly(Lys) in solution. Against expectation,
however, the decrease in uptake bottomed at a poly(Lys):heparin ratio of
about 1:1, and leveled off at about 20% of the initial value in presence
of 10 to 100-fold heparin excess. When the cytotoxicity of MTX-poly(Lys)
was measured in presence of increasing heparin concentrations, it reached
a minimum at a heparin:poly(Lys) ratio of 1:1, and then unexpectedly pro-
ceeded to reappear, as shown in Fig. 11, reaching a maximum at a poly(Lys):
heparin ratio of about 1:100 (Shen and Ryser, 1981c). The fact that cytotoxicity increased even after uptake had leveled off, suggests that heparin
excess does in some way participate in this recovery of cytotoxicity, and
that this participation more than compensates for an 80% decrease in uptake.
This observation made first on two strains of CHO cells (Shen and Ryser,
1981c) was confirmed on a tumor cell line, M5076 (Shen and Ryser, 1983).
Fig. 12 shows that increasing doses of MTX, given either as MTX-poly(Lys)
or as MTX-poly(Lys):heparin complex in presence of heparin excess, cause
identical cytotoxic effects, as revealed by superimposible dose-effect
curves. Evidently this identical pharmacological effect must result from
a different transport process, in which heparin excess enhances the action
of internalized poly(Lys):heparin. The possibility that heparin may have a
cytotoxic effect of its own is excluded by the fact that drug-free complex
is non toxic even in heparin excess, and the fact that leucovorin, an antagonist of MTX, totally prevents the pharmacological effect shown in Fig.
12.

The complex formed between ^{14}C-poly(Lys) and heparin (or poly(Lys) and ^{3}H-heparin) was characterized and studied by Morad et al. (1984), who found a heparin-poly(Lys) stoichiometry of 1.25:1 (by weight) in agreement with the data of Gelman and Blackwell (1973), suggesting that only the sulfate groups of heparin are involved, leaving the carboxyl groups of heparin free.

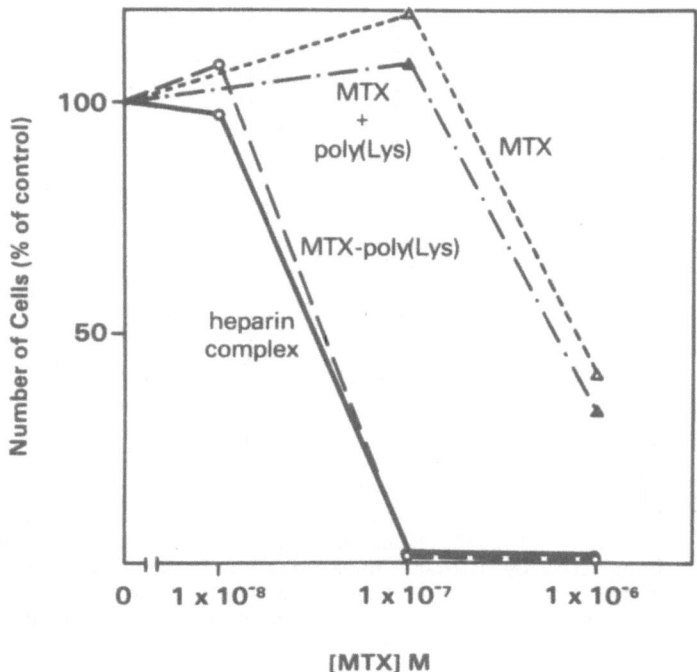

Fig. 12. Cytocidal effect of MTX plus poly(Lys), MTX-poly
(Lys) and MTX-poly(Lys):heparin complex on MTX-
resistant M5076 tumor cells. The complexed and
uncomplexed conjugates are equally effective in
overcoming the MTX transport defect of the tumor
cells. Procedure as in Shen and Ryser, 1981c.

The surface binding as well as the uptake of the complex were measured using either ^{3}H-heparin or ^{14}C-poly(Lys) as a radioactive label. Fig. 13 shows that uptake of complex leveled off with time and with concentration, while in the concentration range tested, uptake of poly(Lys) remained linear as a function of time (Fig. 13, left panel), and as a function of concentration (Fig. 2). Uptake and surface binding of the complex showed saturable kinetics (Fig. 13, right panel), although the saturation concentrations were slightly higher for surface binding. To increase the concentration of complex in the medium, poly(Lys) was increased in presence of heparin excess. The uptake of the complex also differed in its kinetics from that of free heparin (Morad et al., 1984). Heparin uptake increased linearly with concentration, and occured at a very low rate, suggesting a process of fluid-phase endocytosis (Ryser et al., 1983). These data thus demonstrate that the poly(Lys):heparin complex is transported differently than either one of its constituents, and must therefore bind to different surface sites, present in a more limited number than the anionic binding sites of poly(Lys).

116

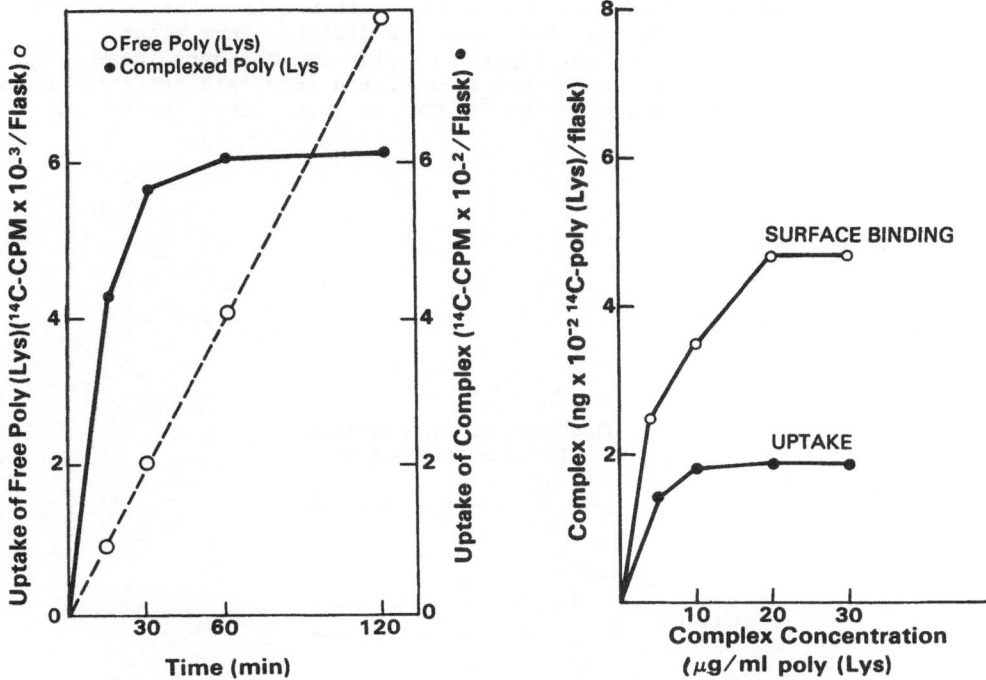

Fig. 13. Cellular uptake of poly(Lys):heparin complex as
a function of time and concentration. Left panel
compares time-dependent uptake of 5 μg/ml [14]C-
poly(Lys) in absence (O) and presence (●) of
a 20-fold heparin excess. Right panel compares
concentration-dependent uptake (●) and surface
binding (O) of ^{14}C-poly(Lys):heparin complexes
formed by adding increasing concentrations of
labelled poly(Lys) to a medium containing 100 μg/
ml heparin. From Morad, et al. 1984.

Specificity of the Binding Site for Poly(Lys):heparin Complexes

Further evidence that the complex interacts with a separate binding
site are: (a) that excess unlabeled heparin does not compete with the binding and uptake of labeled complex; (b) that trypsin-induced cell detachment
modifies the cell surface in such fashion as to cause a marked decrease in
the surface binding of heparin, a slight increase in the binding of complex,
while leaving the binding of poly(Lys) unchanged; (c) that EDTA-induced cell
detachment modifies the cell surface in such fashion as to cause a marked
decrease in the subsequent surface binding of complex, but only a small decrease in the surface binding of poly(Lys), while leaving the surface binding of heparin unchanged (Morad et al., 1984). The specificity of the
surface site with which the complex interacts was further documented by the
fact that (a) replacement of heparin by other polyanions, such as dermatan
sulfate or chondroitin sulfate A and B, yielded complexes that were no
longer competitively displaced by an excess of the respective unlabeled
complexes (Morad, 1984). Even a modified ^{3}H-heparin in which 60% of the
carboxyl groups were reduced to hydroxymethyl groups formed a complex which
was no longer competitively displaced from its binding site; (b) Replacement of heparin with poly(glutamic acid) yielded an MTX-poly(Lys) complex
that was non-cytotoxic (Fig. 11). That the endocytosis of MTX-poly(Lys) and

117

of its heparin complex are different can also be inferred from the fact
that the cellular fates of the two macromolecules differ. Thus after
labeling cells with ^{3}H-MTX-poly(Lys):heparin complex for 24 hours, considerably less small molecular ^{3}H-MTX was re-isolated from cell extracts
than after comparable labelings with the ^{3}H-MTX-poly(Lys) (Shen and Ryser,

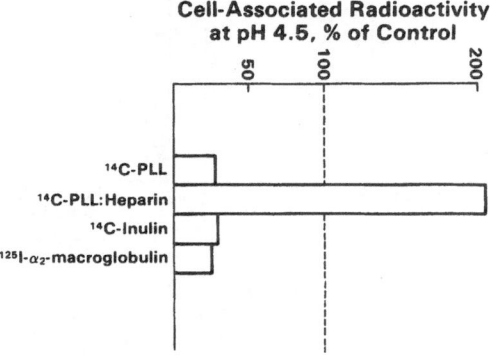

Fig. 14.   Cellular uptake of poly(Lys), poly(Lys):heparin,
           inulin and alpha-2 macroglobulin at pH 4.5. Re-
           sults are expressed in percent of values measured
           at pH 7.0. From Morad, 1984.

1981c). The kinetics of label accumulation over longer periods of times
also differed.

     The fact that the poly(Lys):heparin complex binds to a specific sur-
face site present in limited number and that the ensuing cellular uptake
can be associated with a clear pharmacologic effect satisfies the criteria
that ordinarily define the process of receptor-mediated endocytosis. The
receptor involved has not been characterized, beyond the fact that it is
not fibronectin. Its properties are compatible with those of a surface
proteoglycan (Morad, et al. 1984; Morad, 1984).

Effect of Acid pH on the Uptake of the ^{14}C-Poly(Lys):Heparin Complex

     As mentioned earlier, an acidic endosomal pH will cause the dissociat-
ion of certain ligands from their receptors, and will cause certain viruses

and toxins to cross endosomal membranes and escape undegraded into the cytoplasm (see Fig. 3). It has been shown that short exposure of cells to acid pH may induce a direct passage of diptheria toxin through the plasma membranes (Sandvig and Olsnes, 1980). We found that exposing CHO cells to pH 4.5 in presence of ^{14}C-poly(Lys):heparin markedly increased both the surface binding and the uptake of the complex (Morad et al, 1984). It is particularly significant that when this acid exposure occured in presence of ^{14}C-poly(Lys), of ^{14}C-inulin, or of ^{125}I-alpha 2-macroglobulin, the uptake of these three macromolecules was decreased by 50% or more (Fig. 14). Thus the poly(Lys):heparin complex responds differently to pH than three macromolecules that are suitable models for the three known forms of endocytosis, i.e. fluid-phase, non-specific adsorptive and receptor-mediated described in Fig. 1. This result raises the possibility that at acid pH the poly(Lys):heparin complex might enter cells through a mechanism other than endocytosis, while at neutral pH, it is transported by receptor-mediated endocytosis. This phenomenon has a very sharp pH-dependence between pH 5.0 and pH 4.0. It occurs only in heparin excess and increases linearly when the poly(Lys):heparin ratio increases from 1:1 to 1:20 (Morad, 1984).

The major implications of these findings are (a) that poly(Lys):heparin complexes may offer another example of macromolecule which, like diphtheria toxin and certain viral proteins, undergo a pH-induced conformational change, resulting in their translocation across a cell membrane; (b) that such a translocation could occur at the acid pH of endosomes and lead to the passage of undegraded complex into the cytoplasm. Two additional findings appear to support this hypothesis, namely the fact that the complex is capable of inhibiting DHFR in vitro and thus would be cytocidal if it reached the cytoplasm prior to lysosomal degradation (Shen and Ryser,1981c), and the finding that the amount of small molecular ^{3}H-MTX isolated from exposed cells is smaller following exposure to the complex than following exposure to the simple ^{3}H-MTX-poly(Lys) conjugate (Shen and Ryser, 1981c).

Our data on poly(Lys):heparin complexes fit well into the frame of a conference on polymeric drug carriers. They demonstrate that two polymers upon complexing can give rise to a macromolecular entity that possesses novel properties and is transported into cells differently than its two constituents. While polymers as such are not expected to interact with specific binding sites at the cell surface, our data show that under certain circumstances the complex formed by two polymers may display unexpected specificity in its interaction with cell membranes. Our data further show that among other newly acquired properties, a complex can become sensitive to pH-shifts and may mimic the pH-sensitivity of important functional macromolecules that are known for their unusual membrane interactions at acid pH and their pH-induced conformational changes.

Regardless of their ultimate therapeutic potential, poly(Lys) have proved in our hands to be unique tools to study the cellular fate of drug carriers, and to develop specific drug-carrier linkages for drug-release in different intracellular compartments. Some of our findings will no doubt be relevant to other polymeric systems that may be chosen in the future as drug carriers.

COMMENTS ON DRUG-TARGETING

Poly(Lysine) and other polymeric carriers do not have inherent targeting properties. They can, however, be modified by the addition of specific markers (e.g. galatosyl-, mannosyl-, or mannose phosphate) that will determine their interaction with specific receptors. If such receptors are present selectively on tumor cells or other diseased cells, the potential exists that the modified polymers will be directed selectively to those cells. We have not followed this line of investigation. Our work on drug targeting has instead used natural macromolecules as drug carriers and has followed two different strategies, namely (1) The use of anti-tumor antibodies directed against an antigen expressed at the surface of certain tumor cells (Ballou et al., 1983); (2) The use of a macromolecular complex (in this case an immune complex) known to interact with a specific receptor (Fc-receptor) present at the surface of certain tumor cells (Shen and Ryser, 1984).

In both cases we showed that the macromolecular drug carrier killed selectively the cultured tumor cells bearing the appropriate surface marker. In one case, the cell line used as target was also defective in drug transport (Shen and Ryser, 1984) so that the carrier overcame drug resistance in a tumor cell to which the drug had been selectively targeted.

ACKNOWLEDGEMENTS

This investigation was supported by PHS Grant Numbers CA 14551 (H.J.P.R.) and CA 34798 (W.-C.S.), awarded by the National Cancer Institute, DHHS, USA.

REFERENCES

Ballou, B., Taylor, R.J., Shen, W.-C, Liebert, M., Ryser, H.J.-P., Solter, D., and Hakala, T.R., 1983, Daunomycin targeting to the MH-15 teratocarcinoma using anti-SSEA-1, Fed. Proc. 42: No. 3, p. 685.

Duncan, R., Lloyd, J.B., and Kopecek, J., 1980, Degradation of side chains of N-(2-hydroxypropyl)metacrylamide copolymers by lysosomal enzymes, Biochem. Biophys. Res. Commun., 94:284.

Gellman, R.A., and Blackwell, J., 1973, Heparin-polypeptide interactions in aqueous solution, Arch. Biochem. Biophys., 159:427.

Geuze, H.J., Slot, J.W., Strous, G.J.A.M., Lodish, H.F., and Schwartz, A.L., 1983, Intracellular site of asialoglycoprotein receptor-ligand uncoupling: double-label immunoelectron microscopy during receptor-mediated endocytosis, Cell, 32:277.

Helenius, A.E., Mellman, I.S., Wall, D.A., and Hubbard, A.L., 1983, Endosomes, Trends Biochem. Sci., 8:245.

Morad, N.A., 1984, Membrane binding and transport of poly(lysine):heparin complexes in cultured Chinese hamster ovary cells, Doctoral Dissertation, Boston University.

Morad, N., Ryser, H.J.-P., Shen, W.-C., 1984, Binding and endocytosis of heparin and poly(lysine) are changed when the two molecules are given as a complex to Chinese hamster ovary cells, Biochim. Biophys. Acta, 801:117.

Ryser, H.J.-P. and Shen, W.-C., 1978, Conjugation of methotrexate to poly (L-lysine) increases drug transport and overcomes drug resistance in cultured cells, Proc. Natl. Acad. Sci. U.S.A., 75:3867.

Ryser, H.J.-P., Drummond, I., and Shen, W.-C., 1982, The cellular uptake of horseradish peroxidase and its poly(lysine) conjugate by cultured fibroblasts is qualitatively similar despite a 900-fold difference in rate, J. Cell. Physiol., 113:167.

Ryser, H.J.-P., Morad, N. and Shen, W.-C., 1983, Heparin interaction with cultured cells: possible role of fibronectin in uncoupling surface binding and endocytosis, Cell Biol.:Intnl. Rep., 7:923.

Sandvig, K., and Olsnes, S., 1980, Diphtheria toxin entry into cells is facilitated by low pH, J. Cell Biol., 87:828.

Shen, W.-C., and Ryser, H.J.-P., 1978, Conjugation of poly-L-lysine to albumin and horseradish peroxidase: A novel method of enhancing the cellular uptake of proteins, Proc. Natl. Acad. Sci., U.S.A., 75:1872.

Shen, W.-C., and Ryser, H.J.-P., 1979, Poly (L-lysine) and (D-lysine) conjugates of methotrexate: Different inhibitory effect on drug resistant cells, Mol. Pharmacol., 16:614.

Shen, W.-C., and Ryser, H.J.-P., 1981a, Selective protection against the cytotoxicity of methotrexate and methotrexate-poly(lysine) by thiamine pyrophosphate, heparin and leucovorin, Life Sciences, 28: 1209.

Shen, W.-C., and Ryser, H.J.-P., 1981b, Cis-aconityl spacer between daunomycin and macromolecular carriers: A model of pH-sensitive linkage releasing drug from a lysosomotropic conjugates, Biochem. Biophys. Res. Comm., 102:1048.

Shen, W.-C., and Ryser, H.J.-P., 1981c, Poly(L-lysine) has a different membrane transport when complexed with heparin, Proc. Natl. Acad. Sci. U.S.A., 78:7589.

Shen, W.-C., and Ryser, H.J.-P., 1981d, Conjugation of methotrexate to 5 basic polypeptides: comparison of inhibitory effect on cells defective in drug transport, Fed. Proc., 40: No. 3, Part I, p. 642.

Shen, W.-C., and Ryser, H.J.-P., 1982, The transport and release of poly (basic amino acids)-bound drugs in mammalian cells, Proceedings of the International Union of Pure and Applied Chemistry, 28th Macromolecular Symposium, Amherst, MA p. 368.

Shen, W.-C., and Ryser, H.J.-P., 1983, Poly(Lysine):heparin complex as methotrexate transport carrier in drug resistant cells, Fed. Proc. 42, No. 3, p. 361.

Shen, W.-C., and Ryser, H.J.-P., 1984, A soluble immune complex as drug carrier targeted to Fc-receptor-bearing cells, Proc. Natl. Acad. Sci. U.S.A., 81:1445.

Shen, W.-C., Ryser, H.J.-P., and LaManna, L., 1985, Disulfide spacer between methotrexate and poly(D-lysine), J. Biol. Chem., 260-10905.

Tycko, B., and Maxfield, F.R., 1982, Rapid acidification of endocytotic vesicles containing alpha-2 macroglobulin, Cell, 28:643.

Trouet, A., Masquelier, M., Baurain, R., and Deprez-DeCampeneere, D., 1982, A covalent linkage between daunorubicin and proteins that is stable in serum and reversible by lysosomal hydrolases, as required for a lysosomotropic drug-carrier conjugate: in vitro and in vivo studies, Proc. Natl. Acad. Sci., U.S.A., 79:626.

TARGETING OF COLLOIDAL CARRIERS AND THE ROLE OF SURFACE PROPERTIES

S.S. Davis,[1] S.J. Douglas,[1] L. Illum,[2] P.D.E. Jones,[1]
E. Mak[1] and R.H. Muller[1]

1. Pharmacy Department, University of Nottingham
   Nottingham NG7 2RD
2. Pharmaceutics Department, Royal Danish School of
   Pharmacy, 2 Universitetsparken, Denmark

INTRODUCTION

Colloidal carriers in the form of liposomes, emulsions and microspheres
have been studied as a means of delivering drugs to selected sites in the
body following parenteral administration.[1-3] By far the greatest attention
has been focussed on the intravenous route, although some have mentioned
the advantages of subcutaneous and intraperitoneal administration for
delivery to the lymphatic system and regional lymph nodes.[2]

Passive or natural targeting of drugs can be achieved using colloidal
carriers, by the exploitation of physiological processes. For example,
large particles greater than about $10\,\mu$m in diameter can be trapped readily
and efficiently in the lungs by a process of simple mechanical filtration.[4]
This first example of passive targeting has been suggested for the delivery
of antitumor agents using biodegradable particles made from albumin or
synthetic polymers.[5] Particles less than about $5\,\mu$m in diameter will be
removed by the cells of the reticuloendothelial system. The Kupffer cells
residing in the liver are particularly effective in this respect and a
colloid recognised as foreign will be removed in proportion to blood flow
through the liver.[6] The extraction efficiency can be 90% or greater, with
a half-time of less than one minute.[4] Other cells of the reticuloendothelial system within the spleen and bone marrow as well as the monocytes in
the blood have a much smaller role to play in removing particles simply
because of the dominant role of the Kupffer cells.[7] Particles smaller
than about 100 nm in diameter have the possibility of escaping from the
vascular system through holes or fenestrations in the endothelial cells
lining blood capillaries. In some tissue sites this can result in colloidal
carriers reaching underlying cells provided that the basement membrane is
also incomplete or missing (e.g. the parenchymal cells of the liver and
some (but certainly not all) tumor cells).

The rapid uptake of colloidal carriers by the Kupffer cells provides
a second opportunity for passive targeting. However, efficient uptake by
these cells does not normally lead to delivery of the drug to parenchymal
cells or tumor cells in the liver. Instead, the colloidal carrier is taken
up into the Kupffer cell by a process of phagocytosis and is transported
to the lysosomes of the cell by the usual processes of intracellular traff-

icking.[8] This natural process of drug targeting can be of advantage if indeed the target site is the lysosome, as would be the case in enzyme deficiency diseases or where an invading micro-organism or parasite is resident at this site. Colloidal carriers (liposomes and microsphere systems) are therefore rational approaches for the selective delivery of drugs in the treatment of the tropical diseases (leishmaniasis)[9] and fungal infections (candidiasis).[10] However, if the target site is not the liver, the voracious nature of the Kupffer cell represents a major obstacle that could be difficult to circumvent.[11] One approach would be to block the Kupffer cells with placebo colloid or an agent that forms colloidal particles in situ through interaction with blood components (e.g. dextran sulphate). Certainly, using this concept, liver uptake can be reduced substantially[12] but it does not provide a realistic opportunity for therapeutic applications.

Clearly, not all particles of colloidal size are removed by the cells of the reticuloendothelial system. The red and white cells and platelets in the blood are not recognised as foreign unless they are modified or damaged in some way. Similarly, the natural fat particles, the chylomicrons, are ignored by Kupffer cells but are "captured" by cells high in lipoprotein lipase. The chylomicron copy, Intralipid, formulated from a vegetable oil stabilised by egg phosphatides, is removed only to a limited extent by Kupffer cells.[13] Others, particularly in the liposome field, have explored relationships between Kupffer cell uptake and the nature of incorporated phosphatide using in-vivo and in-vitro models.[14] In such work, physicochemical factors such as particle size and particle charge have been measured and attempts made to correlate these with uptake data. Less classical factors such as surface hydrophobicity, particle shape and rigidity have also been proposed as having relevance.[15]

In our own joint studies we have suggested that the process of particle recognition and uptake by Kupffer cells is determined largely by physicochemical factors and that through a proper understanding of colloidal and interfacial phenomena it should be possible to design colloidal carriers that are largely ignored by the Kupffer and other cell types within the reticuloendothelial system. If this could be achieved, then colloidal particles could be "addressed" appropriately with some form of homing system (e.g. monoclonal antibodies, lectins) to direct them to selected destinations.[16] We are employing polymer microspheres as model carriers. These are different in terms of their particle size, surface charge and surface nature. Polymeric coating agents, in the form of soluble block copolymers with well defined hydrophilic and hydrophobic domains, have been studied in detail with the result that liver uptake can be reduced from greater than 90% to about 25% of administered dose in rabbits. In some situations particles are directed to the bone marrow whereas in others they are found in the general circulation and are not localized at one particular site.

The important stages affecting the fate of an injected colloid are shown in Figure 1. Foreign particles will be coated with blood components almost immediately after injection into the blood. The extent and nature of the coating is determined by the surface nature of the colloidal particle.[17] Certain materials (opsonins) render the particles well recognised by the Kupffer cells. For example, adsorption of immunoglobulin G (IgG) can lead to cascade of the complement system (C_3b). Kupffer cells are believed to have a number of receptors on their surface that can be triggered by opsonized (and non-opsonized) particles. These receptors and mambrane components that mediate the non-specific uptake of colloidal particles, operate independently of one another. The actual process of uptake of a particle by a Kupffer cell is one of deposition and adhesion (rapid) followed by internalization (slow).

The factors affecting the phagocytosis of colloidal particles can be followed in vivo using non-invasive techniques such as gamma scintigraphy, sequential sacrifice and the evaluation of organ levels as well as in-vitro techniques such as phagocytosis by mouse peritoneal macrophages in a stimulated or non-stimulated condition. More recently, viable isolated Kupffer cells in culture have been utilised.

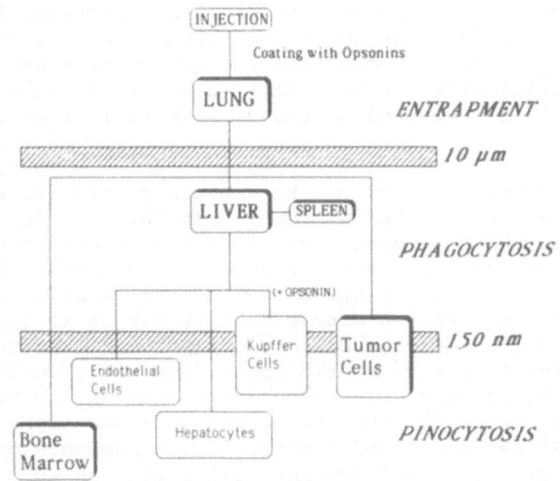

Fig. 1. The biofate of colloidal particles.

It can be deduced from studies in the field of biocompatibility and bioadhesion that a polar (hydrophilic) surface should be poorly coated with blood components. For example, polymeric microspheres having a coating of the non-ionic surface active agent Polysorbate 20 (polyoxyethylene 20) monolaurate sorbitan) adsorbed much less IgG than did uncoated and thus more hydrophobic microspheres (0.83 and 0.03 percent uptake of IgG by weight respectively).[18] Thus a polar, hydrophilic surface should have a dysopsonic effect. This can be created by the adsorption or incorporation of suitable molecules into the particle surface.

The deposition and subsequent adhesion of particles to cells will be determined by both non-specific physicochemical processes as well as specific receptor-mediated biological mechanisms. The theories of colloid science have been applied with success to describe in quantitative terms some of the important aspects of adhesion. For example, if one considers a (negatively) charged particle and a (negatively) charged cell surface, the competitive processes of attraction and repulsion can be described by potential energy diagrams derived by the application of the so-called DLVO theory of colloid stability.[19] Colloidal particles will attract each other (van der Waals interactions) but if the particles are charged there will be (Coulombic) repulsive effects. These competitive forces will vary with particle-cell separation and the balance can be represented in the form of a potential energy diagram. Figure 2 represents a simple version of such a potential energy diagram. As the particles approach each other there can be a net attraction but at closer approach there is a potential energy barrier. The size of this barrier (measured in kT where k is the Boltzman constant and T the absolute temperature) will be determined by repulsive (Coulombic) forces. A colloidal particle undergoing Brownian motion will have a potential (thermal) energy equal to kT. Therefore, if the potential energy barrier is high, adhesion will be minimal. However, if the barrier is small, or if the particle has additional kinetic energy (i.e. flowing in the blood) it may be able to overcome the energy barrier and then will find itself in an attractive energy "well" (i.e. strong adhesion). Considerable energy will be required to remove a particle from such an energy well.

The potential energy diagram can be used in quantitative terms to describe the rapid and effective interaction (deposition and adhesion) of particles to Kupffer cells. Deposition and adhesion can be prevented if the size of the repulsive barrier is increased. This could be achieved by increasing the surface charge on the particle, for example by incorporation of ionizable groups. However, a more effective way is to exploit the colloidal phenomenon of steric stabilisation (polymer excluded volume) and to create a repulsive steric barrier. Space does not permit a detailed treatment of the subject here, but those interested in the theory should consult the book by Napper.[20] The biological considerations have been discussed well by Silberberg[21] and Rutter.[22] The presence of an adsorbed layer of hydrated polymer at the surface will create an anti-adhesive effect[21] by the formation of repulsive steric interactions.

Thus we would predict that an adsorbed molecule that gives a hydrophilic surface and stabilization through steric (excluded volume) considerations should result in a coated particle being largely ignored by opsonins and the Kupffer cells of the liver.

A schematic diagram of such a system is shown in Figure 3. The polymer coating should be well anchored to the particle so that it is not displaced by plasma proteins that have a high affinity for the surface. It is also possible to graft polymers to the surface of particles, thereby giving extremely good anchoring. However, this usually needs to be done during the process of particle preparation. For preformed particles a block copolymer is required that will have domains that will provide both good steric stabilization as well as anchoring. With model polymer microspheres this translates into hydrophilic and hydrophobic domains respectively. Nonionic surfactants of the poloxamer series (polyoxyethylene-polyoxpropylene copolymers) are well recognised as being of low toxicity and indeed are found in certain intravenous pharmaceutical systems.[23] We have investigated a range of these materials by in-vitro and in-vivo experiments.

Colloidal characteristics such as particle size and charge have been measured using conventional methods i.e. laser light scattering (photon correlation spectroscopy) and laser doppler anemometry, respectively. In contrast, the characterisation of surface polarity in a physically meaningful way has been much more difficult and various approaches are still being considered.

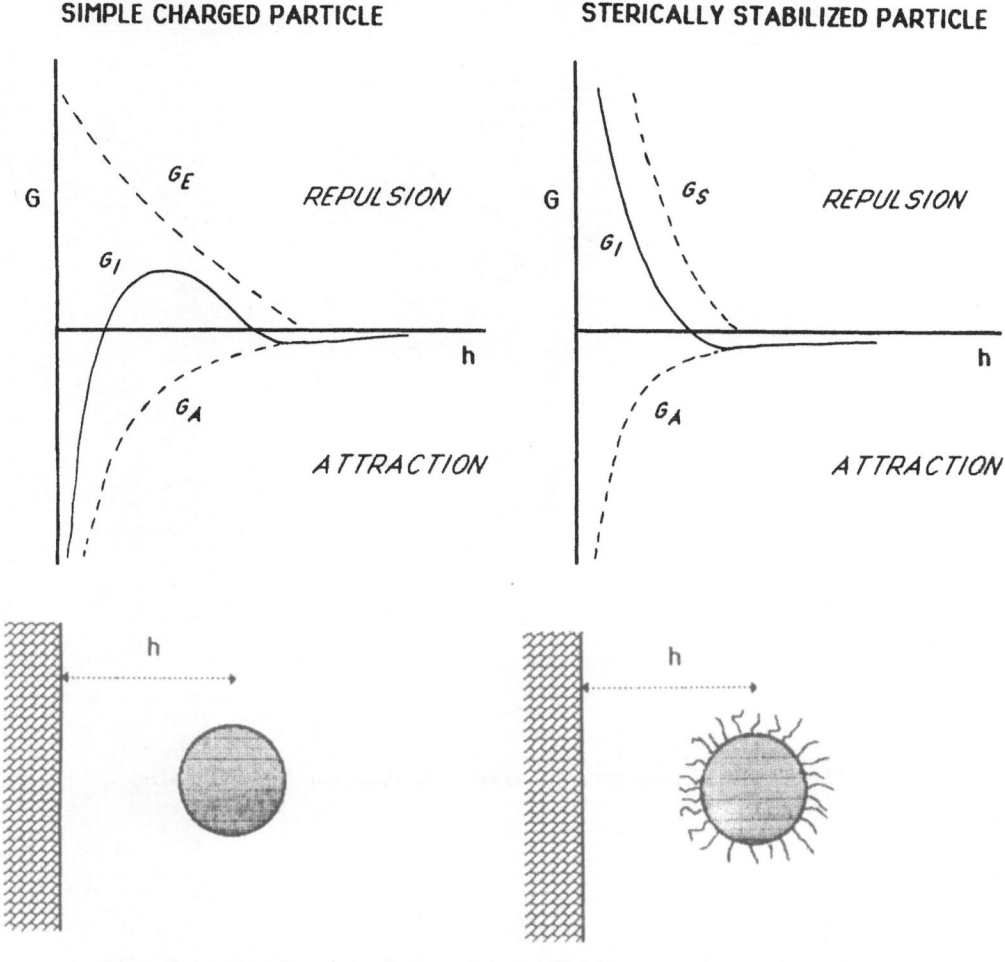

Fig. 2.   Interaction free energy (G) versus separation (h) curves for particle approaching a macroscopic body of same charge.

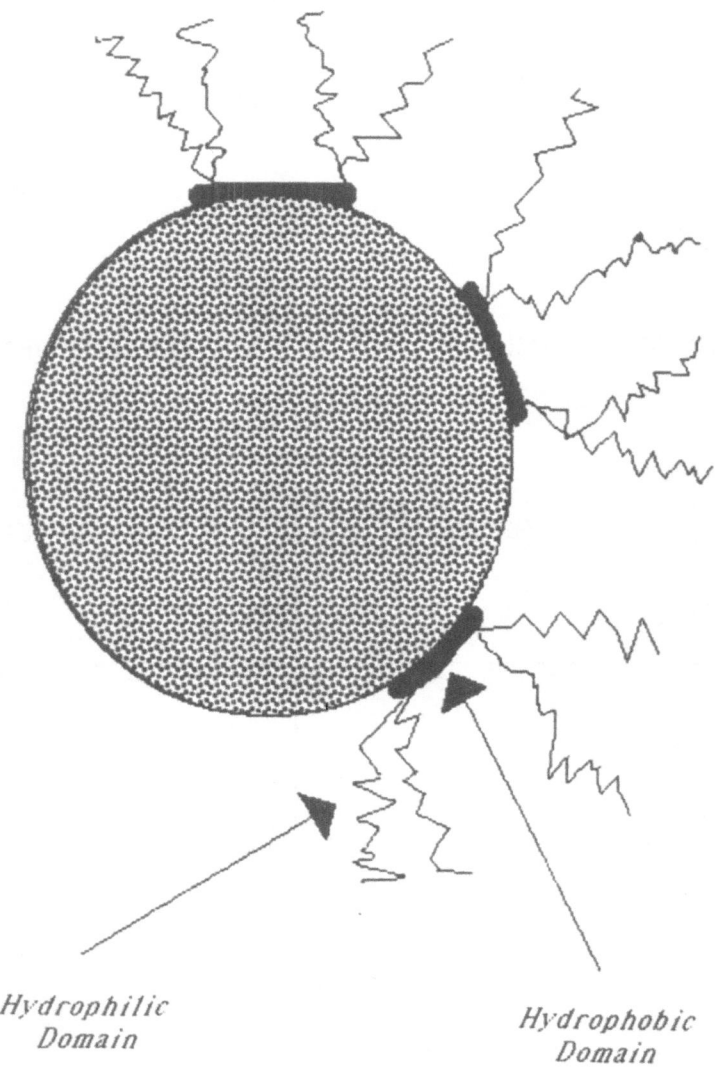

Hydrophilic
Domain

Hydrophobic
Domain

Fig. 3. Surface coating to minimize opsonization and adhesion.

Table 1. The properties of Poloxamers and their adsorption on polystyrene microspheres.

Polox-amer	Mol. weight (average)	Mol. blocks[b] (moles EO[a] PO[a])		Area/molecule[b] (nm^2)	Adsorbed layer thickness[b] (nm)	HLB number	Critical micelle conc. (g/dl x 10^{-3})
181	2000	3	30	2.85	-	3	2.6
182	2500	7	30	3.20	0.9	7	2.0
184	2900	13	30	5.90	1.7	15	1.4
108	5000	46	16	6.51	4.8	31	3.0
237	7700	62	39	-	-	24	-
188	8350	75	30	15.10	8.7	29	5.3
238	10800	97	39	17.52	11.3	28	4.8
407	11500	98	67	-	-	22	-
338	14000	128	54	24.26	13.5	27	7.2

(a) EO, ethylene oxide; PO, propylene oxide
(b) from Kayes and Rawlins[23]

MATERIALS AND METHODS

Coating Agents

The poloxamer series of non-ionic surfactants are ABA block copolymers where A and B refer to hydrophilic polyoxyethylene and hydrophobic polyoxypropylene groups respectively. A variety of these materials has been obtained from commercial sources (Ugine Kuhlman and Atochem) where the two domains have different sizes. The poloxamers listed in Table 1 are presently available to us. The Tetronics (Ugine Kuhlmann) are a range of polyoxyethylene-polyoxypropylene condensates based on ethylenediamine.

Model Particles

Polystyrene Microspheres. A range of polystyrene particles was obtained from a commercial source (Polysciences Inc.). The particles were of different sizes and had different surface groupings (Table 2). Quoted particle sizes were checked by photon correlation spectroscopy (PCS) (Malvern Instruments)[24] or by Coulter Counter. Surface charge (zeta potential) was measured using laser doppler anenometry (LDA)[25] or the new process of Amplitude Weighted Phase Structuration (AWPS)[26] which provides for the simultaneous measurement of charge and particle size for small particles (< 100 nm).

Poly(alkyl 2-cyanoacrylate) particles. Nanoparticles were prepared from butyl 2-cyanoacrylate (Sichel Werke, Hannover, FRG) by a dispersion polymerisation process in an acidic aqueous environment.[27] The influence of physicochemical factors such as temperature, pH of solution, monomer concentration, type and concentration of electrolyte, stirring rate and acidifying agent on the resultant particle size was investigated. Particles of different sizes could be prepared by control of these variables as well as by the use of different stabilisers such as dextrans, β -cyclodextrins and non-ionic surfactants.[27,28] Details of these experiments and the properties of the particles obtained are given in Tables 3 and 4.

Polyhydroxyl stabilisers such as the dextrans can copolymerise with

129

Table 2. Standard microspheres

Microspheres and surface groupings	Size (nm)	Zeta potential[a] (mV)
Polystyrene (PS)	60	-64.0
	170	-77.1
	1000	-72.9
PS-Fluorescent (PSF)	190	-51.4
PS-COOH	190	-82.7
PSF-COOH	280	-83.6
PS-OH	250	-86.0
PS-NH$_2$ (aromatic)	180	-88.6
PS-NH$_2$ (aliphatic)	190	-27.2

(a) pH 7.0; 0.01M phosphate buffer

Table 3. Preparation of poly(butyl 2-cyanoacrylate) nanoparticles. The effect of formulation variables.

Variable	Range	Number average diameter (nm)	Peak molec. wt
Stirring rate	600-3000 rpm	110-122	-
Monomer conc.	1-7% v/v	100-160	1710-2190
pH	1.75-3.5	95-190	1490-1910
Temperature	4-80°C	131-142	-
Electrolyte			
NaCl	0.01-0.25	128-133	-
CaCl$_2$	(mol dm^{-3})	127-136	-
Stabiliser			
Dextran 70	0.05-	105-193	980-1420
Dextran 40	2.50% w/v	105-230	1050-2180
Dextran 10		97-585	900-1700
β-cyclodextrin	0.75-1.75% w/v	2700-3450	-
Poloxamer (0.5%)	variety of stabilisers	61-224	3440-3670
Polysorbate (0.5%)	variety of stabilisers	38-58	1980-2160

Table 4. Preparation of poly(butyl 2-cyanoacrylate) nanoparticles.
The effect of surfactant stabiliser (0.5% w/v concentration).

| Poloxamer | HLB | Molecular weight of block | | Number average diameter |
		Hydrophilic	Hydrophilic	dn
338	27	11264	3132	63
238	28	8536	2262	61
237	24	5456	2262	108
188	29	6600	1740	132
184	15	1144	1740	224

Table 5. Nanoparticles (made of poly(butyl 2-cyanoacrylate) with
different surface charges.

Stabilizer	Zeta potential (mV)
Dextran 70	-19.3
Dextran sulfate	-45.5
DEAE dextran	+33.0

alkylcyanoacrylate monomer to form covalent linkages between the dextran
and the cyanoacrylate polymer, thereby producing an interfacial layer of
dextran firmly grafted to the particle sufrace.[28] In this way it has been
found possible to change the surface characteristics and surface charge
of poly(butyl 2-cyanoacrylate) nanoparticles[29] (Table 5).

Evaluation of surface polarity (hydrophobicity)

While it is recognised that the surface polarity of particles can
have an important influence on the interaction of colloidal particles with
biological environments (e.g. protein adsorption, cell adhesion), rigorous
methods for the evaluation of appropriate parameters have yet to be elucidated satisfactorily. We have examined a number of approaches to include
hydrophobic interaction chromatography,[30] two-phase partition,[31] and adsorption of radiolabeled and fluorescent probes to cell surfaces.[32] As
yet we are still undecided as to the complete suitability of such techniques to provide data that are relevant to the study of biological phenomena but we have been able to separate our model particles into different
categories. Two approaches will be described here.

Two phase partition experiments. Colloidal particles, in particular
microorganisms, can be separated according to the hydrophobic or hydrophilic characteristics of their surfaces by partition methods, using either
classical methods (hydrocarbon/water phases) or polyethylene glycol-dextran
systems.[31,33] Systems based upon dextrans (20, 70 and 500) and polyethylene glycol (PEG, from 2000 to 35000) have been examined. When mixed,
dextran-PEG systems separate into a PEG rich (top) phase and dextran rich

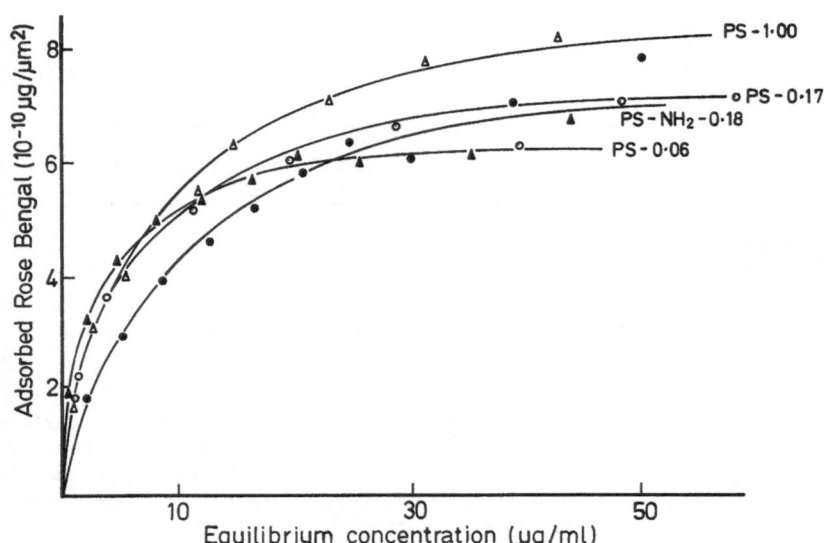

Fig. 4. Model polystyrene particles: Adsorption of rose bengal.
Data are corrected for surface area differences.

(bottom) phase. These separated phases have different polarities and particles and will equilibrate according to their surface characteristics. We have been able to separate model polystyrene particles by following their equilibrium distribution between a PEG-rich phase and the interface between the two phases. Details are shown in Table 6. The system dextran 20 (8%) and PEG 35000 (4.5%) provided a good distribution of particles between the top phase and the interface.

Adsorption of a fluorescent probe. The anionic dye rose bengal is adsorbed to the surface of polymer microspheres according to a classical Langmuir adsorption isotherm[34] and adsorption isotherms for different particles can be obtained by centrifugation techniques. An alternative spectroscopic method is also available since, when adsorbing at the solid surface, the rose bengal molecule undergoes a change in conformation and this change is reflected in a bathochromic shift in the UV spectrum. As a consequence, it is possible to measure the contributions of adsorbed and free solute without the necessity of separation.[34]

Adsorption isotherms for rose bengal onto the various polystyrene particles are shown in Figure 4. The data have been corrected for differences in particle size assuming that the polystyrene particles are rigid spheres. The data can be analysed to give total uptake (plateau region adsorption) as well as the constants derived from an analysis via a Scatchard plot or by an evaluation based on the partition isotherm (Table 7). For the partitioning experiments the uptake of rose bengal onto the surface was measured as a function of nanoparticle concentration and expressed as partition coefficients between particle surface and dispersion medium. The slope of the straight lines obtained by plotting the partition coefficients against surface area is taken as a measure of hydrophobicity; the PS-0.06 particles (slope: 9.124×10^{-11} ml/μm^2) are the most hydrophobic in nature.

Labeling of Microspheres

The rate and extent of uptake of microspheres into organs can be followed by the non-invasive technique of gamma scintigraphy.[4] In order to perform such studies the particles must contain a suitable gamma emitting radiolabel, e.g. iodine-131, or technetium-99m.

Polystyrene microspheres were labeled by iodine-131 using the method of Huh et al.[35] where particles are irradiated in the presence of 131-I sodium iodide. The labeled iodine is incorporated into the surface of the particle and unattached iodide is removed by dialysis or separation on a column.

Poly(butyl 2-cyanoacrylate) nanoparticles proved much more difficult to label. Douglas[36] has examined a variety of approaches for the incorporation of iodine-131 or technetium-99m and he found eventually that dextran[37] labeled with technetium-99m could be used as the stabilising agent for the preparation of nanoparticles.

A solution of dextran 10 (MW 9000) was mixed with a solution of stannous chloride in concentrated hydrochloric acid and then technetium-99m in the form of pertechnetate was added. The labelled nanoparticles were prepared as previously by stirring the monomer with the labelled dextran mixture. The resulting suspension was then passed through a Sepharose CL4B column and the nanoparticle fraction collected. The labelling efficiency was about 15%. If required, coating agents were added to the system prior to injection. A 15-minute equilibration preiod was used.

Table 6. Two-phase partition studies on polystyrene particles[a]

Model Systems (see Table 1)	DEXTRAN 20 (12%) PEG 2000 (9%)		DEXTRAN 500 (6.5%) PEG 2000 (6.5%)	
	Top phase	Inter-face	Top phase	Inter-face
PS-0.17	96	4	100	–
PS-COOH	87	13	100	–
PS-NH$_2$ (arom)	94	6	100	–
PS-OH	86	14	100	–
	DEXTRAN 20 (8%) PEG 35,000 (4.5%)		DEXTRAN 500 (3.75%) PEG 35,000 (2.5%)	
	Top phase	Inter-face	Top phase	Inter-face
PS-0.17	75	25	80	20
PS-COOH	79	21	98	2
PS-NH$_2$ (arom)	74	26	95	5
PS-OH	63	37	88	12

(a) Dextrans and PEG systems dissolved in 0.01M phosphate buffer (pH 7.4) containing 0.05M MaCl. Total volume was 5 ml and particle concentration was 2.5% (25μl used).

Adsorption of Poloxamers

Adsorption of non-ionic surfactants such as the poloxamers and Tetronics on solid surfaces can be described by an isotherm of the Langmuir type.[38] However, the Langmuir equation is derived under the following assumptions.

  i. homogenous adsorption
 ii. monomolecular layer is formed
iii. no solute-solvent interaction or solute-solute interaction in surface or bulk phase
 iv. solvent and solute molecules have equal molecular cross-sectional surface areas

The adsorption of non-ionic surfactants on the surfaces of hydrophobic microspheres will take place under the first two conditions but not the last two. The good fit obtained to the Langmuir equation is due to compensation effects[38] and although constants from the equation can be obtained, their use in interpreting mechanisms is very limited. Consequently, Kronberg et al.,[38-40] have proposed that the Flory-Huggins expression should be used to calculate the chemical potential in bulk solution as well as the surface phase. The Flory-Huggins approach specifically takes into account differences in molecular size between solute and solvent as well as solute-solvent interactions. In deriving their model, Kronberg et al.,[38-40] assumed that the surfactant molecule is adsorbed only with its hydrophobic part in contact with the solid surface. This condition, how-

Table 7. Adsorption of rose bengal onto polystyrene microspheres

	K (ml/μg)	N (μg/mg)	Scratchard analysis Total surface area (μm^2/mg)	Max.amount bound per unit surface area (μg/μm^2)	(slope of initial part of isotherm obtained from partition analysis (ml/μm^2)
PS-0.06	0.40	62.95	9.52×10^{10}	6.61×10^{-10}	9.124×10^{-11}
PS-0.17	0.289	24.13	3.36×10^{10}	7.18×10^{-10}	6.39×10^{-11}
PS-1.00	0.279	4.45	6.349×10^{9}	7.02×10^{-10}	6.023×10^{-11}
PSF-COOH-0.21	0.271	22.50	2.72×10^{10}	8.27×10^{-10}	5.36×10^{-11}
PS-NH$_2$-0.18 (Arom)	0.135	24.32	3.21×10^{10}	7.575×10^{-10}	2.723×10^{-11}
PS-OH-0.25	0.0416	4.13	2.285×10^{10}	1.8099×10^{-10}	1.42×10^{-12}
PSF-0.19	0.12	22.34	3.007×10^{10}	7.428×10^{-10}	2.335×10^{-11}
PS-NH$_2$-0.19 (Aliph)	0.227	20.88	3.00×10^{10}	6.94×10^{-10}	5.29×10^{-11}

ever, may not be true for poloxamers since it is believed that adsorption can also occur to a certain extent via the polyoxyethylene chains which form loops at the surface.[23]

The model predicted that the free energy of adsorption could be split into two contributions

$$\Delta G_{adsorption} = \Delta G_{exchange} + \Delta G_{orientation}$$

The first term ($\Delta G_{exchange}$) represented the exchange of surface-solvent (water) contacts for surface-solute (surfactant) contacts, while the second term ($\Delta G_{orientation}$) represented differences in solute-solvent interaction strength in bulk solution and in the surface phase. When applied to the adsorption of non-ionic surfactants of the nonylphenol-polyethylene oxide series to polystyrene microspheres it was shown that approximately 20% of the adsorption free energy was due to exchange of surface-water contacts for surface-surfactant contacts and 80% was due to orientation of the surfactant molecule at the surface. That is 20% of the adsorption is due to the properties of the surface and 80% to the properties of the surfactant. In this way, adsorption can be likened to micelle formation where unfavourable contacts between water and the hydrocarbon (hydrophobic) regions of the surfactant are replaced by more favourable hydrocarbon-hydrocarbon contacts and water-water contacts (i.e. the hydrophobic interaction). Similar reasoning can be used for both the poloxamer and Tetronic series.

While the properties of the surface (surface polarity) may only make a minor contribution to Δ $G_{adsorption}$ it will be of considerable importance when examining differences between the adsorption of the same surfactant onto different microspheres. Kronberg and Stenius[39] have indicated that surface polarity (Δ E) can be defined as

$$\Delta E = \frac{a^o_1}{k\,T}\,(\gamma_1 - \gamma_2)$$

where γ_1 and γ_2 are the interfacial free energies (tensions) between the surface and pure solvent and solute respectively and a^o_1 is the surface area occupied by one polymer segment.

In principle, Δ E can be obtained from contact angle measurements or by the application of solubility parameter theory, but such approaches are dependent on a number of assumptions and there are additional experimental limitations.

Kronberg and Stenius[39] have presented data for adsorption of a short chain non-ionic surfactant onto four types of microsphere with different polarity as defined by their solubility parameters. An apparent plateau in the adsorption isotherm decreased as the surface polarity increased and they recommended that the polarity of unknown microsphere surfaces could be determined by the adsorption of the same known surfactant onto that surface.

We have yet to examine in detail the adsorption isotherms for the uptake of the various poloxamer and Tetronic systems onto the model particles described in Table 2. However, we would predict from surface polarity and solubility parameter standpoints that the uptake of a given poloxamer to a polystyrene microsphere should be more effective than the uptake of the same poloxamer to a poly(butyl 2-cyanoacrylate) particle stabilised by dextran. In this respect, Kayes and Rawlins[23] have presented data for the adsorption of poloxamers onto polystyrene microspheres of average diameter 312 nm possessing only surface carboxyl groups. Their values, where available, are given in Table 1.

Mouse Peritoneal Macrophages

The interaction of colloidal particles with macrophages can be measured using in vitro models such as mouse peritoneal macrophages[41-43] or isolated Kupffer cells.[43] Test particles in the form of 5.25 μm polystyrene microspheres were incubated at 37°C with peritoneal macrophages obtained from female MFI mice. Uptake of particles by macrophages was determined by microscopic counting.

In-vivo Experiments

Female New Zealand white rabbits (weight range 3-5 kg) were divided into groups of 3 and were injected intravenously via the marginal ear vein with radiolabelled particles. The total volume of the injection was in the range of 0.5 - 1.0 ml and the total number of particles was about 10^{13}. Particles were coated with various polymers by incubation in a 2% aqueous solution (usually for 24 h). Control experiments were performed by injection of the polymer solution via one ear and the particles via the contralateral ear.

The distribution of the labeled microspheres was followed by placing the rabbits on a gamma camera (Maxi Camera II, General Electric Company) (40 cm field of view) fitted with an appropriate collimator. Dynamic images (20 sec per frame) were recorded for 15 min and then static frames were

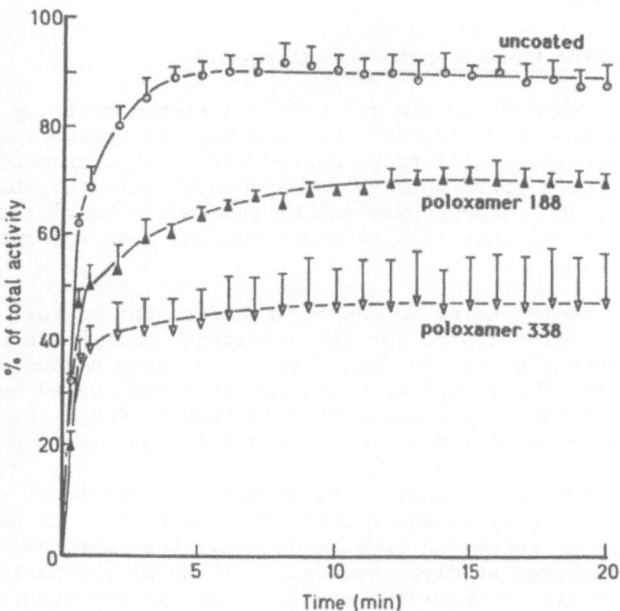

Fig. 5. Uptake of 60 nm polystyrene particles in liver/spleen
of rabbit: Effect of surface coatings.

recorded at suitable times for periods up to 11 days.[4] The data were recorded and processed by a dedicated computer system. Regions of interest
were created around the liver and spleen, heart and lung, whole body and,
when appropriate, the left hind leg. After a period of 8 to 11 days the
animals were sacrificed and the organs removed. The total activity in the
different organ sites and in the carcass were determined using a large
sample well-type gamma counter.

Some initial experiments have also been conducted in rats. Animals
were injected via the penal vein with coated (poloxamer 338) polystyrene
microspheres (60 nm diameter) surface-labeled[35] with iodine-125. Uncoated
microspheres of the same size were used as controls. Animals were sacrificed one hour after injection, the livers were flushed free of blood by
perfusing with buffer containing 1 mM EGTA, and the radioactivity in the
liver, blood, spleen and femurs measured.

RESULTS AND DISCUSSION

In-vitro Studies Using Mouse Peritoneal Macrophages

Data for the uptake of coated and uncoated microspheres by mouse peritoneal macrophages are shown in Table 8. The highest uptake was found for
the uncoated microspheres while those coated with the poloxamers and
Tetronic 908 had a much lower uptake. As a general rule the ability of the
coating agent to reduce phagocytosis can be related to the molecular weight
of the coating agent and the relative contributions from hydrophilic and
hydrophobic domains.

The effect of adding serum to the media before phagocytosis experiments
were conducted, was investigated for the uncoated microspheres and those
coated with Poloxamer 338 and 188 (Table 8). The serum had no significant
effect on the uptake of uncoated microspheres or those coated by Poloxamer
338. However, the marked suppression of phagocytosis found for Poloxamer
188 in the absence of serum was no longer found in the presence of serum.

The results demonstrate clearly the effect of a non-ionic coating
agent on the uptake of polystyrene microspheres by peritoneal macrophages.
Even in the absence of serum (no opsonic coating) the macrophages ingested
the uncoated microspheres avidly. However, coating of the particles with
a polymer layer capable of providing a steric barrier for stabilisation
greatly reduced particle uptake. Such results support our proposal that
inhibition of particle uptake (recognition) can be achieved by exploitation
of physicochemical principles. Indeed on examination of the literature
we find other examples where other hydrophilic coating agents have been reported to reduce the phagocytosis of colloidal particles and cells.[44] In
some cases the coating materials were of biological origin (biopolymers)[45,46]
but would be expected to provide a similar dysopsonic and steric stabilisation effect as the poloxamers and Tetronics. Mention can also be made of
the recent work of Davis and Hansrani[47] who found that poloxamer 338 and
188 employed as emulsifying agents were able to suppress the uptake of fat
emulsion particles by mouse peritoneal macrophages.

The adsorbed layer of polymer can be displaced from the surface of the
particle if the hydrophobic anchoring group has a lower affinity for the
surface than adsorbing blood components.

In-vivo Studies in Rabbit

Curves for the uptake of polystyrene microspheres (60 nm diameter) in
the liver/spleen region of the rabbit are shown in Figure 5. The dramatic

Table 8. The uptake of polystyrene microspheres by mouse peritoneal macrophage.

Coating material	Serum	% uptake in 60 min compared to control
none	no	100[a]
	yes	100
Poloxamer 188	no	22
	yes	90
338	no	14
	yes	23
238	yes	42
407	yes	33
Tetronic 908	yes	30

(a) control.
(b) includes contribution from spleen and blood pool.
(c) 11 days.

Table 9. Uptake of polystyrene microspheres by the liver (spleen) and the effect of coating.

System	Particle size (nm)	Uptake as % of control	
		20 min[b]	8 days
Uncoated[a]	60	100	100
	1240	–	100[c]
Coated, Poloxamer 188	60	78	104
Poloxamer 338	60	51	51
	1240	–	80[c]
Tetronic 908	60	38	21

effect of coating the particles with a non-ionic polymeric agent is well demonstrated. Total uptake at 20 min can be reduced from 90% of administered dose to less than 30%.[48-50] The kinetics of uptake can be evaluated using a simple one compartment exponential model. Interestingly, while the coating agent reduces the extent of uptake (Table 9) it does little to change the rate of uptake (Figure 5). If larger particles are used in such experiments then the coating effect is very much reduced and the larger proportion of the dose is still found in the liver at sacrifice. Scintiscans taken 3 hours after the administration of uncoated and coated microspheres are shown in Figure 6. A distinct liver image can be seen for the uncoated particles while uptake in the bone marrow is shown for the particles coated with poloxamer 338. Coating the particles with Tetronic 908 leads to suppression of liver/spleen uptake but no general localisation at other sites.

As discussed in the introduction, cells of the reticuloendothelial system are found not only in the liver and spleen but elsewhere in the body. However, normally the voracious uptake of particles by the Kupffer cells means that other sites have little access to the administered dose. Coating the particles with poloxamer 338 apparently suppresses uptake by Kupffer cells and thereby allows significant uptake by the bone marrow. Thus the deposition/recognition processes relevant to Kupffer cell uptake are not the same as for the cells of the bone marrow. The residual levels in the liver could well be associated with particles taken up by endothelial cells and hepatocytes[51] together with activity in the blood pool associated with circulating microspheres. Presently we are investigating the possibilities of separating liver cells by enzyme perfusion as a method of ascertaining the distribution of administered particles to the different cell types of the liver (and the possible selective effect of surface coatings)[14] as well as being a means of providing isolated Kupffer cells[43] in culture for studies on particle-cell interactions and their control by physicochemical approaches.

Yoshida[52] has reported recently electron microscope studies on endocytic uptake of colloidal particles in the rabbit. Polystyrene microspheres in the liver sinusoids could be removed by the endothelial cells by formation of bristle coated vesicles if the microspheres were less than 100 nm in size. Uptake of particles in the bone marrow by endothelial cells was by bristle-coated vesicles and multiparticulate-pinocytic vesicles. Particles as large as 2.02 μm were sequestered in this way.

The fact that particles coated with poloxamer 338 find their way to the bone marrow, while those coated by Tetronic 908 seem to be largely ignored by the cells of the RES, opens up possibilities for targeting, not only to the different cell types of the RES but also to other sites. For example, a monoclonal antibody could be linked to a particle coated with a steric stabilizer (or to a stabilizer grafted to the particle surface) to explore concepts in active targeting, particularly to tumor sites.

Extended scintigraphic imaging and examination of organ levels at sacrifice (8 or 11 days after dosing) demonstrated that the inhibitory effects of coating agents on liver uptake seen at short time periods were maintained for much longer periods of time for poloxamer 338 and Tetronic 908 but not for poloxamer 188. The transient effect for the last coating agent can be explained by the gradual displacement of the coating material by blood components. This is similar to the effect seen in the experiments with mouse peritoneal macrophage where added serum led to uptake values similar to those for controls (uncoated microspheres). Apparently the hydrophobic domain for poloxamer 188 is not sufficiently well anchored to the particle surface to prevent displacement. Thus, the relative affinities of blood components and coating agents will be of importance unless the

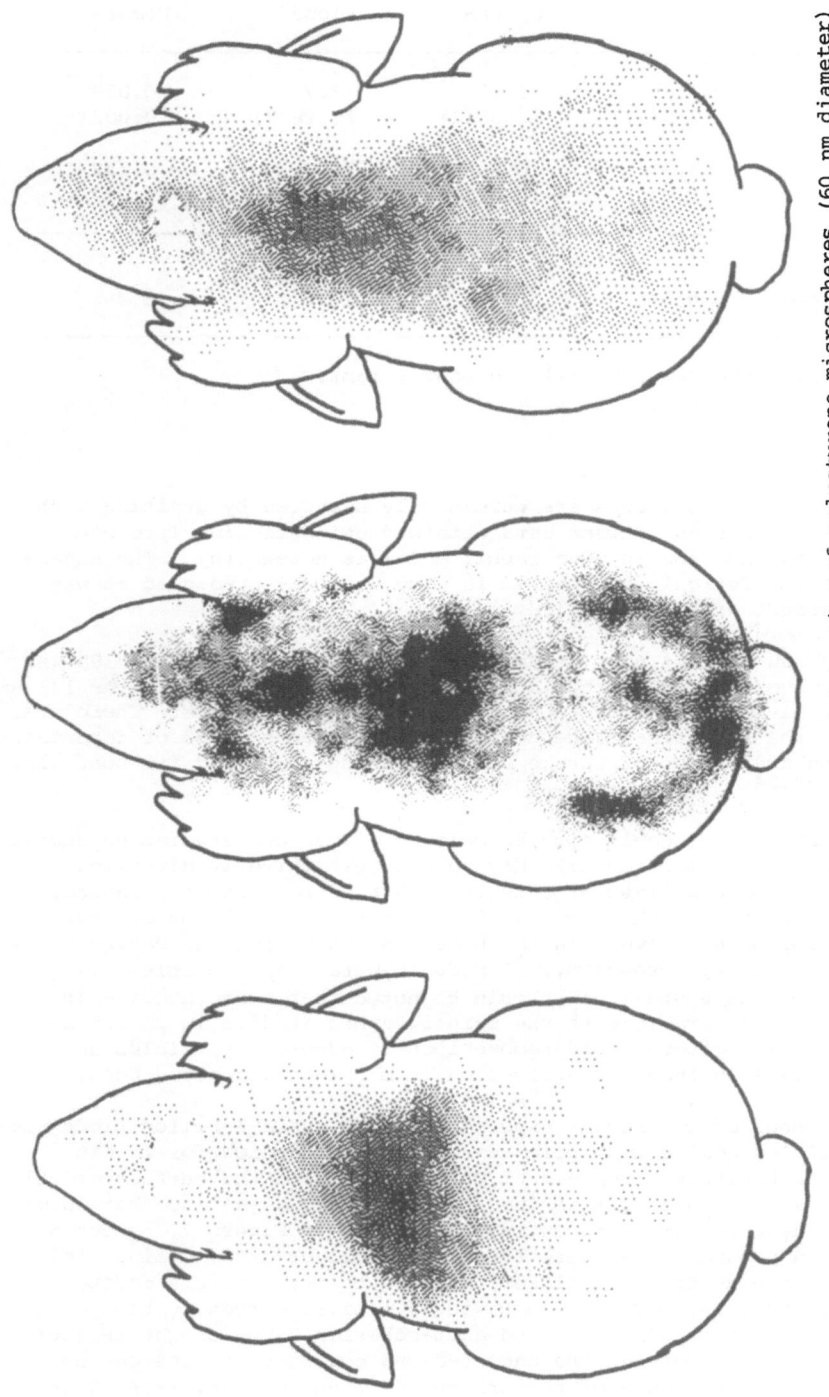

Fig. 6. Scintiscans of rabbits 3 hours after administration of polystyrene microspheres (60 nm diameter) Left, uncoated; Middle, coated with poloxamer 338; Right, coated with Tetronic 908.

Table 10. Uptake of polystyrene microspheres (60 nm in diameter) in the organs of the rat following IV injection.

System	Liver	Spleen	Blood[a]	Femurs
Uncoated control	47.4[b] ±2.6(7)	1.05 ±0.65(4)	3.7 ±0.28(7)	0.059 ±0.002(3)
Coated with poloxamer 338	3.5 ±1.1(2)	0.39 ±0.015(2)	39.2 ±5.4(2)	0.142 ±0.037(2)
Ratio coated/uncoated	0.073	0.36	10.6	2.4

(a) Blood volume taken as 6.5 ml/100 g body weight.
(b) % uptake at one hour

stabilising hydrophilic groups are permanently anchored by grafting techniques. The correlation between data obtained using the in-vitro mouse peritoneal system and the in vivo rabbit model is noteworthy. The superior effect of the Tetronic 908 system in vivo was well predicted as was the limited effect of poloxamer 188.

Recently Leu et al[53] have described how in the rat poloxamer 188 was effective in diverting polymethylmethacrylate microspheres (diameter 131 nm) away from the liver (30% reduction) for periods up to 7 days. Therefore, the nature of the surface and the coating agent appears to be of importance in determining the extent of uptake of the coating agent and its possible displacement.[40,54]

The animal species could also be relevant. Our own studies conducted in the rat show that poloxamer 338 may be more effective at diverting particles away from the liver of this species than for rabbit. Indeed, for the coated particles the liver activity is about 7% of the control (Table 10). The higher levels in the blood and bones for the coated particles are particularly noteworthy. A reduced uptake in the spleen is also found for the coated system. It should be noted that these studies in rat are not directly comparable with the scintigraphic studies in rabbit because the perfusion method could remove loosely adherent particles and scintigraphic data include a contribution from spleen and blood pool.

Finally, some of our recent experiments with gamma-labelled biodegradable nanoparticles coated with poloxamer 338 by incubation for 15 min (studies on the kinetics of uptake of poloxamer 338 to the surface of colloidal particles show that equilibrium is reached rapidly[31] and that 15 min should be an adequate time for coating) are shown in Figure 7. About 50% of the uncoated nanoparticles were found in the liver after 3 min. This indicates that the dextran stabilizer, grafted to the particle during the polymerization process, may well be having a similar effect to the poloxamers in terms of dysopsonization and anti-adhesion. The slight reduction in liver uptake found between the uncoated and coated particles can be associated with the presence of the grafted dextran 10 stabilizer at the interface that may well reduce the uptake of the poloxamer to the particle surface. The decrease in total activity with time seen for both systems is evidence for the biodegradation of poly(alkyl 2-cyanoacrylate) nanoparticles.

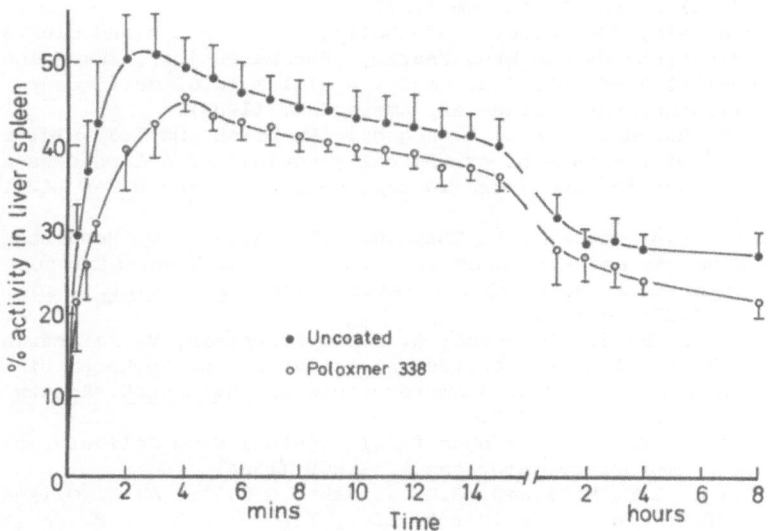

Fig. 7. Uptake of poly(alkylcyanoacrylate) nanoparticles in liver/spleen of rabbit: Effect of surface coatings.

● - uncoated   o - coated with poloxamer 338

REFERENCES

1. S.S. Davis, L. Illum, J.G. McVie and E. Tomlinson, eds., "Microspheres
    and Drug Therapy: Pharmaceutical, Immunological and Medical Aspects",
    Elsevier, Amsterdam (1984).
2. J.N. Weinstein and L.D. Leserman, Liposomes as drug carriers in cancer
    chemotherapy, Pharmac. Ther. 24:207 (1984).
3. G. Gregoriadis, Targeting of drugs, Nature 265:407 (1977).
4. L. Illum, S.S. Davis, C.G. Wilson, M. Frier, J.G. Hardy and N.W. Thomas,
    Blood clearance and organ deposition of intravenously administered
    colloidal particles: the effect of particle size, nature and shape,
    Int. J. Pharm. 12:135 (1982).
5. T. Yoshioka, M. Hashida, S. Muranishi and H. Sezaki, Specific delivery
    of mitomycin C to the liver, spleen and lung: nano- and micro-
    spherical carriers of gelatin, Int. J. Pharm. 8:131 (1981).
6. E. Wisse, Ultrastructure and function of Kupffer cells and other sinus-
    oidal cells in the liver, in: "Kupffer Cells and Other Sinusoidal
    Cells", E. Wisse and D.L. Knook, eds., Elsevier/North Holland
    Biomedical Press, Amsterdam (1977).
7. J.W.B. Bradfield, The reticuloendothelial system and blood clearance,
    in: "Microspheres and Drug Therapy: Pharmaceutical, Immunological
    and Medical Aspects", S.S. Davis, L. Illum, J.G. McVie and
    E. Tomlinson, eds., Elsevier, Amsterdam, (1984).
8. G. Poste, O. Bucana, A. Raz, P. Bugeski, R. Kirsh and I.J. Fidler,
    Analysis of the fate of systemically administered liposomes and
    implications for their use in drug delivery, Cancer Res. 42:1412
    (1982).
9. C.R. Alving, E.A. Steck, W.L. Chapman, V.B. Waits, L.D. Hendricks,
    G.M. Schwartz and W.L. Hanson, Therapy of leishmaniasis: superior
    efficacies of liposome-encapsulated drugs. Proc. Natl. Acad. Sci.
    U.S.A. 75:2959 (1978).
10. R.L. Hopfer, K. Mills, R. Mehta, G. Lopez-Berestein, V. Fainstein and
    R.L. Juliano, In-vitro antifungal activities of Amphotericin B
    and liposome encapsulated Amphotericin B, Antimicrob. Agents
    Chemother. 25:387 (1984).
11. G. Poste and R. Kirsh, Site specific (targeted) drug delivery in
    cancer chemotherapy, Biotechnol. 1:869 (1984).
12. R.T. Proffitt, L.E. Williams, C.A. Presant, G.W. Tin, J.A. Oliana,
    R.C. Gamble and J.D. Baldeschweiler, Liposomal blockade of the
    reticuloendothelial system. Improved tumor imaging with small
    unilamellar vesicles, Science 220:502 (1983).
13. I. Fraser, H. Pearson, V. Bowry and P.R.F. Bell, The intravenous Intra-
    lipid tolerance test, J. Leuk. Biol. 36:347 (1984).
14. F. Roerdink, J. Regts, B. Van Leeuwen and G. Scherphof, Intrahepatic
    uptake and processing of intravenously injected small unilamellar
    phospholipid vesicles in rat, Biochim. Biophys. Acta 770:195 (1984).
15. C. Capo, F. Garrouste, A.M. Benoliel, P. Bongrand and R. Depieds,
    Non-specific binding to macrophages: evaluation of the influence
    of medium-range electrostatic repulsion and short-range hydrophobic
    interaction, Immunol. Commun. 10:35 (1981).
16. L. Illum, P.D.E. Jones, J. Kreuter, R.W. Baldwin and S.S. Davis, Ad-
    sorption of monoclonal antibodies to polyhexylcyanoacrylate nano-
    particles and subsequent immunospecific binding to tumour cells
    in vitro, Int. J. Pharm. 17:65 (1983).
17. T.M. Saba, Physiology and physiopathology of the reticuloendothelial
    system, Arch. Intern. Med. 126:1031 (1970).
18. A. Rembaum, J. Ugelstad, J.T. Kemshead, M. Chang and G. Richards, Cell
    labelling and separation by means of mono-disperse magnetic and
    non-magnetic microspheres, in: "Microspheres in Drug Therapy:
    Pharmaceutical, Immunological and Medical Aspects", S.S. Davis,
    L. Illum, J.G. McVie and E. Tomlinson, eds., Elsevier, Amsterdam,
    (1984).

19. P.R. Rutter and B. Vincent, The adhesion of micro-organisms to sur-
    faces: physicochemical aspects, in: "Microbial adhesion to sur-
    faces", R.C.W. Berkeley, J.M. Lynch, J. Melling, P.R. Rutter and
    B. Vincent, eds., Ellis Horwood, Chichester, (1980).

20. D.H. Napper "Polymeric Stabilisation of Colloidal Dispersions",
    Academic Press, London (1983).

21. A. Silberberg, The role of membrane-bound macromolecules and macro-
    molecules in solution in cell/cell encounters in flowing blood,
    Ann. N.Y. Acad. Sci. 416:83 (1984).

22. P.R. Rutter, The physical chemistry of the adhesion of bacteria and
    other cells, in: "Cell Adhesion and Motility", A.S.G. Curtis and
    J.D. Pitts, eds., Cambridge University Press, Cambridge, (1980).

23. J.B. Kayes and D.A. Rawlins, Adsorption characteristics of certain
    polyoxyethylene-polyoxypropylene block copolymers on polystyrene
    latex, Colloid Polymer Sci. 257:622 (1979).

24. J.M. Roe and B.W. Barry, Photon correlation spectroscopy of pharmaceut-
    ical systems (sodium dodecyl sulphate, sodium deosycholate and
    chloropromazine hydrochloride mcielles and polystyrene latices),
    Int. J. Pharm. 14:159 (1983).

25. A.W. Preece and N.P. Luckman, A laser doppler cytopherometer for
    measurement of electrophoretic mobility of bioparticles, Phys.
    Med. Biol. 26:11 (1981).

26. K. Schatzel and J. Merz, Measurement of small electrophoretic mobil-
    ities by light scattering and analysis of the amplitude weighted
    phase structure function, J. Chem. Phys. 81:2482 (1984).

27. S.J. Douglas, L. Illum, S.S. Davis and J. Kreuter, Particle size and
    size distribution of poly(butyl 2-cyanoacrylate) nanoparticles.
    Influence of physicochemical factors, J. Colloid Interface Sci.
    101:149 (1984).

28. S.J. Douglas, L. Illum and S.S. Davis, Particle size and size distrib-
    ution of poly(butyl 2-cyanoacrylate) nanoparticles. Influence of
    added stabilisers, J. Colloid Interface Sci. 103:154 (1985).

29. S.J. Douglas, L. Illum and S.S. Davis, Poly(butyl 2-cyanoacrylate)
    nanoparticles with differing surface charges, J. Controlled
    Release, in press.

30. G. Halperin, M. Tauber-Finkelstein and S. shaltiel, Hydrophobic chroma-
    tography of cells: Adsorption and resolution on homologous series
    of alkylagaroses, J. Chrom. 317:103 (1984).

31. V.P. Shanbhag and C.G. Axelsson, Hydrophobic interaction determined by
    partition in aqueous two-phase systems. Partition of proteins in
    systems containing fatty acid esters of poly(ethylene glycol),
    Eur. J. Biochem. 60:17 (1975).

32. T. Malmqvist, Bacterial hydrophobicity measured as partition of palm-
    itic acid between the two immiscible phases of cell surface and
    buffer, Acta Path. Microbiol. Immunol. Scand., Sect. B, 91:69
    (1983).

33. M. Rosenberg, D. Gutnick and E. Rosenberg, Adherence of bacteria to
    hydrocarbons: A simple method for measuring cell-surface hydro-
    phobicity, FEMS Lett. 9:29 (1980).

34. L. Illum, M.A. Khan, E. Mak and S.S. Davis, Evaluation of the carrier
    capacity for poly(butyl 2-cyanoacrylate) nanoparticles, In prep-
    aration.

35. Y. Huh, G.W. Donaldson and F.T. Johnson, A radiation-induced bonding
    of iodine at the surface of uniform polystyrene particles, Radiat.
    Res. 60:42 (1974).

36. S.J. Douglas, The preparation and characterisation of poly(alkyl) cyano-
    acrylate) nanoparticles, PhD Thesis, University of Nottingham (1986).

37. E. Henze, H.R. Schelbert, J.D. Collins, A. Hajaj, J.R. Barrio and
    L.K. Benet, Lymphoscintigraphy with Tc-99m labeled dextran, J.
    Nucl. Med. 23:923 (1982).

38. B. Kronberg, Thermodynamics of adsorption of nonionic surfactants on latexes, J. Colloid Interface Sci. 96:55 (1983).

39. B. Kronberg and P. Stenius, The effect of surface polarity on the adsorption of non-ionic surfactants. I. Thermodynamic considerations, J. Colloid Interface Sci. 102:410 (1984).

40. B. Kronberg, P. Stenius and G. Igeborn, The effect of surface polarity on the adsorption of non-ionic surfactants. II. Adsorption to poly(methylmethacrylate) latex, J. Colloid Interface Sci. 102:418 (1984).

41. F. Roerdink, N.M. Wassaf, E.C. Richardson and C.R. Alving, Effects of negatively charged lipids on phagocytosis of liposomes opsonized by complement, Biochim. Biophys. Acta, 734:33 (1983).

42. T.P. Stossel, R.J. Mason, J. Hartwig and M. Vaughan, Quantitative studies of phagocytosis by polymorphonuclear leukocytes. Use of emulsions to measure the initial rate of phagocytosis, J. Clin. Invest. 51:615 (1972).

43. D.L. Knook and E.C. Sleyster, Preparation and characterisation of Kupffer cells from rat and mouse liver, in: "Kupffer Cells and Other Sinusoidal Cells", E. Wisse and D.L. Knook, eds., Elsevier/North Holland Biomedical Press, (1977).

44. C.J. van Oss, D.R. Absolom and H.W. Neumann, Interaction of phagocytes with other blood cells and with pathogenic and nonpathogenic microbes, Ann. N.Y. Acad. Sci. 416:332 (1984).

45. E. Whitnack, A.L. Bisno and E.H. Beachey, Hyaluronate capsule prevents attachment of group A streptococci to mouse peritoneal macrophages, Infect. Immun. 31:985 (1981).

46. R.J. Grasso, R. Ganguly and J.F. Breen, Inhibition of yeast phagocytosis in macrophage cultures treated with slime polysaccharide purified from Pseudomonas aeruginosa, J. Leuk. Biol. 36:771 (1984).

47. S.S. Davis and P.J. Hansrani, The effect of surface characteristics on the phagocytosis of lipid particles, Int. J. Pharm. 23:69 (1985).

48. L. Illum and S.S. Davis, Effect of non-ionic surfactants on the fate and deposition of polystyrene microspheres following intravenous administration, J. Pharm. Sci. 72:1086 (1983).

49. L. Illum and S.S. Davis, The organ uptake of intravenously administered colloidal particles can be altered using a non-ionic surfactant (Poloxamer 338), FEBS Lett. 167:79 (1984).

50. L. Illum and S.S. Davis, The kinetics of uptake and organ distribution of colloidal drug carrier particles delivered to rabbits, Proc. Second European Congress on Biopharmaceutics and Pharmacokinetics, Salamanca, Spain, Vol. II, 97 (1984).

51. D.P. Praaning-van Dalen, A. Brouwer and D.L. Knook, Clearance capacity of rat liver Kupffer, endothelial and parenchymal cells, Gastroenterol. 81:1036 (1981).

52. K. Yoshida, H. Nagata and H. Hoshi, Uptake of carbon and polystyrene particles by the sinusoidal endothelium of rabbit bone marrow and liver and rat bone marrow, with special reference to multiparticle-pinocytosis, Arch. Histol. Jap. 47:303 (1984).

53. D. Leu, B. Manthey, J. Kreuter, P. Speiser and R.P. DeLuca, Distribution and elimination of coated polymethyl [2-C14] methacrylate nanoparticles after intravenous injections in rats, J. Pharm. Sci. 73:1433 (1984).

54. W. Norde, Adsorption of proteins at solid surfaces, in: "Adhesion and adsorption of polymers", L. Lee, ed., Plenum Press, New York, (1978).

TARGETABLE NANOPARTICLES

P. Couvreur,[1] V. Lenaerts, A. Ibrahim, L. Grislain,
L. Van Snick, P. Guiot, C. Verdun and F. Brasseur

[1]Laboratoire de Pharmacie Galénique, Université de Paris XI
  92290 Chatenay-Malabry, France

Laboratoire de Pharmacie Galénique, Université Catholique de
Louvain, 1200 Brussels, Belgium

INTRODUCTION

The concept of drug-carriers has been developed as an alternative to
a tissue distribution dependant on the molecular action of the drug. Class-
ically, an intravenously administered drug is distributed in the body as
a function of intrinsic properties of the molecule. A pharmacologically
active concentration is reached in the diseased tissue at the expense of
massive distribution in the rest of the body. For many drugs, e.g.
cytostatics or antileishmaniasis agents, this poor specificity raises a
toxicological problem which represents a real obstacle to an effective
therapy. An improvement to this situation could be brought about by the
use of a carrier directing the drug to the target tissue.

Entrapment of cytotoxic drugs inside endocytosable carriers such as
liposomes improves drug specificity and reduces toxicity towards non-
diseased cells.[1,2] Work in this field has resulted in the development of
polyacrylamide nanocapsules.[3] Polyacrylamide nanocapsules also may be
useful in promoting cellular uptake via endocytosis for compounds that do
not gain access to lysosomes.[4]

Due to their polymeric nature, these small capsules (diameter of about
200 nm) may be more stable than liposomes in biological fluids and during
storage. Furthermore, they can entrap various molecules in a stable and
reproducible way. However, this new lysosomotropic carrier is unlikely to
be digested by lysosomal enzymes, thus restricting its clinical use. With
these considerations in mind, biodegradable nanoparticles made by polymerization of various alkyl-cyanoacrylate monomers were developed recently.[5,6]
These polymers were chosen because of their biodegradability.[7,8] Their described use in surgery[9] was, from a toxicological point of view, an advantage.

Furthermore, the anionic polymerization of cyanoacrylates (Fig. 1) in
aqueous media requires no energy source for its initiation, thus avoiding
the alteration of drug molecules that can occur when such means as γ or
UV radiations or elevated temperature are needed. The aim of this paper
is to describe the in-vivo distribution of nanoparticles and discuss the

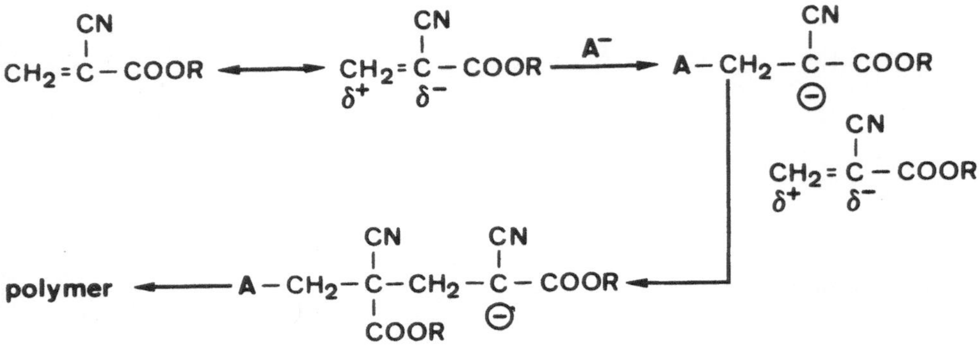

CN
|
CH_2=C−COOR ⟷ CH_2=C−COOR $\xrightarrow{A^-}$ A−CH_2−C−COOR
δ^+ δ^- $\ominus$

CN
|
CH_2=C−COOR
δ^+ δ^-

CN   CN
| |
polymer ⟵ A−CH_2−C−CH_2−C−COOR ⟵
| $\ominus$
COOR

Fig. 1. Polymerization mechanism of cyanoacrylate
monomers.

possibility of modifying tissue distribution and the pharmacokinetics of
both doxorubicin and dactinomycin after linkage to nanoparticles and to
magnetized nanoparticles. Interaction with macrophages in culture will
also be discussed.

PREPARATION AND CHARACTERIZATION OF NANOPARTICLES

Nanoparticles were prepared by emulsion polymerization of alkylcyanoacrylate monomers following an anionic initiation mechanism.

Drug-free Nanoparticles

These were prepared from two different monomers: isobutyl- and hexylcyanoacrylate. Monomer (100 μl) was added under continuous mechanical
stirring to 10 ml of an aqueous polymerization medium generally containing
glucose (5%), dextran 70 (1%) and citric acid (0.5%). After polymerization
for 4 h (isobutylcyanoacrylate) and 24 h (hexylcyanoacrylate), an homogeneous milky suspension of nanoparticles was obtained. The nanoparticles
show a spherical shape and an average diameter of 0.2 μm under scanning
electron microscopy. by simply modifying the acidity of the polymerization
medium, it is however, possible to obtain larger or smaller particles.
After freeze-fracturing and examination by electron microscopy, the inner
structure of the carrier appears highly porous, developing a large specific
area favourable to sorption processes. No continuous, limiting envelope
surrounding the particle could be identified.

Molecular weights of polyalkylcyanoacrylate polymers forming nanoparticles were determined by a GLC-HPLC method.[10] In comparison to other
particulate carriers described in the literature, polyalkylcyanoacrylate
nanoparticles seem to be an original colloidal drug delivery system because
they are built by numerous small oligomers (MW, 600-3000) rather than by
one or a few macromolecules. Such oligomeric systems should be more easily
cleared from the body and more capable of avoiding overload of the reticuloendothelial system (RES).

Drug-loaded Nanoparticles

Drug-loaded nanoparticles were prepared as above after dissolution of
dactinomycin (50 μg/ml) or doxorubicin (100 to 750 μg/ml) in the polymer-

Table 1. Particle size and drug adsorption rate before and after freeze-drying of doxorubicin (DOX-NP) and dactinomycin-(DACT-NP) loaded polyisobutylcyanoacrylate nanoparticles.

	DOX-NP		DACT-NP	
	Size (nm)	Adsorption rate (%)	Size (nm)	Adsorption rate (%)
Before freeze-drying	180 ± 10	89 ± 5	215 ± 10	95 ± 3
After freeze-	190 ± 10	88 ± 5	242 ± 10	95 ± 3

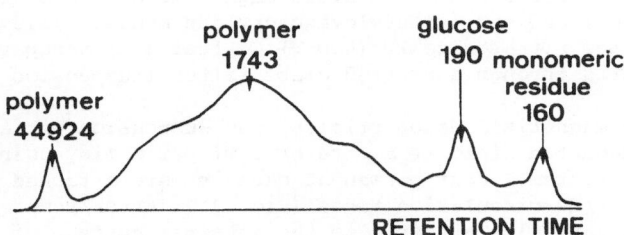

Fig. 2. GLC chromatographic profiles of polyisobutyl-cyanoacrylate nanoparticles loaded with doxorubicin.

ization medium.[11] After addition of isobutylcyanoacrylate and complete
polymerization of the monomer, drug-loaded nanoparticle samples were
freeze-dried for 90 h under vacuum (6×10^{-2} m bar). Resuspension of the
solid nanoparticle formulation was made by simple addition of distilled
water (5 ml) to the vials. Determination of doxorubicin linked to nanoparticles was carried out by fluorimetry after ultracentrifugation. For
the determination of dactinomycin linked to nanoparticles, 20 μl of ^{3}H
dactinomycin (specific activity: 322 GBq/mmol) 18.5 MBq/ml was added to
the polymerization medium before adding the monomer. After polymerization,
the suspension was centrifuged at 20,000 rpm and radioactivity measured by
scintillation counting in both sediment (linked dactinomycin) and supernatant (free dactinomycin).

As shown in Table 1, nanoparticles are efficient for the adsorption
of drugs. Furthermore, comparative size measurements showed no significant modification of the carrier dimensions after freeze-drying. While
dactinomycin did not seem to induce any change in nanoparticle molecular
weights, the polyalkylcyanoacrylate chromatographic profile was greatly
modified after binding of doxorubicin (Fig. 2). Indeed, the increase in
the molecular weight of the polymer as well as that of the molecular weight
dispersity was significant. Compared to unloaded nanoparticles, the great
difference is, however, the advent of a bimodal distribution of molecular
weights with a peak corresponding to a (high) molecular weight of about
45,000 (Fig. 2). This could be due to the intervention of doxorubicin as
initiator (by its free amine function) of the anionic polymerization of
the cyanoacrylic monomer. The drug could also be covalently linked to the
beginning of the polymer chain. Studies are now in progress to test this
hypothesis.

Magnetically Responsive Nanoparticles

These were prepared by anionic polymerization of the monomer in the
presence of ultrafine magnetic particles of between 0.01 to 0.05 μm diameter.[12]

After 1 g of glucose and 1 g citric acid had been dissolved in 100 ml
of distilled water, 0.7 g of magnetite particles were dispersed by ultrasonic treatment over 15 min. The suspension was passed through a fritted
glass filter (pore size 9-15 μm) to avoid magnetite agglomerates. ^{3}H-
dactinomycin (2 ml) and ^{14}C-isobutylcyanoacrylate monomer (1.5 ml) were
added and dispersed ultrasonically (400 W). After 3 h, nanoparticles were
formed and filtered through a fritted glass filter (suspension A).

To separate magnetized nanoparticles, the suspension was allowed to
flow through a magnetic field at a rate of 1 ml per 3 min, using a pumping
circulation tube system. Four permanent magnets were attached to the external surface of the circulation tubes (Fig. 3). After removal of the
magnets, the nanoparticles attached to the internal surface of the tubes
were washed with 100 ml of an aqueous solution containing NaCl (0.7%) and
$CaCl_2 \cdot 2H_2O$ (0.2%). This magnetically responsive particle suspension was
finely resuspended by ultrasonic treatment of 15 min at 400 W and filtered
through fritted glass (suspension B).

The determination of the ^{3}H-dactinomycin content in the nanoparticles
was performed by centrifugation of suspensions A and B for 1 h at 20,000
rpm. The measurement of ^{3}H in the sediment and supernatant was made by
liquid scintillation counting as described above. ^{14}C in suspensions A
and B was measured to determine the percentage of magnetized nanoparticles.

Magnetized nanoparticles had a diameter of 0.22 μ which was highly
reproducible. Moreover, comparison between ^{14}C radioactivity obtained with

nanoparticles suspension A and B showed that 93% of the prepared carrier was magnetically responsive. Furthermore, these magnetized particles adsorbed 80% of ^{3}H-dactinomycin (suspension B). With constant-flow conditions to test the magnetic susceptibility of the nanoparticles, we found that 28% w/w magnetite was necessary to obtain a high percentage of responsive magnetic nanoparticles.

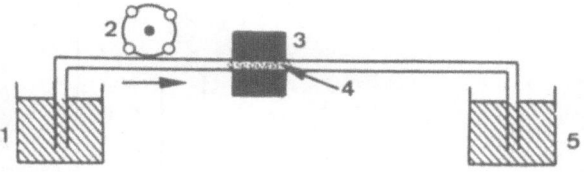

1. initial suspension
2. peristaltic slow speed pump (3 ml / min)
3. magnet 3500 Gauss
4. magnetized nanoparticles
5. non-magnetized and/or low magnetically
                                      responsive particles

Fig. 3. Scheme of constant flow apparatus for the separation of magnetised nanoparticles. From reference 12.

INTERACTION OF DACTINOMYCIN LOADED NANOPARTICLES WITH MACROPHAGES IN CULTURE

Cell Cultures

In an initial experiment, we have compared the uptake of ^{3}H-dactinomycin D by mouse peritoneal macrophages when the drug is free in the culture medium or adsorbed on polybutylcyanoacrylate nanoparticles, the external concentration of ^{3}H-dactinomycin being the same in both cases.[13] It is shown in Figure 4 that dactinomycin uptake is three times greater when this compound is presented to cells by way of the particulate vector and when 20% fetal calf serum was present in the culture medium (Fig. 4; blocks 1 and 3).

On the other hand, one can assume that the dissociation between dactinomycin and the particles in the culture medium is not an instantaneous process since the penetration of the medium into the internal lattice of the particles would be rather slow. Indeed, the butyl moieties may confer to the polymeric lattice a partial hydrophobic character. In conclusion, one can reasonably assume that when dactinomycin is linked to the particles, the drug uptake reflects the uptake of particles.

In the experiment shown in Figure 5 and which is correlated to the time effect on the interaction, the number of dactinomycin molecules associated per mg of cell protein can be converted into the number of particles

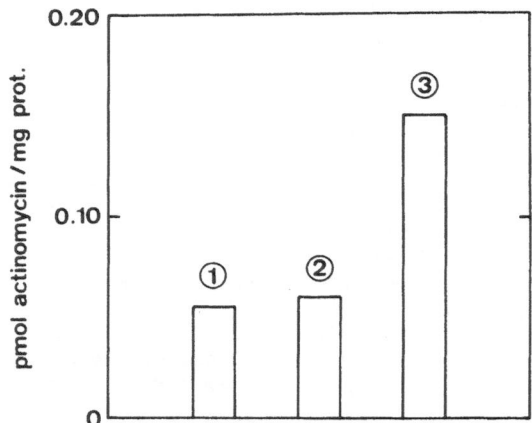

Fig. 4. Uptake of ^{3}H-actinomycin D by mouse peritoneal macrophages. Block 1, free actinomycin was presented during one hour to the cells maintained at 37°C and cultured in a culture medium containing 20% serum. Concentration of actinomycin was 1.3 pmole.ml^{-1}; Block 2, same experimental conditions as in block 1 except that the culture medium did not contain serum and that the actinomycin was adsorbed on nanoparticles; Block 3, same experimental conditions as in block 1 except that actinomycin was absorbed on nanoparticles and presented to cells in the presence of serum. From reference 13.

associated per cell since we know that one particle contains on the average 120 drug molecules and that one cell corresponds to 83.3 pg of cell protein. A quantitative estimation of the interaction is then possible and Table 2 summarizes the data obtained. We also see from Fig. 5 that a plateau is reached which corresponds to roughly 120 nanoparticles per cell. This last number represents an equivalent volume of 0.34% of the cell volume itself.

Temperature effect is illustrated in Fig. 6. The association of the particles with macrophages is nearly non existent near 0°C and is rapidly

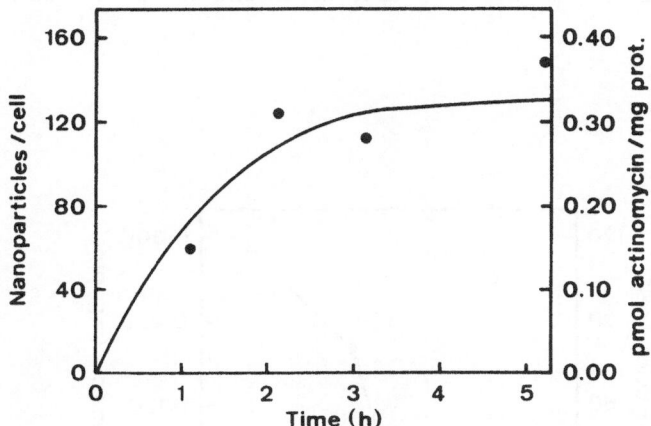

Fig. 5. Time dependance of the association between polybutylcyanoacrylate nanoparticles and mouse peritoneal
macrophages. The external particle concentration was
1.2×10^{10} per ml and about 200,000 cells per ml were
used and maintained at 37°C for different times. From
reference 13.

Table 2. Interaction between polybutylcyanoacrylate nanoparticles and mouse peritoneal macrophages[a].

Parameters	
Number of particles taken up by one cell in one hour	60
Corresponding volume represented by this number	$0.66 \mu m^3$
% of the cell volume represented by this number[b]	0.17%

(a) The cells were maintained at 37°C and the external nanoparticle
   concentration was 1.2×10^{10} ml^{-1}.
(b) The volume represented by a mouse peritoneal macrophage is taken
   as $395 \mu m^3$.

increasing when the temperature is raised up to 20°C, which suggest that the association is cell-energy dependent. Since the polymeric nature of the particles excludes fusion with the plasma membrane, the pattern of Fig. 6 suggests that endocytosis may play a role in the uptake process.

Additional evidence for endocytosis arises from the fact that the uptake of nanoparticles is protein source-dependent as we can see by the

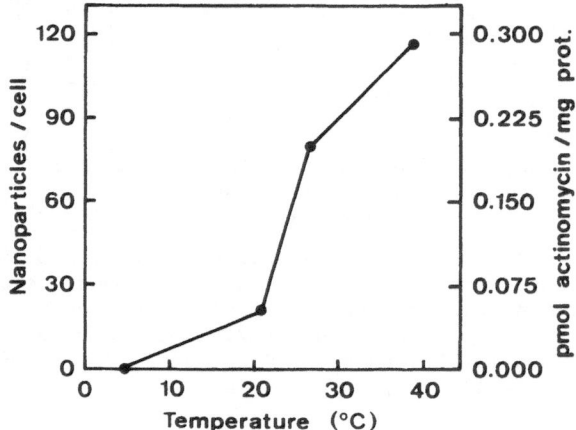

Fig. 6. Effect of temperature on the interaction between polybutylcyanoacrylate nanoparticles and mouse peritoneal macrophages. The external particle concentration was 1.2 x 10^{10} per ml and about 200,000 cells per ml were used and incubated for 2 h at different temperatures. From reference 13.

comparison of blocks 2 and 3 in Fig. 4 and that the uptake of the particulate vector seems to enhance the microvilii formation by cells. Indeed, scanning electron microscopic examination of macrophages interacting with the nanoparticles shows that the particles are taken up by means of microvilii (Fig. 7) and that the number of microvilii developed is greater than for cells in the absence of nanoparticles.

Biological Degradation in Lysosomal Extracts

Because nanoparticles are picked up by cells through phagocytosis, it is likely that they are digested by lysosomal enzymes. Therefore, degradation of nanoparticles has been tested in presence of lysosomal extracts (tritosomes). ^{14}C-polyhexylcyanoacrylate nanoparticles were incubated at pH 5.6 in the presence of tritosomes. Samples taken at various time intervals were centrifuged and the radioactivity was evaluated in the supernatent (soluble polymer) and the pellet (insoluble polymer), allowing the calculation of the percentage of polymer solubilization. Fig. 8 shows that in an acid medium (pH 5.6), tritosomes have raised the degradation rate of the nanoparticles, comparatively to that measured in absence of the biological extract. The reaction tends to slow down after about 1 h. Upon add-

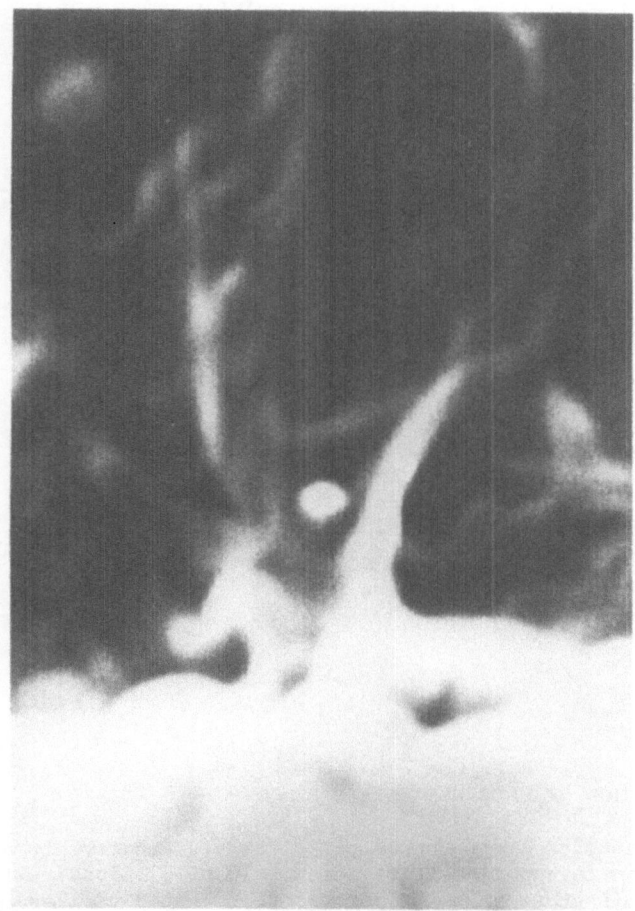

Fig. 7. Scanning electron microscopic appearance of a macrophage interacting with the nanoparticles. The cell body and the microvilii surrounding the particles can be seen. At the center of the field two microvilii enveloping a particle is 200 nm. The spherical structures taken up by the microvilii are related to nanoparticles since they are not observed for macrophages in the absence of particles. From reference 13.

ition of fresh tritosomes (after 2 h), the kinetics increase markedly indicating that the previous observed plateau might be due to a relative instability of the enzyme involved in nanoparticle solubilization. It is also reasonable to assume that nanoparticles administered in vivo are degradated, mainly by an enzymatic mechanism, into water soluble oligomers.

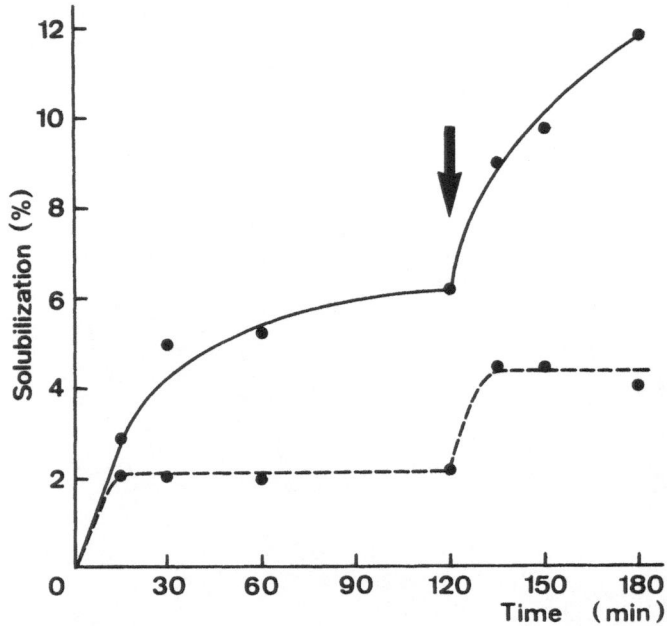

Fig. 8. Solubilization of polyisobutylcyanoacrylate nanoparticles in the presence (——) and in the absence (---) of tritosomes. The arrow indicates addition of a new tritosomes aliquot.

TISSUE DISTRIBUTION OF FREE NANOPARTICLES

The extravascular distribution of polyalkylcyanoacrylate nanoparticles is rapid after intravenous administration (Fig. 9). Blood half-life is about 5 min. This shows that nanoparticles are rapidly taken up from the circulating blood by various organs. For this reason, the body-distribution of radioactive nanoparticles in mice has been studied at different times after injection.[14] An autoradiographic study (Fig. 10) shows that 5 min after intravenous administration most of the particles are found in the organs of the reticuloendothelial system, mainly the liver. For the longer time intervals, the liver still retains the major part of nanoparticles as compared to other organs. However, total liver radioactivity gradually decreases and is excreted via the faeces and urine. Excretion of the nanoparticles is complete after 7 days.

The situation can thus be summarized by the assertion that the liver acts as a reservoir towards nanoparticles, conditioning their rapid first-phase disappearance from the blood and their second-phase release in the body under a degraded and excretable form. Owing to the high proportion of the carrier taken up by the liver, it is also probable that this organ represents the major site of metabolism for these polymeric particles. It was therefore very important to determine the distribution of nanoparticles between the different liver cell types in vivo. Thus, radioactive nano-

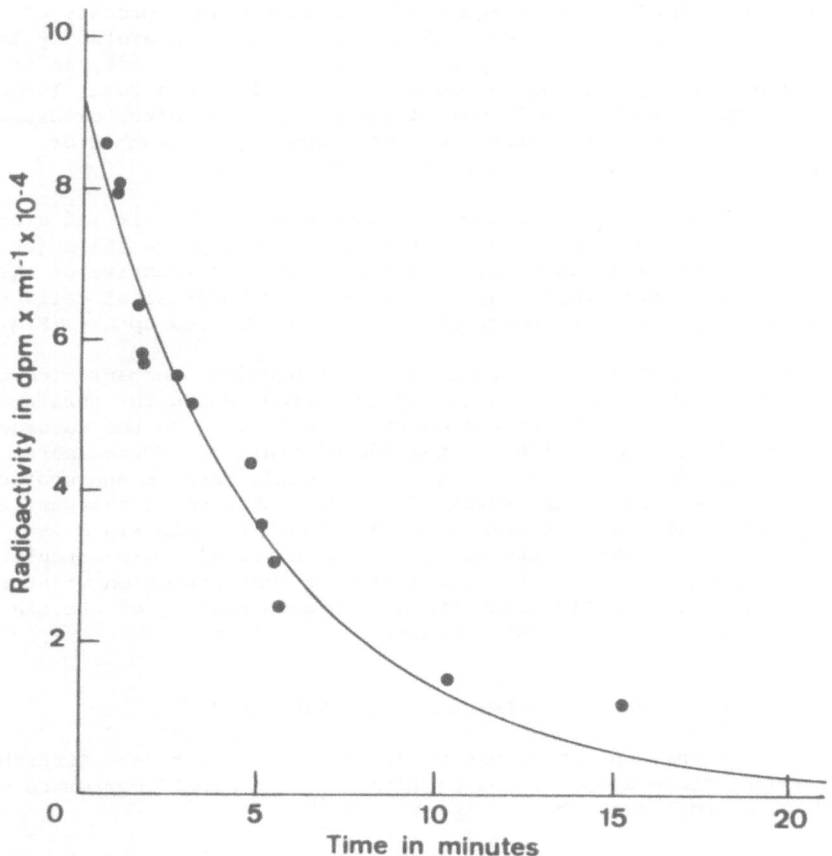

Fig. 9. Disappearance of radioactivity from blood after
         injection of ^{14}C-labelled polyisobutylcyanoacryl-
         ate nanoparticles into mice. From reference 14.

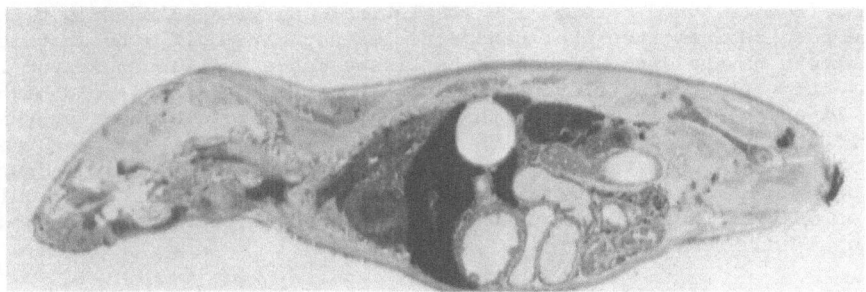

Fig. 10. Whole body autoradiography of a mouse 5 min after
         intravenous administration of ^{14}C-labelled poly-
         hexylcyanoacrylate nanoparticles.

particles were injected[15] intravenously (40 mg polymer/kg; 3.2 x 10^6 to 6.4 x 10^6 dpm/kg) to anaesthetised rats. At time intervals the liver was perfused firstly with a Hank's medium (8°C) and then with a pronase or collagenase solution in Hank's medium (8°C). Cells were separated by the cold temperature method developed by Nagelkerke.[16] Radioactivity in the cell preparations, nanoparticles samples, blood samples, and total liver fraction were determined by liquid scintillation counting after oxidization of the samples. Protein content was determined by the method of Lowry,[17] using bovine serum albumin as a standard.

Intravenously administered nanoparticles are found distributed among the various compartments of the liver. The Kupffer cells are the major liver site for internalization of the nanoparticles, irrespective of their size, and following a very rapid capture mechanism. Endothelial cells show a slower and lower uptake, and parenchymal cells a very low uptake (Fig. 11).

After intraperitoneal administration of radioactive nanoparticles to rats, radioactivity was found to be solely localized inside the peritoneal space (Fig. 12). No diffusion of the carrier was visible on the autoradiographic pictures 1 hour, or 1 month after administration. Furthermore, the granular aspect of the radioactive areas could result from an aggregation of the particles in the peritoneal cavity (Fig. 12). Absence of nanoparticle resorption could be of interest when it is desirable to maintain a drug for a long time in a close cavity without diffusion in the systemic circulation. As an example, one can mention the intraarticular administration of steroids in the treatment of rheumatoid arthritis or the chemotherapy of certain cancers with attendant pleural extravasation.

TISSUE DISTRIBUTION OF DACTINOMYCIN-LOADED NANOPARTICLES

The aim of the present study was to determine if the tissue distribution of a nanoparticle-associated drug (dactinomycin) could correlate with the distribution profile of the carrier itself.[18]

Two groups of 23 rats were injected by puncture in the femoral vein. Each group received either free ^{3}H-dactinomycin associated with polybutylcyanoacrylate nanoparticles. At 0.5, 3 and 24 h, 11, 6 and 6 rats respectively were sacrificed in each of the two groups. Triplicate samples of blood and fresh tissues including spleen, small intestine, muscle, kidneys, liver and lungs were taken for ^{3}H analysis.

Table 3 shows the concentrations of ^{3}H-dactinomycin found in blood, spleen, small intestine, kidneys, liver and lungs 24 h after administration of either free or bound forms. The statistical analysis (Student's t-test) of these results indicates that nanoparticles notably modify the distribution pattern of the drug. Indeed, the tissue concentrations observed in rats injected with polybutylcyanoacrylate nanoparticles are 64-fold higher for the liver, 44-fold higher for the spleen and 4.7-fold higher for the lungs compared to rats injected with the free form. Likewise, after the same time, the muscle retains 5.6-fold more dactinomycin D when the drug is bound to the nanoparticles. Blood concentrations after intravenous administration of both free and nanoparticle-associated dactinomycin are shown in Tables 4 and 5.

These results provide evidence that nanoparticles of polybutylcyanoacrylate can be successfully used to modify the tissue distribution of dactinomycin. It is noteworthy that the carrier greatly increases the uptake of the drug by the tissues rich in reticuloendothelial cells. Although the mechanism of this uptake is not yet completely elucidated, it could be that nanoparticles gain access to the cells via endocytosis.

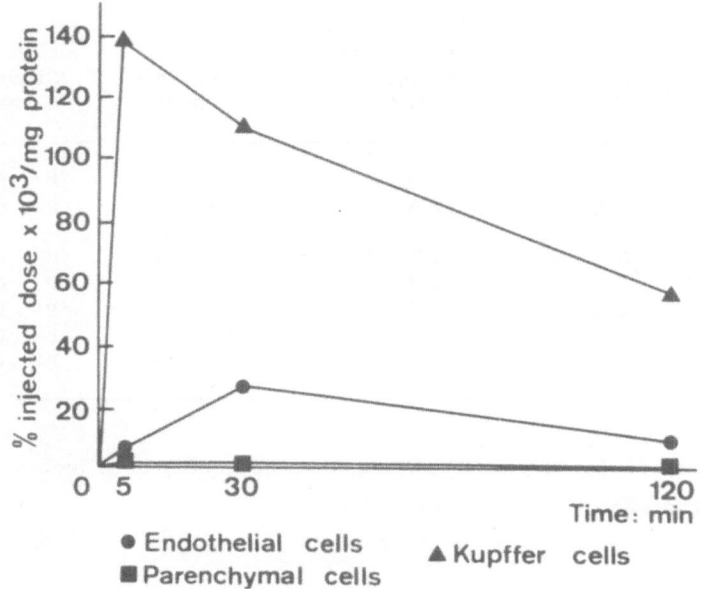

Fig. 11. Percentage (x 10^3) of the injected dose of
radioactive nanoparticles per mg protein in
different cell types of the liver and in
total liver as function of time after intravenous administration. From reference 15.

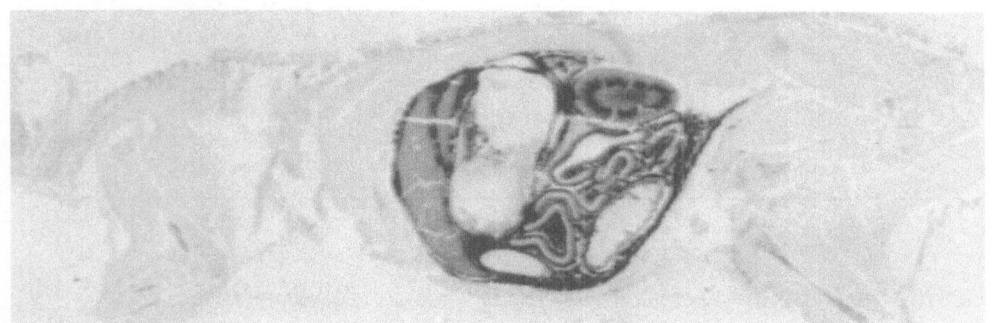

Fig. 12. Whole body autoradiography of a rat 1 h after
intraperitoneal administration of ^{14}C-labelled
polyhexylcyanoacrylate nanoparticles.

Table 3. Tissue content of ^{3}H-dactinomycin 24 h after injection (% of injected dose per g of wet tissue)[a]

Organ	Free ^{3}H-dactinomycin (F)[b]	^{3}H-dactinomycin-loaded nanoparticles (PBN)[b]	Ratios PBN/F	Student's t-test
Blood	0.2 ± 0.4	0.3 ± 0.0	1.39	P ≤ 0.05
Spleen	1.1 ± 1.0	48.7 ± 20.9	44.31	P 0.002
Small Intest.	0.9 ± 0.2	1.9 ± 0.3	1.91	P 0.001
Muscle	0.1 ± 0.0	0.7 ± 0.1	5.61	P 0.001
Kidneys	4.2 ± 0.4	5.0 ± 2.3	1.19	P > 0.1
Liver	0.7 ± 0.2	44.8 ± 8.5	64.07	P ≤ 0.001
Lungs	0.9 ± 0.3	4.4 ± 2.3	4.69	P 0.001

[a] Drug concentration in μCi/g of wet organ weight and per injected mCi.

[b] Results are the mean ± S.E. from 6 animals.

Table 4. Blood concentrations after intravenous administration of free ^{3}H-dactinomycin or ^{3}H-dactinomycin-loaded polyisobutylcyanoacrylate nanoparticles.

Time (min)	Free ^{3}H-dactinomycin (% of injected dose)	^{3}H-dactinomycin bound to nanoparticles (% of injected dose)
1	80.2 ± 5.2	77.8 ± 6.6
2	35.3 ± 3.3	61.5 ± 3.6
3	21.2 ± 1.6	49.2 ± 1.6
5	15.2 ± 0.6	32.0 ± 0.6
10	14.2 ± 0.7	10.9 ± 4.4
15	13.8 ± 0.4	10.5 ± 0.6
20	13.5 ± 0.6	11.4 ± 0.6
30	12.7 ± 0.5	7.8 ± 0.7
40	12.0 ± 0.9	7.1 ± 0.4
60	10.8 ± 1.7	5.8 ± 0.1

TISSUE DISTRIBUTION OF DACTINOMYCIN-LOADED MAGNETIZED NANOPARTICLES

Although nanoparticles can profoundly modify the pattern of tissue distribution of various drugs, excessive accumulation by the reticuloendothelial system remains a problem. Therefore we developed nanoparticles with magnetic properties since, magnetic guidance of intravascular particles by an externally placed electromagnet has proved feasable.

Mice kidneys were chosen for targeting the magnetized drug-carrier,

Table 5. Blood concentrations after intravenous administration of free ^{3}H-dactinomycin or ^{3}H-dactinomycin-loaded polyhexylcyanoacrylate nanoparticles.

Time (min)	Free ^{3}H-dactinomycin (% of injected dose)	^{3}H-dactinomycin bound to nanoparticles (% of injected dose)
1	80.2 ± 5.2	92.2 ± 5.4
2	35.3 ± 3.3	84.0 ± 4.0
5	15.2 ± 0.6	65.3 ± 7.8
10	14.2 ± 0.7	46.9 ± 5.4
15	13.8 ± 0.4	37.3 ± 3.8
20	13.5 ± 0.6	32.1 ± 3.5
40	12.0 ± 0.9	24.6 ± 2.2
120	–	14.0 ± 3.2

because of their accessibility and because of ease of comparing the drug concentration in a magnet-bearing kidney with the paired kidney.

A magnet was placed on the left kidney of each of ten mice with the right kidney being used as a reference (Fig. 13). The mice were then intravenously injected with 0.3 ml of nanoparticle suspension B. After 10 min they were killed. Each kidney was isolated and separately homogenized with 0.5 ml of distilled water. The homogenized tissue suspension (100 μl) was treated with tissue oxidizer and its radioactivity determined. To test the possibility of avoiding excessive accumulation of the carrier in the liver, the same experiment was carried out in 8 mice with a magnet on each kidney. Kidney and liver radioactivity was determined by tissue counting as described above.

Ten min after intravenous administration of magnetized nanoparticles loaded with ^{3}H-dactinomycin, an average three times higher radioactive concentration was found in the kidney bearing the magnet compared with the control. The mean radioactive concentration per gram tissue was equivalent to 440,245 dpm for the left kidney and to 155,581 dpm for the right kidney. This difference was significant ($P < 0.005$; Student's t = 2.88).

In another experiment, nanoparticles loaded with ^{3}H-dactinomycin were injected into mice bearing a magnet on each kidney; the mean radioactivity found in these organs was three times higher than in the kidneys of control mice without magnet (Fig. 14). At the same time, a one third reduction of radioactivity was found in the liver of mice with magnets on their kidneys compared with the controls.

These results demonstrate the possibility of localizing a high concentration of drug in a desired target by using nanoparticles as a magnetized drug-carrier. Furthermore, the possibility of reducing accumulation in the reticuloendothelial system was demonstrated.

The fixation of magnetite on a colloidal drug carrier could allow therapeutic levels of drugs to be attained at a desired target with smaller doses, allowing side effects, due to accumulation of the drug in other tissues, to be reduced.

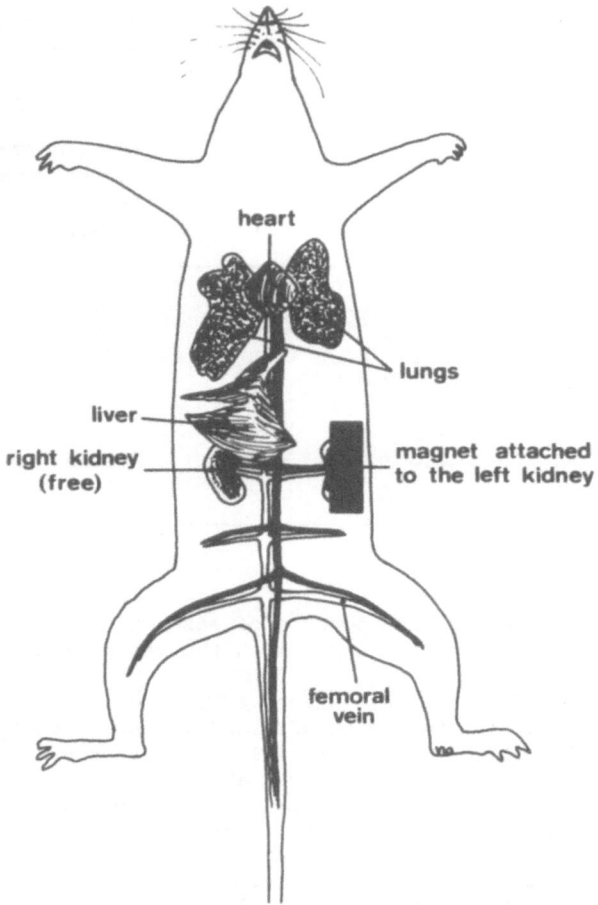

Fig. 13. Scheme of chirgically treated mouse before intravenous administration of magnetized nanoparticles.

CONCLUSIONS

The future of polyalkylcyanoacrylate nanoparticles cannot be dissociated from that of others colloidal carriers. Indeed, the main disadvantage of intravascular particulate carriers is their marked tendency to concentrate in the liver, especially in Kupffer cells, leading to a decrease in the availability of the adsorbed drugs to other tissues. In some cases, this property can be of interest, for example, in the treatment of infectious diseases of the liver. However, modification in the tissue distribution and pharmacokinetic parameters of a drug can be directly exploited in the reduction of drug side-effects and toxicity. Furthermore, better drug specificity could be achieved by the development of magnetically responsive nanoparticles.

ACKNOWLEDGEMENTS

This work was supported by the SOPAR S.A. Company and by the Institut

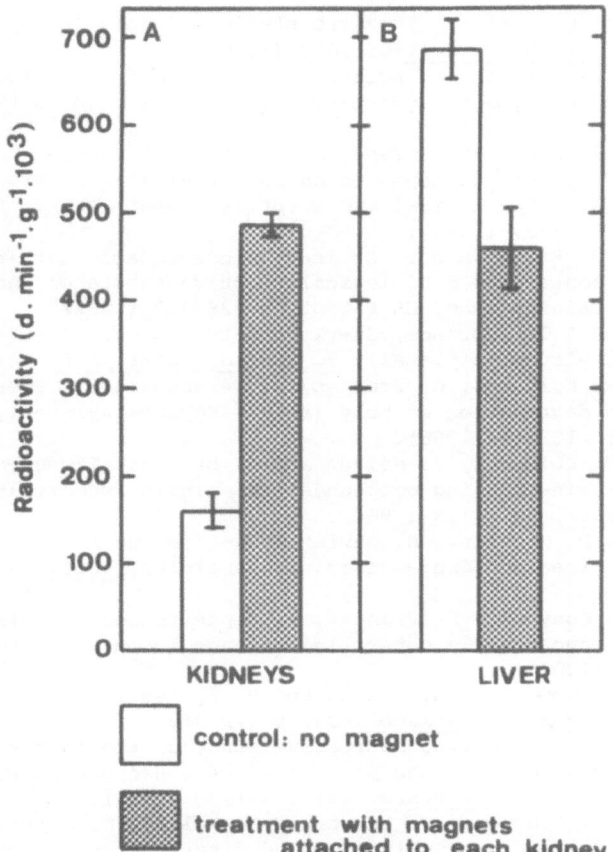

Fig. 14. Kidney (A) and liver (B) distribution of radioactivity at 10 min after intravenous injection
of magnetic polyisobutylcyanoacrylate nanoparticles. Open columns, controls (no magnet);
shaded columns, treatment with magnets attached
to each kidney. Each column represents mean
radioactive concentraion (d. min^{-1} g^{-1}) in the
liver or in the kidney of 8 animals. Total
radioactivity injected was 6,840,000 d. min $^{-1}$
g^{-1}. From reference 12.

pour l'Encouragement de la Recherche Scientifique dans l'Industrie et
l'Agriculture (I.R.S.I.A.) The authors wish to thank Mr. van Diest for
the drawings.

REFERENCES

1. G. Gregoriadis, C.P. Swain, E.J. Wills and A.S. Tavill, Drug-carrier
    potential of liposomes in cancer chemotherapy, Lancet 1:1313
    (1974).
2. M.J. Kosloski, F. Rosen, R.J. Milholland and D. Papahadjopoulos, Effect
    of lipid vesicle (liposome) encapsulation of methotrexate on its
    chemotherapeutic efficacy in solid rodent tumors, Cancer Res.
    38:2848 (1978).

3. J. Kreuter and P. Speiser, In-vitro studies of poly (methylmetacrylate) adjuvants, J. Pharm. Sci. 65:1624 (1976).
4. P. Couvreur, P. Tulkens, M. Roland, A. Trouet and P. Speiser, Nano-capsules a new type of lysosomotropic carrier, Febs Lett. 84:323 (1977).
5. P. Couvreur, B. Kante, M. Roland, P. Guiot, P. Baudhuin and P. Speiser, Polycyanoacrylate nanocapsules as potential lysosomotropic carriers: preparation, morphological and sorptive properties, J. Pharm. Pharmacol. 31:331 (1979).
6. P. Couvreur, M. Roland and P. Speiser, Biodegradable submicroscopic particles containing a biologically active substance and compos-itions containing them, US Patent, 4 329 332 (1982).
7. W.R. Vezin and A.T. Florence, In-vitro heterogenous degradation of poly (n-alkylcyanoacrylates), J. Biomed. Mater. Res. 14:93 (1980).
8. F. Leonard, R. Kulkarni, G. Brandes, J. Nelson and J. Cameron, Syn-thesis and degradation of poly (alkyl-αcyanoacrylates), J. Appl. Polym. Sci. 10:259 (1966).
9. F. Leonard, R. Kulkarni, J. Nelson and G. Brandes, Tissue-adhesives and hemostasis-inducing compounds: the alkylcyanoacrylates, J. Biomed. Mater. Res. 1:3 (1967).
10. L. van Snick, P. Couvreur, D. Christiaens-Leyh and M. Roland, Molecular weights of free and drug-loaded nanoparticles, Pharm. Res. 1:36 (1985).
11. C. Verdun, P. Couvreur, H. Vranckx, V. Lenaerts and M. Roland, Develop-ment of a nanoparticle controlled release formulation for human use, submitted (1985).
12. A. Ibrahim, P. Couvreur, M. Roland and P. Speiser, New magnetic drug carrier, J. Pharm. Pharmacol. 35:59 (1983).
13. P. Guiot and P. Couvreur, Quantitative study of the interaction between polybutylcyanoacrylate nanoparticles and mouse peritoneal macro-phages in culture, J. Pharm. Belg. 38:130 (1983).
14. L. Grislain, P. Couvreur, V. Lenaerts, M. Roland, D. Deprez-Decampeneere and P. Speiser, Pharmacokinetics and distribution of a biodegradable drug-carrier, Int. J. Pharm. 15:334 (1983).
15. V. Lenaerts, J.F. Nagelkerke, T.J.C. van Berkel, P. Couvreur, L. Grislain, M. Roland and P. Speiser, In-vivo uptake of polyiso-butylcyanoacrylate nano-particles by rat liver Kupffer, Endothelial and Parenchymal Cells, J. Pharm. Sci. 73:980 (1984).
16. J.F. Nagelkerke, K.P. Barto and T.J.C. van Berkel, in: "Sinusoidal Liver Cells", D.L. Knook and E. Wisse, eds., Elsevier North-Holland Biomedical Press, Amsterdam (1983).
17. O.H. Lowry, N.J. Rosebrough, A.L. Farr and M.J. Randall, Protein measurement with folin-phenol reagent, J. Biol. Chem. 193:265 (1951).
18. B. Kante, P. Couvreur, V. Lenaerts, P. Guiot, M. Roland, P. Baudhuin and P. Speiser, Tissue distribution of ^{3}H-actinomycin D adsorbed on polybutylcyanoacrylate nanoparticles, Int. J. Pharm. 7:45 (1980).

POLYMERS AS MATRICES FOR DRUG RELEASE

PAOLO FERRUTI

Facolta d'Ingegneria, Universita di Brescia, Viale Europa
n. 39, 25060 Brescia, Italy

INTRODUCTION

The pharmaceutical applications of synthetic polymers may be broadly
divided into two main sections: the physical incorporation of active molecules into a polymeric matrix, from which they may be subsequently released
either by diffusion processes, or by erosion and the synthesis of pharmacologically active polymers. Pharmacologically active polymers may be
classified as follows:

1. Polymers which are pharmacologically active per se, their activity
dependant on macromolecularity. The corresponding monomers, or non-macromolecular models, are inactive.
2. Polymers whose activity depends on moieties structurally related to
well known non-macromolecular drugs, linked to the macromolecular backbone
with covalent bonds which are not expected to be cleaved in order to exert
activity.
3. Polymers which are able to give rise to non-macromolecular active
substances after administration. These may be further divided into two
main categories: a) Polymers in which residues of active molecules are
constituents of the main backbone. Consequently, the active molecules
are released by degradation of the whole macromolecule; b) Polymers in
which residues of active molecules are linked as side substituents to a
polymeric or oligomeric structure with covalent bonds cleavable in body
environments. The release of active molecules does not necessarily involve
a degradation of the whole produce, which may or may not take place as a
separate process.

Leaving apart the physical combinations of drugs with polymers, which
will not be dealt with here, it is apparent that the concept of "matrix"
as a part distinct from the active moieties applied to categories 2 and 3b,
which are formally similar, apart for the cleavability of the drug-matrix
bonds. In categories 1 and 3a these parts merge into a unique structure.
Therefore, polymeric drugs of cat. 2 and 3b are typically conceived as
formed by two distinct parts, the matrix, and the active moieties which are
covalently linked to the matrix either directly, or through an intermediate
spacing arm. It may be observed that the majority of polymeric drugs
studied so far belong to these categories, and especially so to category
3b.

Since this article will be mainly concerned with synthetic aspects, the behavioural differences between polymeric and non-polymeric substances in living organisms will not be considered in detail. They have been reviewed elsewhere (see for instance References 1-3). It may be sufficient to recall here the main results obtainable by binding drugs to polymeric or oligomeric carriers: (a) sustained activity with respect to the free drug; (b) increased adsorption by oral and, possibly, intradermal administration (in the case of oligomeric matrices); (c) preferential localization at the level of given target cells.

GENERAL CONSIDERATIONS ON THE DESIGN OF POLYMER-DRUG DERIVATIVES

The design of a polymer-drug derivative obviously involves two main aspects: the nature of the matrix, and the nature of the linkage between the drug, and the matrix.

From a synthetic point of view, two main routes for the preparation of polymer-drug derivatives can be envisaged. The first one involves the preparation and polymerization of a polymerizable derivative of the drug, as, for instance, an ester of the drug itself with acrylic acid:

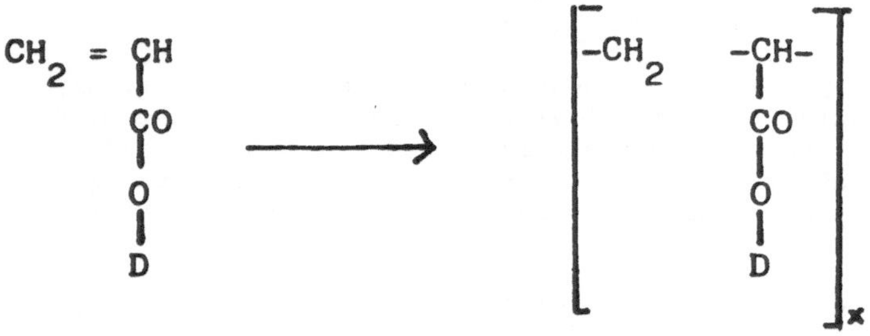

$$CH_2 = \underset{\underset{\underset{\underset{D}{|}}{O}}{\underset{|}{CO}}}{CH} \longrightarrow \left[-CH_2 \quad \underset{\underset{\underset{\underset{D}{|}}{O}}{\underset{|}{CO}}}{-CH-} \right]_x \qquad D = \text{drug residue}$$

This method can be utilized only if the drug molecule does not contain other chemical functions able to interfere with the polymerization process. One such function, which is often present in drugs, is for example a carbon-carbon double bond, which obviously interferes with radical polymerizations. The second method involves the formation of a suitable chemical bond between the drug and a presynthesized matrix. In the most general sense, multifunctional matrices can be prepared, containing functional groups able to react selectively with other functional groups present in the drug molecule. For instance, matrices containing activated ester (or amido groups) may react with hydroxylated or aminated drugs as follows:

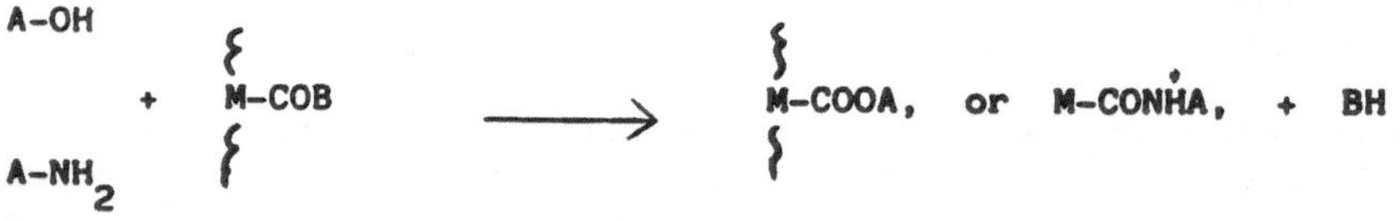

$$\begin{matrix} A-OH \\ \\ A-NH_2 \end{matrix} \quad + \quad M\text{-}COB \quad \longrightarrow \quad M\text{-}COOA, \text{ or } M\text{-}CONHA, \quad + \quad BH$$

A = drug residue

B = leaving group residue

M- = matrix

166

Table 1. Activated derivatives of poly)acrylic acid), poly(methacrylic acid), and related polymeric acids.

System	Structure Unit(Ref)	System	Structure Unit(Ref)
1	N-Acryloylimidazole(7)	7	N-Methacryloilimidazole(12)
2	1-Acriloylbenzotriazole(8)	8	1-Methacryloxybenzotriazole(9)
3	1-Acryloxybenzotriazole(9)	9	N-Methacryloxysuccinimide(10)
4	N-Acryloxysuccinimide(10)	10	2,4,5-trichlorophenylmethacry-late(9)
5	2,4,5-trichlorephenylacrylate (9)	11	Hydroxysuccinimides of poly (N-methacryloylω- amino acids)*(13)
6	4-Nitrophenylacrylate(11)		

System	Structure Unit (Ref)	System	Structure Unit (Ref)
12	1-Benzotriazolides of poly (N-methacryloyl-ω-amino acids)**(14)	14	4-Nitrophenylesters of (N-methacryloyl)oligo-peptides(15,16)
13	1-(4-methacyloxy)benzoyl-benzotriazole(14)		

*n=1,2,3,5,10
**n=1,5

A similar scheme may be adapted for the preparation of hydrazone, urethane or carbonate derivatives of drugs. Alternatively, this reaction scheme may be reversed by preparing, for instance, activated derivatives of carboxyl bearing drugs and causing them to react with hydroxylated matrices:

$$A-COOH \longrightarrow A-COB \quad + \quad HO-M \longrightarrow A-COOM \quad + \quad HB$$

were A, B, and M have the same meaning as in the previous case.

By comparing the two methods, it is apparent that the second one has two significant advantages: first, no interactions are expected to occur with many chemical functions which may be present in the drug molecule, as, for instance, unsaturated structures; secondly, a single "mother" polymer may be utilized to bind covalently a number of active substances. Thus, the synthesis and polymerization of a polymerizable derivative of the drug does not have to be studied in every case.

Most high molecular weight matrices so far studied to prepare polymeric drugs belong to three main families: polyvinylic polymers, poly-
saccharides, and poly-α-aminoacids, including some proteins. Other
structures can be utilized as oligomeric matrices. It may be observed that
in this case an acceptable loading of the active substances to be delivered
can be achieved also if the matrices carry active moieties only at their
ends, while, as a rule, high molecular weight matrices must carry the drug
moieties as side substituents. The following sections will be mainly devoted to polyvinylic matrices and oligomeric matrices.

POLYVINYLIC MATRICES

     Polyvinylic matrices are characterized by a macromolecular backbone
formally deriving from the polymerization of carbon-carbon double bonds,
as in the following very common cases:

$$\left[-CH_2 - CH \right]_x \quad ; \quad \left[-CH_2 - \underset{|}{\overset{CH_3}{C}} \right]_x$$

These matrices have been by far the most studied as drug carriers, owing
to their versatility and to the considerable background of knowledge on
multifunctional polymers of this structure. In the most general sense, a
drug-carrying matrix contains drug-binding units and solubilizing units,[4]
the latter, as a rule, having the function of imparting water solubility
to the resulting polymeric drugs. The binding units may bind the drug
molecules either directly, or through an intermediate spacing arm. In
addition, targeting units may be present.[1-6] In the case of polyvinylic
matrices, binding units, solubilizing units, and spacers usually constitute different steps in designing the corresponding polymeric drugs, and
will be dealt with separately.

Binding and Coupling Units

     The term "binding unit" indicates a unit to which the drug moiety is
attached, either starting from a polymerizable derivative of the drug, or
as a result of a coupling reaction with a pre-synthesized matrix (see
above). In the latter case, the units involved in the coupling reaction
may be called "coupling units". This section will be mainly devoted to
coupling units.

     A further distinction can be made, since the coupling units may
either react directly with the proper drug molecules, or simply provide
a site of attachment for the same molecules after further activation. In
the former case, the coupling units may be also referred to as "activated
units". Many activated units described in the literature correspond to
activated esters or amides of polymeric acids, and especially so of poly
(acrylic) and poly(methacrylic acid), or related polymeric carboxylic
acids. They are reported in Table 1. These derivatives offer many ad-

$$\left[-CH_2 - \underset{\underset{COCl}{|}}{CH} \right]_x \quad ; \quad \left[-CH_2 - \underset{\underset{COCl}{|}}{\overset{CH_3}{C}} \right]_x$$

169

vantages over the more traditional poly(acryloyl chloride) and poly(meth-acryloyl chloride), since they are more selective and more easy to handle, being less moisture-sensitive; moreover, they do not give rise to aggressive by-products on reaction with aminated or hydroxylated compounds, and these reactions can be performed in solvents which in general are not compatible with acyl chlorides, such as for instance sulphoxides, amides, and, in some cases, alcohols.[17,18]

In all cases with the exceptions of poly(N-acryloylimidasole) (system 1) and N-methacryloylimidazole (system 7), the corresponding monomers can be prepared and polymerized, or copolymerized with a number of co-monomers including hydrophilic ones; hence, the introduction of solubilizing units in the resulting matrices is usually not a problem. Poly(N-acryloylimidazole) can be prepared only by reaction of poly(acrylic acid) with N,N-carbonyldiimidazole[7] since every attempt to prepare monomeric N-acryloylimidazole has failed so far. N-Methacryloyl imidazole can be prepared, but it does not homopolymerize, although it can be copolymerized with several other monomers.[12]

In principle, all the above derivatives may lead to ester or amido bonds between the drug and the matrix, according to the general reaction scheme reported in the preceeding section. Their reactivities depend both on the nature of the leaving groups, and the structure of the matrix. The following observations can be made: As a rule, methacryloyl derivatives are much less reactive than acryloyl derivatives, when no spacers are in-serted between the main backbone, and the activated function. Thus, for instance, poly(N-acryloxysuccinimide) (system 3) reacts rapidly and quant-itatively with aliphatic amines at room temperature yielding essentially homopolymeric poly(acrylamide)s, while poly(N-methacryloxy succinimide) (system 9) requires prolonged heating at 100°C to give similar results. When a spacer is inserted, however, the reactivities are practically the same, at least as far as the derivatives of Table 1 are concerned.

Other conditions being equal, the nature of the leaving groups is obviously the most important factor governing reactivities. In this respect, the leaving groups can be roughly arranged as follows:

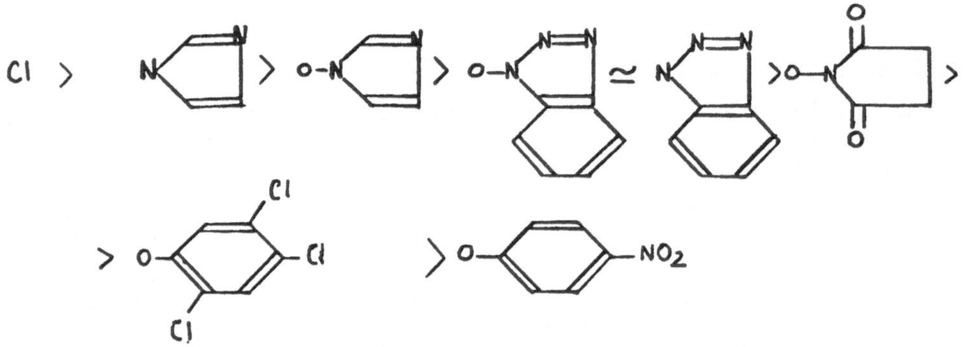

Thus, all derivatives listed in Table 1 may react without catalysts with aliphatic or cycloaliphatic primary or secondary amines giving amido link-ages; moreover, many of them react cleanly with hydrazine giving the corr-esponding hydrazides, as in the following example[19]:

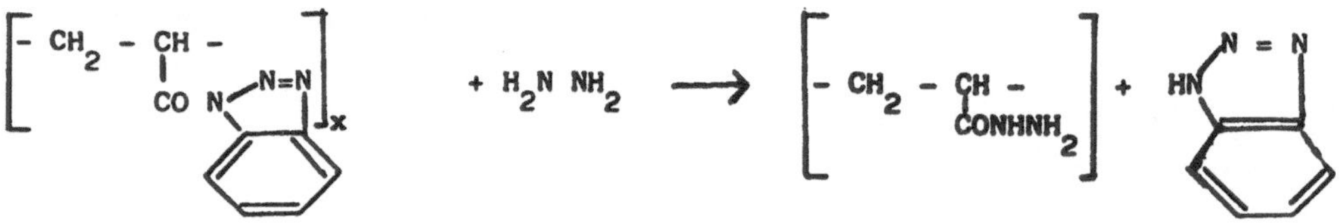

Table 2. Activated derivatives of polymeric carbamic acids

System	Structure Unit (Ref)	System	Structure Unit (Ref)
1	N(N'-vinyl)carbamoyloxy-imidazole(9)	5	N(N'-vinyl)carbamoyloxy-succinimide(9)
2	1(N-vinyl)carbamoyloxy-benzotriazole(9)	6	2,4,5-trichlorophenyl(N-vinyl)carbamate(9)
3	1(N-vinyl)carbamoyloxy-4-methylbenzotriazole(9)	7	N(N'-isopropenyl)carbamoy-loxysuccinimide(9)
4	1(N-vinyl)carbamoyloxy-5-methoxybenzotriazole(9)	8	N(2-methacryloxy)ethyl-carbamoyloxysuccinimide(20)

171

In our experience, the best results in both types of reactions are given by the N-hydroxysuccinimide esters and the benzotriazolides. In presence of hydroxylated co-units, the 4-nitrophenylesters can be used with advantage.

The reaction of the same derivatives with alcohols, phenols, and aromatic amines, is usually much slower, and is only feasible in practice, with the most reactive derivatives. In many cases, nearly quantitative reactions are obtained only by heating at 50-70°C in the presence of catalysts, such as triethylamine or, better, 4-dimethylaminopyridine. In our experience, benzotriazolides offer a good compromise between reactivity and easiness in handling. Another group of derivatives comprises activated esters of polymeric carbamic acids. These are reported in Table 2. When applicable, the above considerations are valid also for these derivatives, apart from the fact that, the leaving group being the same, they seem to be less reactive than the corresponding derivatives of the previous series.

In many instances, the acylation reaction may show considerable selectivity. For example, polymeric benzotriazolides, as well as N-hydroxysuccinimide esters, may react in the cold with hydroxylated amines giving pure, uncrosslinked hydroxylated acrylamides.[17,18]

Good selectivity can also be obtained, in the case of poly(N-acryloylbenzotriazole), with compounds bearing both primary and secondary hydroxyls, the former being more reactive.[21]

Some activated matrices contain anhydride groups. These are reported in Table 3. On reaction with hydroxylated or aminated compounds, they give rise to free carboxyl groups, as in the following example:

Table 3. Activated matrices containing anhydride groups.

System	Structure Unit(Ref)	System	Structure Unit(Ref)
1	Maleic anhydride (22)	3	Copolymer between maleic anhydride and divinylether (DIVEMA)*(24)
2	Acrylic anhydride (cyclopolymerized)(23)		

*The anhydride group on the pyran ring is thought to react first. The structure is rather idealized, since other types of repeating units may be present.

Table 4. Miscellaneous activated matrices.

System	Unit	Structure	Bond Obtain-able	Ref.
1	Acryloylhydrazide	$-CH_2$ $-CH-$ CO NH-NH$_2$	a α	19 25
2	ϵ-methacrylaminoca proylhydrazide	CH$_3$ $-CH_2$ $-C-$ C=O NH (CH$_2$)$_5$ C-NH-NH$_2$ O	a α	19
3	4-methacryloxybenzoyl-hydrazide	CH$_3$ $-CH_2$ $-C-$ C=O O (ring) C-NH-NH$_2$ O	a α	19
4	Allylglycidylether	$-CH_2$ $-CH-$ CH$_2$ $-CH-CH_2$ (epoxide)	b β	26
5	Glycidylmethacrylate	CH$_3$ $-CH_2$ $-C-$ COOCH$_2$ $-CH-CH_2$ (epoxide)	b β	27
6	Glycerylmethacrylate	CH$_3$ $-CH_2$ $-C-$ CO O-CH$_2$-CH-CH$_2$OH OH	c γ	27
7	Vinylmercaptan	$-CH_2$ $-CH-$ SH	d δ	26
8	2(chlorocarbonyloxy) ethylmethacrylate	CH$_3$ $-CH_2$ $-C-$ C=O O-CH$_2$-CH$_2$-O-C-Cl O	e ϵ	28

a, α : hydrazone:M $\sim$ CO - NM - N = C $<$; b, β : 2-hydroxyester, 2-hydroxysulphide, 2-hydroxyamino etc., for example $M\sim CH_2-CH-CH_2-O-C-$ with OH and O ; c, γ : acetal: $M\sim CH-CH_2$ with O O and C ; d, δ : sulphide: $M\sim S - C <$; e, ϵ : carbonate $-O-C-O-$ with O

Table 5. Some examples of coupling units requiring activation to give matrix drug linkages.*

System	Unit	Structure
1	Vinylalcohol	$-CH_2-\overset{}{\underset{OH}{CH}}-$
2	Vinylamine	$-CH_2-\overset{}{\underset{NH_2}{CH}}-$
3	Acrylic acid	$-CH_2-\overset{}{\underset{COOH}{CH}}-$
4	Methacrylic acid	$-CH_2-\overset{CH_3}{\underset{COOH}{C}}-$
5	2-Hydroxyethylmethacrylate	$-CH_2-\overset{CH_3}{\underset{C=O}{C}}-$ $O-CH_2-CH_2-OH$
6	N-(2-Hydroxy)propylmethacryl-amide	$-CH_2-\overset{CH_3}{\underset{CO}{C}}-$ $NH-CH_2-\overset{}{\underset{OH}{CH}}-CH_3$

* In most cases, the molecule to be attached is activated via acylchloride
  groups, or activated ester and amido groups (see Table 1); coupling
  agents, e.g., carbodiimides, are also frequently used.

These groups can act as solubilizing groups, if ionized; however, the whole
polymer-drug derivative behaves as a poly-anion. This, in principle, may
alter the activity of the resulting products. In fact, poly-anions are
often pharmacologically active per se. This is especially true for DIVEMA
(system 3), which is a well known anticancer drug.[24]

Miscellaneous activated matrices, leading to polymer-drug linkages
are reported in Table 4.

Binding units requiring some kind of activation are reported in Table
5. It may be noted that practically all the cases we have found in the
literature deal with polymeric alcohols, acids and amines. As a consequence, the linkages to be obtained are of ester and amido types.

Solubilizing Units

Many polymeric derivatives of drugs, in the form of homopolymers,
would be practically water-insoluble, since the polyvinylic backbone is
hydrophobic, as are many drug moieties. On the other hand, as a rule, a
hydrolytic cleavage of the drug-matrix bonds is expected to occur, and this

174

Table 6. Solubilizing units.

System	Structure Unit(Ref)	System	Structure Unit(Ref)
1	 N-vinylpyrroiidone (29, 30, 31)	7	 2-sulphoxymethylethylacrylate (31)
2	 2-vinylpyridine-N-oxide (30, 32)	8	 acrylamide(26, 30, 31)
3	 3-vinyl-A-methylpyridine- -N-oxide(30,32)	9	 methacrilamide (30)
4	 N-acryloylmorpholine(21)	10	
5	 2(hydroxyethyl)methacrylate (33, 34)		N(2-hydroxypropyl)-metha- crylamide(34)
6	 2-sulphoxymethylethylmeth- acrylate(30, 31, 34)		

*Selected references referring to polymeric drugs containing these units.

requires at least a certain degree of hydrophilicity. Moreover, water solubility is obviously of advantage in terms of administration, dosage, and migrability within the body.

Water solubility can be achieved by introducing units derived from suitable hydrophilic monomers. These units should be biocompatible and, preferably, pharmacologically inactive per se. Some solubilizing units are reported in Table 6.

These units can be introduced by copolymerization of the corresponding monomers either with a polymerizable derivative of the drug, or with an activated monomer, since as pointed out above, many of these can be copolymerized with hydrophilic monomers. Also several monomers corresponding to the binding units of Table 5 can be used as solubilizing units. In these cases, it is sufficient to utilize them only partially in the coupling reaction with drugs, leaving some of them free.

Spacers

In many cases, direct attachment of the drug moieties to a polyvinylic backbone, even if the polymer-drug linkage is hydrolyzable per se, leads to poor pharmacological results. The main backbone, in fact, may exert a shielding effect, rendering it difficult for the cleavage of adjacent bonds. This is even more pronounced when a steric hindrance is present, as, for instance, in derivatives of poly(methacrylic acid) (see also above, when dealing with activated units).

For this reason, a spacer is often inserted between the main chain, and the drug moieties. A spacer may be defined as a group of atoms, usually in the form of a short chain, bringing the drug moieties some distance from the main chain. Even when the polymer-drug linkage is not supposed to be hydrolyzed to exert activity (case 2 of Introduction), a spacer may facilitate the interaction of the active moieties with the receptors.

Many spacers are already present as a structural part of binding units, as, for instance, Systems 11 - 14 of Table 1, system 8 of Table 2, system 3 of Table 3, systems 2 - 7 and 9 of Table 4, systems 5 and 6 of Table 5. Other spacers are introduced as an intermediate step, usually preparing a drug-spacer derivative, functionalized at one end, and then binding it to the matrix by one of the methods outlined above. It would appear that a good spacer should have at least a certain degree of hydrophilicity since a hydrophilic spacer in aqueous media would be hydrated, and, consequently, extended thus exposing the drug moieties. Hydrophobic spacers, such as long alkyl chains, may even hinder the hydrolytic cleavage of the polymer-drug bonds by creating a hydrophobic microenvironment. This was the case, for instance, of some cyclophosphamide/DIVEMA (system 3 of Table 3) derivatives.[35]

The use of proper spacers may also enable the cleavage of the drug-polymer bonds to be modulated in such a way that it takes place only at preferred sites. For instance, spacers of oligopeptidic structure (see system 14 of Table 1) can be hydrolyzed within lysosomes, but are otherwise stable. The release of the drug, as a consequence, is necessarily preceded by uptake into cells.[15,16]

Some Remarks on Polyvinylic Polymeric Drugs

The examples of polymer drug combinations with a polyvinylic structure described so far are very numerous, and to report them is beyond the scope of this article. Only a few general observations will be made here.

a. The main drawback of most polyvinylic matrices is their undegradability within the organism. It is known, in fact, that a high polymer is little or not adsorbed through the gastrointestinal tract, but is practically not excreted through the usual excretion routes if injected. To be eliminated, its molecular weight must be reduced below a certain threshold, which may vary according to the polymer's structure.[1,2] As a consequence, polymeric drugs of polyvinylic structure should be used, in principle, only by oral administration in order to avoid any long-term residence in the body of macromolecular residues. However, very few of the polymeric drugs reported in the literature are said to be active after oral administration.[21] In other cases, activity has not been observed after oral administration, but only after intravenous administration.[25] As a matter of fact, this way of administration seems to ensure in most cases the maximum hope of activity. Two ways can be envisaged to overcome the problem of elimination of the macromolecular residues. The first, and more obvious one, is to reduce the molecular size of the matrix. The second one, especially useful when high molecular weights are needed, is to introduce weak links in the main backbone. The latter approach is discussed by Lloyd in this book.

b. Most of the drug-matrix linkages reported are of ester, amido-urethane-, carbonate-, hydrazone- type. Hence, they are potentially hydrolyzable in aqueous systems. This does not mean that they are actually hydrolyzed at a useful rate after administration to living organisms. On this argument, no comprehensive studies seem to have been published so far, but only scattered hints are available. Therefore, the following rules are only aimed to provide a rough guide, to which a number of exceptions may be found: (i) Amidic bonds, (-CONH- or -CONR-), if not of activated type, are probably little cleaved anywhere in the absence of spacers. When a spacer is present, they may be cleaved inside lysosomes, but are stable in the gastrointestinal tract and possibly also in the blood stream. The same is probably true for urea bonds (-NH CONH-); (ii) Ester bonds (-COO-) are cleavable in the body fluids in the presence of spacers. Without spacers, esters of hydroxylated drugs with poly(acrylic acid) are poorly cleavable, and those with poly(methacrylic acid) are essentially uncleavable. Esters of carboxylated drugs with hydroxylated matrices are probably cleavable in many instances; (iii) Esters of hydroxylated drugs with polymeric carbamic acids (-NH COO-), and polymeric carbonates ($O-\overset{\text{O}}{\underset{}{C}}-O$ -), may be considered cleavable; (iv) Polymeric hydrazones (-CONH N=), and Schiff bases (-C = N-) are cleavable.

The above rules refer to water soluble products in the absence of heavy steric hindrances around the bond to be hydrolyzed. It must be emphasized, however, that the rate of cleavage must be checked in every case, since in biological environments it may vary considerably even for products of similar structures. Moreover, a given rate of release may, or may not, be pharmacologically significant according to the activity of the drug released.

OLIGOMETIC MATRICES

As pointed out above the results expected from oligomeric matrices (i.e. matrices whose molecular weights range from a few hundreds to a few thousands) may be different from those obtainable from high-molecular weight ones. Oligomeric matrices may act as vehiculating agents to facilitate adsorption of drugs through the gastrointestinal tract, and, possibly, through the skin. In the mean time, prolonged activity may be obtained, and toxic side reactions reduced, as in the case of high molecular weight matrices.

Leaving apart a number of oligomeric matrices of different structures proposed by several workers,[37-41] these considerations are especially true in the case of poly(ethylenglycol) (PEG)[18] and its homologue poly(propylene glycol) (PPG). These have good biocompatibility, and are commercially available as fractions of well defined molecular weights. They have amphiphilic properties, and therefore may be expected to act as vehiculating agents across physiological barriers.

Table 7. Functionalization of poly(ethyleneglycol)s

System	Reactive function introduced	Reagent(s) used	Ref.
1,1	Cl, Br	SoBr$_2$, SOCl$_2$	41-43
2	-OSO$_2$—⟨benzene⟩—CH$_3$(T$_s$), OSO$_2$CH$_3$(M$_s$)	TsCl, MsCl	44,45
3	-R-NCO	OCN-R-NCO	46-48
4	-OOC-Cl	COCl$_2$	43
5	-OCH$_2$-CH-CH$_2$ (epoxide)	ClCH$_2$CH-CH$_2$ (epoxide)	49
6	-OC-(imidazole)	N,N-'carbonyldiimidazole	50
7	-C-CH$_2$CH$_2$-C-N(imidazole)	10 a) + imidazole, DCC	51,52
8	-C-CH$_2$CH$_2$COO- (succinimide)	10 a) + HON(succinimide) DCC	53
9	-C-CH$_2$CH$_2$-N(benzotriazole)	10 a) + 1H-benzotriazole, DCC	51,52
10.a	-OCO-CH$_2$CH$_2$COOH	succinic anh.	51,52
10.b	-OCO(CH$_2$)$_3$COOH	glutaric anh.	51,52
11	-O-CH$_2$COOH	BrCH$_2$COOEt+alkoxide	41
12	-OCH$_2$ CHO	MnO$_2$	53
13	-OC-O—⟨benzene⟩—NO$_2$	Cl C O—⟨benzene⟩—NO$_2$	54

Both PEG and PEG monomethylethers can be used as such as carriers for carboxyl-bearing drugs. However, a number of other end-functions have been introduced, with the purpose of obtaining either drug-binding, or also protein-binding derivatives. Some of these are reported in Table 7, together with the drug-matrix linkages which can be obtained.

The question of the cleavability of these linkages deserves some further comments. While in the case of polyvinylic matrices it was doubtful if many ester derivatives could be hydrolyzed in the body without spacers, PEG esters and urethanes are.[55] Probably, PEG carbonates and amides are also cleavable, but at a slower rate.

Some examples of PEG - drug combinations studied by this author are given in Table 8. The case of 4-isobutylphenyl-2-propionic acid (IBUPRO-FEN), (system 1 of Table 8), is a good example of the results obtainable.[56]

Two types of derivatives were prepared, in which poly(ethyleneglycol)s were esterified at both ends, or at one end only. Their molecular weights ranged from 400 to 2400. Their pharmacological activity was determined in rats after oral administration, and compared with that of the free drug. It was found that in all cases a more sustained activity could be achieved. Plasma levels of drug remained within therapeutic limits much longer. Moreover, in the lower molecular weight compounds, the initial activity was higher than that of equivalent doses of the free drug. This means that the bioavailability was considerably increased, probably because of a vehiculating effect of the matrix across the gastrointestinal tract.

These results were confirmed by 3α, 7β -dihydroxy - 5 - β - cholan - 24-oic acid (ursodeoxy cholic acid) derivatives.[56] In the case of nicotinic acid derivatives, the partition profiles in a three-phase partition

Table 8. Some examples of PEG - drug combinations

System	Drug linked	Structure	Ref.
1	Ibuprofen ('-isobutyl-phenyl-2-propionic acid)		55
2	Ursodeoxycholic acid		56
3	Nicotinic acid		57

test simulating skin adsorption have been studied.[58] It was concluded that PPG's are more efficient vehicles than PEG's, and that increasing the molecular weight of the matrix above about five hundred has an adverse effect on adsorption properties.

REFERENCES

1. J. Drobnik and F. Rypacek, Soluble synthetic polymers in biological systems, Advanc. Polym. Sci. 57:1 (1984).
2. R. Duncan and J. Kopecek, Soluble synthetic polymers as potential drug carriers, Advanc. Polym. Sci. 57:53 (1984).
3. J. Pitha, Polymer-cell surface interactions and drug targeting in: "Targeted Drugs", E. Goldberg, ed., John Wiley and Sons, New York (1983).
4. P. Ferruti, Macromolecular drugs acting as precursors of non-macromolecular active substances. Preliminary considerations, Pharmacol. Res. Commun. 7:1 (1975).
5. P. Ferruti, Macromolecular Drugs, Il Farmaco, Ed. Sci. 3:220 (1977).
6. H. Ringsdorf, Synthetic Polymeric Drugs, Mid. Macromolecular Monogr. 5:197 (1978).
7. P. Ferruti and F. Vaccaroni, Polymeric acrylic and methacrylic esters and amines by reaction of poly(acrylic acid) and poly(methacrylic acid) with N,N'-carbonyldiimidazole and alcohols or amines, J. Pol. Sci. 13:2859 (1975).
8. P. Ferruti, A. Fere and G. Cottica, Poly-1-acryloylbenzotriazole as polyester and polyacrylamide precursor, J. Polym. Sci. 12:553 (1974).
9. H.G. Batz, G. Franzmann and H. Ringsdorf, Model reactions for synthesis of pharmacologically active polymers by way of monomeric and polymeric reactive esters, Angew. Chem. 12:1103 (1972).
10. P. Ferruti, A. Bettelli and A. Fere, High polymers of acrylic and methacrylic esters of N-hydroxysuccinimide as polyacrylamide and polymethacrylamide precursors, Polymer. 13:462 (1972).
11. C.P. Su and H. Morawetz, Reactivity of polymer substituents. Aminolysis of p-nitrophenylester residues attached to various polymer backbones, J. Polym. Sci. 15:185 (1977).
12. P. Ferruti and G. Cottica, 1-Methacryloylimidazole as methacrylating agent, J. Polym. Sci. 12:2453 (1974).
13. H.G. Batz and J. Koldehoff, Monomere und Polymere Succinimidoester von co-Methacryloylaminosäuren, ihre Darstellung und inhre Reaktion mit Aminen, Makromol. Chem. 177:683 (1976).
14. P. Ferruti, F. Vaccaroni and M.C. Tanzi, Synthesis and exchange reactions of some polymeric benzotriazolides, J. Polym. Sci., 16:1435 (1978).
15. R. Duncan and J.B. Lloyd, Degradation of side chains of N-(2-hydroxypropyl) methacylamide copolymers by lysosomal enzymes, Biochem. Biophys. Res. Commun. 94:284 (1980).
16. R. Duncan, H.C. Cable, J.B. Lloyd, P. Rejmanova and J. Kopecek, Polymers containing enzymatically degradable bonds, Makromol. Chem, 184:1997 (1983).
17. P. Ferruti, Functionalization of polymers in: "Reactions on Polymers", J.A. Moore ed., D. Reidel Publishing Co., Boston (1973).
18. P. Ferruti, A.S. Angeloni, G. Scapini and M.C. Tanzi, New oligomers and polymers as drug carriers in: "Recent Advances in Drug Delivery Systems", J.M. Anderson and S.W. Kim, eds., Plenum Press, New York, (1984).
19. P. Ferruti, M.C. Tanzi and F. Vaccaroni, Polymeric hydrazides by reaction of hydrazine with polymeric benzotriazolides, J. Polym. Sci. 17:277 (1979).
20. G. Franzamann and H. Ringsdorf, Pharmakologisch aktive Polymere, 12: Depotformen von Chlorambucil durch kovalente Bindung on Polymere, Makromol. Chem., 177:2547 (1976).

21. N. Ghedini, P. Ferruti, V. Andrisano and G. Scapini, Synthesis of a high molecular weight polymeric derivative of $3\alpha,7\beta$-dihydroxy-5β-cholan-24-oic acid (ursodeoxycholic acid), Synth. Comm. 13:707 (1983).

22. An example of Polymer/drug (dopamine) combination was given by G. Reinish at the 26th minisymposium on "Polymers in Medicine and Biology", Prague, July 9-12, 1984.

23. T. Hirano, W. Klesse and H. Ringsdorf, Polymeric derivatives of activated cyclophosphamide as drug delivery systems in antitumor chemotherapy, Makromol. Chem. 180:1125 (1979).

24. P.P. Umrigar, S. Ohasshi and G.B. Butler, Synthesis and properties of alternating copolymers of potential antitumor activity containing 5-fluorouracil, J. Polym. Sci. 17:351 (1979).

25. F. Ascoli, G. Casini, M. Ferappi and E. Tubaro, A polymeric nitrofuran derivative with prolonged antibacterial action, J. Med. Chem. 10: 97 (1967).

26. J. Pitha, S. Zawadzki and B.A. Hughes, Carriers for drugs and enzymes based on copolymers of allylglycidylether with acrylamide, Makromol. Chem. 183:781 (1982).

27. J.C. Brosse, J.C. Soutif and G. Pinazzi, Synthesis and modification of some new acrylate polymers. Fixation of active compounds, Proceedings I.U.P.A.C. 28th Macromol. Symposium, Amherst, Mass, USA, p.383 (1982).

28. C. Pinazzi, J.C. Rabadeaux and A. Pleurdeau, Synthèse et polymérization de polyméthacrylates porteurs de la quinine. Etude comparée de la toxicité et de l'immunogénicité des formes libres et polymériques, Makromol. Chem. 179:1699 (1978).

29. M. Tahan, Y. Calderon and A. Zilkha, Synthesis of some polymer models of potentially biologically active compounds, Israel J. Chem. 12: 785 (1973).

30. G. Domke, I. Lüderwald, M. Medina, R. Mantoya, L. Bravoluga, H. Ringsdorf, E. Schmidt, A.M. Silva, J. Soto and G. Walter, The prolongation of the action of pharmaceutical preparations Chemical radioprotection with polymers, Polym. Prepr. Am. Chem. Soc. 16:494 (1975).

31. G. Batz, H. Ringsdorf and H. Ritter, Pharmacologically active polymers; 7. Cyclophosphamide and steroid hormone-containing polymers as potential anticancer agents, Makromol. Chem. 175:2229 (1974).

32. H.G. Batz, H. Daniel, G. Franzmann, J. Koldehoff, H. Herz, H. Ringsdorf and K. Stokhaus, Pharmakologisch aktive Polymere, Arzneim. Forsch, 27:1884 (1977).

33. G. Pinazzi, J.P. Benoit, J.C. Rabadeux and A. Pleurdeau, Synthèse de polymères porteurs d'un antivitamine K, la Phénindione, Eur. Polym. J. 15:1069 (1979).

34. P. Molz, H. Ringsdorf, G. Abel and P.J. Cox, Synthesis and first in vitro cytotoxicity studies of bis (2- chloroethyl) amino groups containing polymers. Pharmacologically active polymers 22, Int. J. Biol. Macromol. 2:245 (1980).

35. T. Hirano, H. Ringsdorf and D.S. Zaharko, Antitumor activity of monomeric and polymeric cyclophosphamide derivatives compared with in vitro hydrolysis, Cancer Res. 40:2263 (1980).

36. G. Bauduin, D. Bondon, J. Martel, Y. Pietrasanta and B. Pucci, Study of telomers with potential pharmacological activity. 2. Telomers of acrylic acid and grafting of hydroxylated compounds, Makromol. Chem. 182:773 (1981).

37. G. Bauduin, J.M. Bessière, D. Bondon, J. Martel and Y. Pietrasanta, Recherche de télomères à activité pharmacologique potentielle, 4. Réactions sur les telomères de l'alcool vinylique, Makromol. Chem. 182:3397 (1981).

38. G. Bauduin, J.M. Bessière, D. Bondon, J. Martel and Y. Pietrasanta,

Recherche de télomères à activité pharmacologique potentielle. 5. Réactions sur les telomères de l'acide acrylique, Makromol. Chem. 183:3491 (1982).

39. G.P. Pinazzi, A. Menil, J.C. Rabadeux and A. Pleurdeau, Polyiso-prenes and polybutadiene derivatives of potential biomedical inter-est. Part II. J. Polym. Sci. 52:1 (1975).

40. G.P. Pinazzi, A. Menil, J.C. Rabadeux and A. Pleudeau, Polyisoprenes and polybutadiene derivatives of testosterone, Polymer, 12:447 (1974).

41. A.F. Bückmann, M. Morr and G. Johansson, Functionalization of poly (ethylene glycol) and monomethoxy-poly (ethylene glycol), Makromol. Chem. 182:1379 (1981).

42. A. Chaabouni, P. Hubert, E. Dellacherie and J. Neel, Synthèse de poly(oxyéthylène)s rendus biospécifiques par fixation de stéroides en extrémités de chaîns. Utilisation d'un poly-(oxyethylene) substitué par des groupes estradiol pour extraire l'isomérase 5 $\longrightarrow$ 4 oxo-3 steroide par partage d'affinité, Makromol. Chem. 179: 1135 (1978).

43. J.C. Galin, P. Rempp and J. Parrod, Preparation de chaînes macromol-éculaires dotées d'extrémités functionnelles réactives, C.R. Acad. Sc. Paris, 260:5558 (1965).

44. J.M. Harris, N.H. Hundley, T.G. Shannow and E.C. Stuck, Poly(ethylene glycols) as soluble, recoverable, phase-transfer catalysts, J. Org. Chem. 47:4789 (1982).

45. J.M. Harris and J. Milton, Laboratory synthesis of polyethylene glycol derivatives, J. Makromol. Sci. 25:325 (1985).

46. A. Okamoto, K. Toyoshima and I. Mita, Kinetic study on reactions between polymer chain-ends II, Eur. Polym. J. 19:341 (1983).

47. S. Zalipsky, C. Gilon and A. Zilkha, Attachment of drugs to poly-ethylene glycols, Eur. Polym. J. 19:1177 (1983).

48. F. Brandstetter, H. Scott and E. Bayer, New polymer protecting group in oligonucleotide synthesis. 2. Hydroxyethyl-phenyl-thio-ether of polyethylene glycol, Tetrahedron Lett., 31:2705 (1974).

49. J. Pitha, K. Kocilek and M.G. Caron, Detergents linked to poly-saccharides: preparation and effects on membranes and cells, Eur. J. Biochem. 94:11 (1979).

50. L. Tondelli, M. Laus and A.S. Angeloni, Poly(ethylene glycol) imi-dazolyl formates as oligomeric drug-binding matrices, J. Controll. Release 1:251 (1985).

51. P. Ferruti, M.C. Tanzi, L. Rusconi and R. Cecchi, Succinic half-esters of poly(ethylene glycol)s and their benzotriazole and imi-dazole derivatives as oligomeric drug-binding matrices, Makromol. Chem. 182:2183 (1981).

52. L. Ruscono, M.C. Taniz, C. Zambelli and P. Ferruti, Activated deriv-atives of succinic and glutaric half- - esters of polypropylene glycols, and their exchange reactions with hydroxy- and amino-compounds, Polymer 23:1689 (1982).

53. E. Boccu', R. Largajolli and F.M. Veronese, Coupling of monomethoxy polyethylene glycols to proteins via active esters, Z. Natur. Forsch. C. Biosci. 38C:94 (1983).

54. F.M. Veronese, R. Largajolli, E. Boccu', C.A. Benassi and O. Schiavon, Appl. Biochem. Biotechnol., in press (1986).

55. R. Cecchi, L. Rusconi, M.C. Tanzi, F. Danusso and P. Ferruti, Syn-thesis and pharmacological evaluation of poly(oxyethylene) derivat-ives of 4-isobuthylphenyl-2- -propionic acid (Ibuprofen) J. Med. Chem. 24:622 (1981).

56. N. Ghedini, P. Ferruti, V. Andrisano, M.R. Cesaroni and G. Scapini, Oligomeric derivatives of 3 α, 7 β - -dihydroxy-5 β -cholan-24-oic acid (ursodeoxycholic acid), Synth. Comm. 13:701 (1983).

57. N. Ghedini, V. Andrisano, V. Zecchi, A. Tartarini, G.G. Scapini and P. Ferruti, J. Controll. Release, in press (1986).

LIPOSOMES IN VIVO: A RELATIONSHIP BETWEEN STABILITY AND CLEARANCE?

Gregory Gregoriadis and Judith Senior

Medical Research Council Group, Academic Department of
Medicine, Royal Free Hospital School of Medicine
Pond Street, London NW3 2QG, U.K.

INTRODUCTION

Quantitative retention of entrapped drugs by liposomes en route to
their destination is usually an essential prerequisite for the successful
use of the carrier. Furthermore, for cell targets other than those of the
reticuloendothelial system (RES) (by which liposomes are taken up rapidly),
there is a need for prolonging the presence of liposomes in the circulation.[1,2] Recent developments in this laboratory show that both liposomal
permeability to entrapped solutes in the presence of blood and clearance
from the circulation can be controlled by simple adjustments of the structural characteristics of the vesicles.

RETENTION OF ENTRAPPED SOLUTES BY LIPOSOMES IN THE PRESENCE OF BLOOD PLASMA

Following findings[3-5] that plasma high density lipoproteins (HDL) de-
stabilize liposomes by removing phospholipid molecules, it was reasoned[6-8]
that packing of the lipid bilayers with excess cholesterol and/or the inclusion of phospholipids with a high liquid-crystalline phase transition
temperature, may inhibit or abolish HDL action. Thus, when quenched carboxyfluorescein (CF) was used as a water soluble marker[7,9,10] (for rationale of its use see Reference 8) it was found that, for compositions of
equimolar cholesterol and total phospholipid, CF loss in the presence of
blood plasma was reduced to an extent dependant on the nature of the phospholipid used. For example,[7] by increasing the proportion of sphingomyelin
(SM) relative to egg phosphatidylcholine (PC) in cholesterol-rich small
unilamellar vesicles (SUV), it was possible to achieve increasing solute
retention. Retention values (expressed as % CF latency) were shown[7] to
reach 100 when SM constituted at least 77% of the total phospholipid. By
using equimolar distearoyl phosphatidylcholine (DSPC) and cholesterol,
liposomal stability was further improved with CF latency retaining its
initial value (100%) for at least 48h.[7] Work[5,10,11] with liposomes containing radio-labelled phospholipids showed, in addition, that lipid compositions which lead to enhanced solute retention in the presence of blood,
do indeed reduce or abolish altogether phospholipid loss to HDL. For instance, increase of the proportion of SM in liposomes is paralleled with
a decrease in the elution of the radiolabelled lipid together with HDL.[11]
Evidence supporting the involvement of HDL in phospholipid removal has also
come from experiments[12] in which PC liposomes devoid of cholesterol were

Table 1. Solute retention by cholesterol-rich liposomes in the presence of blood plasma.

Solute	Retention (%)
Carboxyfluorescein	95.0 (30 min)
Sucrose	95.9 (30 min)
Inulin	92.3 (30 min)
Bleomycin	92.6 (2h)
Melphalan	67.6 (10 min)
Vincristine	48.9 (10 min)

Small unilamellar liposomes composed of equimolar PC and cholesterol and containing the appropriate solute were incubated in the presence of blood plasma at 37°C. Values denote % of the solute that was still entrapped at the end of the incubation period (shown in parentheses) (Modified from References 11, 13 and 16).

Table 2. Solute retention by cholesterol-rich liposomes in the presence of blood plasma.

Liposomes	Solute retention (%)	
	[III]In-bleomycin	CF
SUV		
PC, Chol	92.6 (2h)	
	51.2 (24h)	61.7 (20h)
DSPC, Chol	99.5 (24h)	100.0 (24h)
DSPC, DPPE, Chol	97.4 (24h)	
REV		
PC, Chol		90.5 (10min)
		46.6 (24h)
DSPC, Chol		98.5 (24h)

Small unilamellar (SUV) or reverse phase evaporation (REV) vesicles composed of equimolar total phospholipid and cholesterol (Chol) and containing [III]In-labelled bleomycin or 0.2M CF were incubated in plasma at 37°C. At time intervals (shown in parenthesis) samples were chromatographed. Values denote solute recovered with the liposomes fraction and is expressed as % of total recovered. Size (nm ± SD) of SUV liposomes was 64. 0 ± 5.9. REV were filtered through a nucleopore filter of 0.4 µm pore diameter (modified from references 15 and 17).

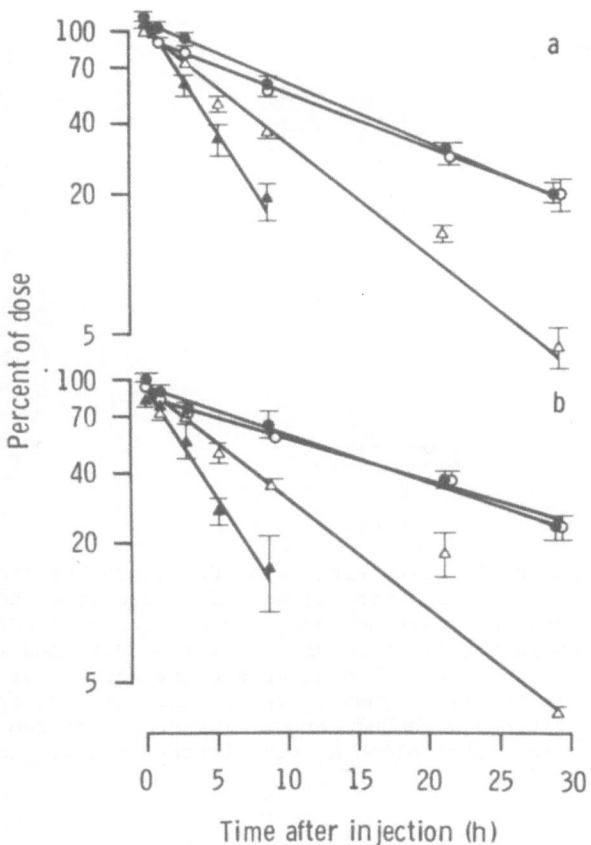

Fig. 1.   Effect of dose on the clearance of cholesterol-
           rich DSPC liposomes from the circulation. Mice
           in groups of 4 or 5 were injected intravenously
           with 0.2 ml SUV composed of equimolar DSPC and
           cholesterol and containing 0,2M carboxyfluorescein
           and tracer [14C]DSPC.   Injected doses (mg phospho-
           lipid per kg body weight) were 71.4 (O), 40.7 (●),
           21.4 (△) and 7.4 (▲). Values are % ± S.D. of in-
           jected quenched carboxyfluorescein (a) or 14C radio-
           activity (b) per total blood at time intervals after
           injection (from Reference 15).

incubated with plasma from lipoprotein-deficient mice: there was virtually
no loss of PC. When purified HDL, low and intermediate density (LDL and
IDL) or very low density (VLDL) mouse lipoproteins were added to lipoprotein-deficient mouse plasma in increasing amounts to cover physiological
levels of the lipoproteins, it was found that only HDL had a vesicle de-
stabilizing effect.[12]

     Contrary to the behaviour of CF and other polar solutes[13-15] (they are
retained quantitatively by cholesterol-rich liposomes in the presence of
plasma), solutes of lipophilic nature entrapped in similar liposomes tend
to leak.[16] Nonetheless, as with polar solutes, excess cholesterol in the
liposomal membranes and the correct choice of their phospholipid content
influence lipophilic solute retention by the vesicles, albeit less effectively.[16] In this respect, only a fraction of entrapped vincristine (about

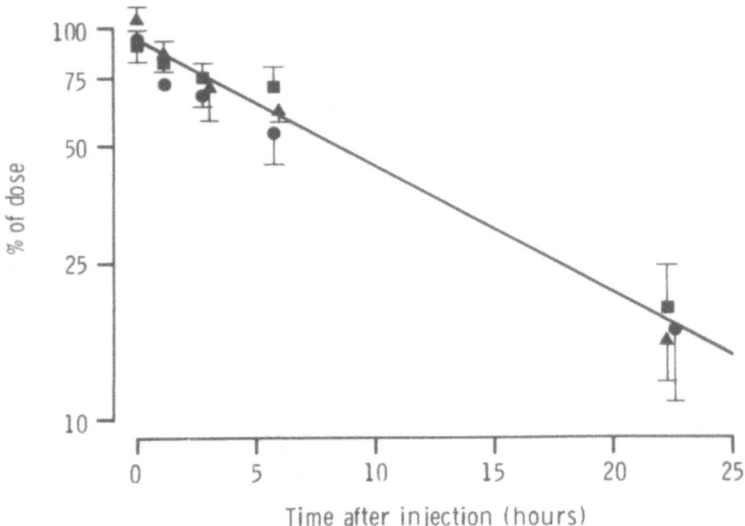

Fig. 2. Clearance of cholesterol-rich DSPC liposomes from
the circulation. Mice (5 animals) were injected
intravenously with SUV composed of equimolar DSPC
and cholesterol. Liposomes contained 0.2M and tracer
^{III}In-bleomycin in the aqueous phase and tracer [^{14}C]
DSPC in the lipid phase. Values are % $\pm$ S.D. of
administered quenched carboxyfluorescein or radioactivity per total blood at time intervals after injection (from Reference 15).

49%) or melphalan (about 68%) is retained in the presence of plasma by
cholesterol-rich PC SUV. These were able to retain over 90% of CF and
other water soluble solutes under similar conditions (Table 1).

RETENTION OF DRUGS BY CIRCULATING LIPOSOMES AND LIPOSOME CLEARANCE

On the basis of radioactive lipid label and latent CF measurements
and of values of lipid label transfer (if any) to HDL, we have shown
(e.g. Table 2 in conjunction with Fig. 2) that those liposomal lipid compositions which contribute to quantitative drug retention in the presence of
plasma in vitro at 37°C, exert a similar action in the blood circulation
of injected animals.[7-10,16-18] In the case of very stable liposomes, the
lipid and CF markers are cleared from the circulation at the same rate[7-10]
and, as discussed elsewhere,[12] there is virtually no transfer of phospholipid to HDL. Additional evidence that such liposomes retain their structural integrity in the blood is the persistence of highly quenched CF in
blood samples taken several hours after injection.[17] Moreover, clearance
rates of small unilamellar vesicles are linear[7,18] and depend on the amount
of administered lipid. In the case of cholesterol-rich DSPC liposomes, for
instance, half-lives were 12h (2 and 1.2 mg), 7h (0.6 mg) and 3h (0.2 mg
administered phospholipid per mouse) (Fig. 1).

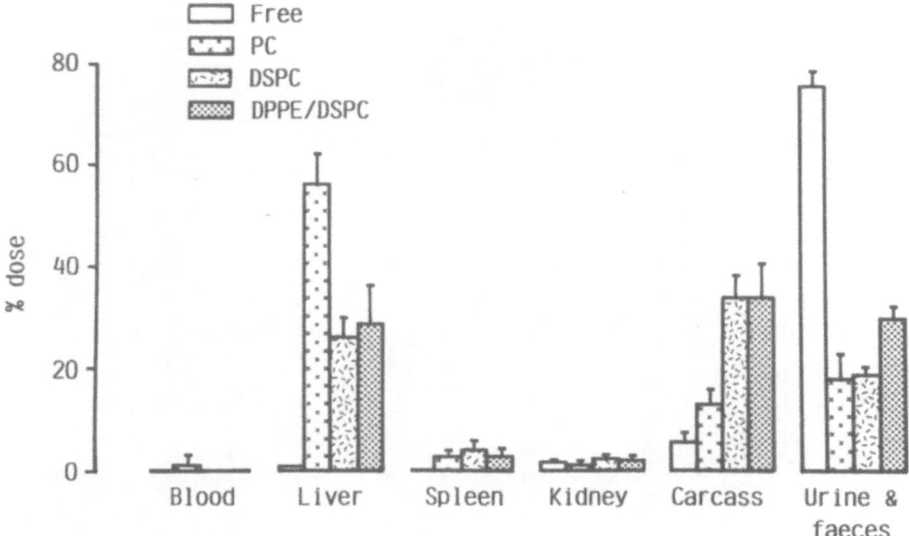

Fig. 3. Tissue distribution of cholesterol-rich liposomes
after intravenous injection. Mice in groups of 6
were injected with [III]In-bleomycin, free or entrapped
in SUV composed of equimolar PC and cholesterol,
equimolar DSPC and cholesterol, or DSPC, DPPE and
cholesterol (0.8:0.2:1 molar ratios). At 26h (free
and PC SUV) or 72h (other SUV) after injection, animals were killed, and radioactivity was measured in
blood, liver, spleen, kidney, the carcass and urine
and faeces. Values are % of dose per total blood,
total tissues, or collected 26-72 urine and faeces
(from Reference 15).

It has become apparent recently[17,19] that a direct relationship exists
between solute (CF) retention by liposomes in the presence of blood and
rate of vesicle clearance from the circulation. Thus, rates of clearance
for several liposomal preparations shown to be increasingly stable in
plasma at 37°C,[7] exhibit after intravenous injection increasingly longer
half-lives. These range from about 2h (cholesterol-rich PC liposomes) to
about 11h (cholesterol-rich SM liposomes).[7] This relationship is also
apparent[7] in the case of two preparations of DSPC, one without and the
other with (equimolar) cholesterol. Suggestions by others[20] that excess
cholesterol in vesicles inhibits vesicle uptake by the liver are in conflict
with observations made[7,9,18,21] with a number of cholesterol-rich preparations which differed only in phospholipid composition and which exhibited
widely different half-lives (0.1 to 20h). A plausible explanation for this
direct relationship observed[17,19] between rates of clearance of liposomes
and their solute retention in the presence of plasma could be that as
retention of a given solute is directly related to the extent of membrane
lipid packing, it would also be similarly related to the difficulty with
which blood elements (for instance opsonins) responsible for the removal
of foreign matter from the circulation may adsorb or insert themselves onto
the vesicles.

According to more recent work,[15] even those liposomes which exhibit
long half-lives do not escape the RES to reach intravascular or extravasc-

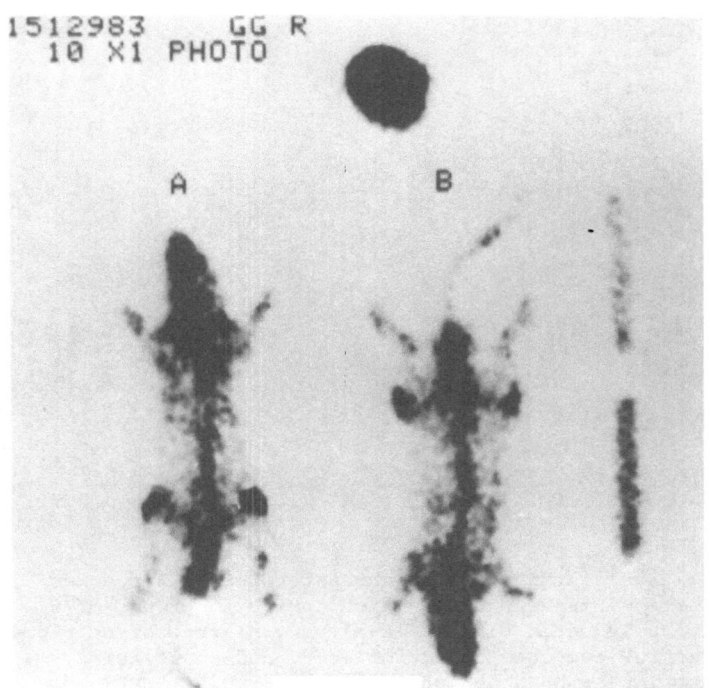

Fig. 4. Scans of carcasses of rats injected with radiolabelled cholesterol-rich SUV. Two rats (A and B)
were injected intravenously with SUV composed of
equimolar DSPC and cholesterol and containing [111]In-
bleomycin. At 72h after injection, the rats were
killed and, after the removal of all tissues including the stomach and large and small intestines,
scanned. Radioactivity distribution (% of total in
the carcass) in the bones of various areas in the
two rats (A and B) was, respectively, head, 14.5 and
13.9; thorax, 23.2 and 26.8; top of legs and pelvis,
28.9 and 29.6; tail, 4.3 and 2.8. Radioactivity (%
of dose for each animal) in total blood was 3.2(A)
and 1.2(B), and in skeletal muscle 0.1 and 0.1 per
g tissue. Carcass radioactivity was 25.1% (A) and
26.1% (B) of the injected dose. Rats are placed in
the Figure with head (A) and tail (B) at the top
(from Reference 15).

ular compartments. For instance, intravenously injected liposomes which
have a long half-life and, in addition, retain practically all their [111]In-
labelled bleomycin content (Fig. 2), are taken up by the hepatic and splenic
moieties of the RES only to a relatively small extent (28-30% of the dose)
(Fig. 3). Much of the remainder of the dose is found in the bones (Fig. 4),
presumably the macrophages of the bone marrow.

Attempts to obtain liposomes with long half-lives in the circulation
must, on the basis of what has been already discussed for SUV, be based on
preparations that retain their solute contents quantitatively in the
presence of plasma. Unfortunately, however, although the direct relationship between stability (solute retention) and half-life shown to exist for

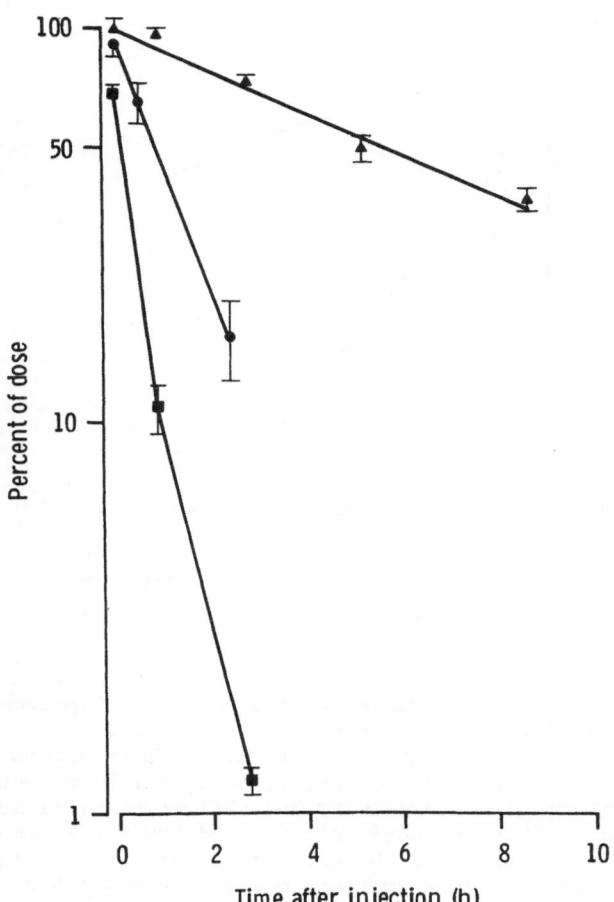

Fig. 5. Effect of size of cholesterol-rich DSPC liposomes
on their clearance from the circulation. Mice in
groups of 4 or 5 were injected intravenously with
SUV (▲), REV passed twice through nucleopore
filter of 0.2 μm diameter (●) and REV passed twice
through nucleopore filter of 0.4 μm diameter (■).
All liposomes were composed of equimolar DSPC and
cholesterol and contained 0.2M carboxyfluorescein.
Values are percentage ± S.D. of the injected quenched
carboxyfluorescein per total blood at time intervals
after injection (from Reference 15).

SUV also applies to larger vesicles, half-lives achieved even for the most
stable large liposomes are only modestly long. For instance, according
to Fig. 5, half-lives of 60 and 20 min were observed for the very stable
(Table 2) DSPC REV liposomes filtered through polycarbonate membranes of
0.2 and 0.4 μm pore diameter respectively. It would therefore seem that
vesicle size overrides membrane permeability in determining vesicle clearance. On the other hand, the relationship between reduced membrane permeability and prolonged vesicle celarance established for small liposomes
also exists for large vesicles. Thus, the half-life of large liposomes
(0.4 μm diameter) composed of equimolar DSPC and cholesterol was much
greater than for liposomes of similar size composed of equimolar PC and

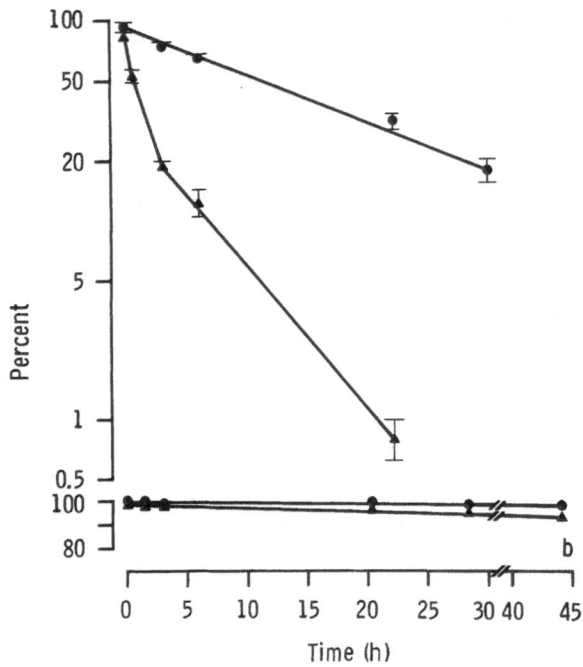

Fig. 6.   The effect of vesicle surface charge on liposome
          stability in the presence of blood plasma and
          clearance from the circulation. Mice in groups of
          4 or 5 were injected intravenously (a) with neutral
          SUV (●) composed of DSPC and cholesterol (1:1 molar
          ratio) or with negatively charged SUV (▲) composed
          of DSPC, DSPA and cholesterol (0.75:0.25:1, molar
          ratios). Values at time intervals are % $\pm$ S.D. of
          the injected quenched carboxyfluorescein per total
          blood. Neutral and negatively charged liposomes as
          above were incubated in the presence of 5 volumes
          of mouse blood plasma at 37°C (b). Both types of
          liposomes contained 0.2M carboxyfluorescein. Values
          at time intervals are percentage $\pm$ S.D. latent
          carboxyfluorescein expressed as percentage of
          latency in the initial preparations (➘ 98%) (from
          Reference 15).

cholesterol (not shown). As with SUV, permeability of large liposomes to
CF in the presence of blood plasma was greater with the latter lipid composition than with the former (Table 2). Large liposomes can accommodate
much more drug per lipid unit weight and are, in addition, easier to prepare. Therefore, prolongation of their circulation time would be an important advance towards their use in drug targeting in vivo.

     Another type of liposome for which pronounced stability does not
parallel prolonged circulation time is negatively charged (cholesterol-
rich) small unilamellar vesicles composed of equimolar DSPC and cholesterol.

Although permeability for these liposomes to CF in plasma is very low and almost identical to that for neutral SUV (about 90% latency at 45h; Fig. 6). It is thus again reasonable to conclude that surface charge overrides bilayer permeability in determining rate of vesicle clearance. Earlier work from this laboratory[22] has shown that neutral liposomes become negatively charged in the presence of blood plasma, whereas the charge of originally negative liposomes remains negative. It was also proposed[22] that acquisition of a negative charge by uncharged liposomes was the result of α_2-macroglobulin adsorption on their surface. It is of interest that similar findings using latex microspheres have been reported by others[23] who attributed the change in charge to the adsorption of a plasma component. As with negative liposomes, negatively charged latex microspheres exhibited a shorter half-life than those with increasingly positive charge.[23] The same authors [23] have, on the basis of previous observations,[24] attempted to explain the way by which adsorbed plasma components influence particle clearance rates depending on the original surface charge: the surface properties (e.g. surface charge) of the particles may alter the steric arrangement of the adsorbed plasma components to a form specifically recognized by the RES. This could also apply to liposomes, in view of the role that α_2-macroglobulin is believed to play as a phagocytosis promoting factor.[25]

ACKNOWLEDGEMENTS

Work presented was supported by grants from the Medical Research Council, British Council, the Wellcome Trust and British Technology Group. We thank Mrs. B. Thompson for excellent secretarial assistance.

REFERENCES

1. G. Gregoriadis, Targeting of Drugs: Implications in Medicine, Lancet, 2:241 (1981).
2. D.A. Tyrrell and B.E. Ryman, Liposomes: Bags of potential, Essays Biochem., 16:49 (1980).
3. L. Krupp, A.V. Chobanian and I.P. Brecher, The in vivo transformation of phospholipid vesicles to a particle resembling HDL in the rat, Biochem. Biophys. Res. Commun., 72:1251 (1976).
4. G. Scherphof, F. Roerdink, M. Waite and I. Parks, Disintegration of phosphaticylcholine liposomes in plasma as a result of interaction with high density lipoproteins, Biochim. Biophys. Acta, 542:296 (1978).
5. C. Kirby, J. Clarke and G. Gregoriadis, Cholesterol content of small unilamellar liposomes controls phospholipid loss to high density lipoproteins in the presence of serum, FEBS Lett., 111:324 (1980).
6. G. Gregoriadis and C. Davis, Stability of liposomes in vivo and in vitro is promoted by their cholesterol content and the presence of blood cells, Biochem. Biophys. Res. Commun., 89:1287 (1979).
7. J. Senior and G. Gregoriadis, Is half-life of circulating small unilamellar liposomes determined by changes in their permeability? FEBS Lett., 145:109 (1982).
8. C. Kirby, J. Clarke and G. Gregoriadis, Effect of the cholesterol content of small unilamellar liposomes on their stability in vivo and in vitro, Biochem. J., 186:591 (1980).
9. J. Senior and G. Gregoriadis, Stability of small unilamellar liposomes in serum and clearance from the circulation: the effect of the phospholipid and cholesterol components, Life Sci., 30:2123 (1982).
10. C. Kirby and G. Gregoriadis, The effect of the cholesterol content of small unilamellar liposomes on the fate of their lipid components in vivo, Life Sci., 27:2223 (1980).

11. G. Gregoriadis, C. Kirby, J. Senior and B. Wolff, Fate of liposomes in vivo: Control leading to targeting, in: "Receptor-Mediated Targeting of Drugs" G. Gregoriadis, G. Poste, J. Senior and A. Trouet, eds., Plenum Publishing Company, New York (1984).

12. J. Senior, G. Gregoriadis and K. Mitropoulos, Stability and clearance of small unilamellar liposomes: Studies with normal and lipoprotein-deficient mice, Biochim. Biophys. Acta, 760:111 (1983).

13. C. Kirby and G. Gregoriadis, Plasma-induced release of solutes from small unilamellar liposomes is associated with pore formation in the bilayers, Biochem. J., 199:251 (1981).

14. B. Wolff and G. Gregoriadis, The use of monoclonal anti-Thy_1, IgG_1 for the targeting of liposomes to AKR-A cells in vitro and in vivo, Biochim. Biophys. Acta, 802:259 (1984).

15. J. Senior, J.C.W. Crawley and G. Gregoriadis, Tissue distribution of liposomes exhibiting long half-lives in the circulation after intravenous injection, Biochim. Biophys. Acta, 839:1 (1985).

16. C. Kirby and G. Gregoriadis, The effect of lipid composition of small unilamellar liposomes containing melphalan and vincristine on drug clearance after injection into mice, Biochem. Pharmacol., 32:609 (1983).

17. J. Senior, Fate and behaviour of liposomes in vivo: A review of controlling factors, Crit. Rev. Ther. Drug Carrier Systems, In press (1986).

18. G. Gregoriadis and J. Senior, The phospholipid component of small unilamellar liposomes controls the rate of clearance of entrapped solutes from the circulation, FEBS Lett., 119:43 (1980).

19. G. Gregoriadis, Targeting of drugs with molecules, cells and liposomes, TIPS, 4:304 (1983).

20. H.M. Patel, N.S. Tuzel and B.E. Ryman, Inhibitory effect of cholesterol on the uptake of liposomes by liver and spleen, Biochim. Biophys. Acta, 761:142 (1984).

21. K.J. Hwang, K.-F.S. Luk and P.L. Beaumier, Hepatic uptake and degradation of unilamellar sphingomyelin cholesterol liposomes: A kinetic study, Proc. Natl. Acad. Sci. USA, 77:4030 (1980).

22. C.D.V. Black, and G. Gregoriadis, Interaction of liposomes with blood plasma proteins, Biochem. Soc. Trans., 4:253 (1976).

23. D.J. Wilkins and P.A. Myers, Studies on the relationship between the electrophoretic properties of colloids and their blood clearance and organ distribution in the rat, Br. J. Exp. Pathol., 47:568 (1966).

24. D.J. Wilkins and A.D. Bangham, The effect of some metal ions on in vitro phagocytosis, J. Reticuloendothelial Soc., 1:233 (1964).

25. C.S.J. Allen, Ph.D. Thesis, University of Illinois, (1974).

LIPOSOMES AS DRUG CARRIERS TO LIVER MACROPHAGES: FUNDAMENTAL AND THERAPEUTIC ASPECTS

Frits Roerdink[1], Joke Regts[1], Toos Daemen[1], Irma Bakker-Woudenberg[2] and Gerrit Scherphof[1]

[1]Laboratory of Physiological Chemistry, University of Groningen, Bloemsingel 10, 9712 KZ Groningen, and
[2]Department of Clinical Microbiology, Erasmus University Rotterdam, P.O. Box 1738, 3000 DR Rotterdam, The Netherlands

INTRODUCTION

It is well established that after intravenous injection large-size liposomes are rapidly cleared from the blood and taken up by cells belonging to the reticulo-endothelial system (RES). Particularly, fixed macrophages in liver (Kupffer cells) and spleen are actively involved in the uptake of the vesicles.[1-5] Uptake occurs by way of endocytosis followed by intralysosomal degradation of liposomal lipids and release of entrapped substances.[5] The natural affinity of liposomes for macrophages has been exploited in the application of liposomes as a drug delivery system to this cell type. For example, liposomes have been used as carriers of antimicrobial agents in the treatment of intracellular infections such as experimental leishmaniasis,[6] candidiasis[7,8] or listeriosis.[9] In these infections the microorganisms are lodged in the lysosomes of tissue macrophages, precisely the site where the liposomes end up after intravenous injection. Thus, encapsulation of relevant antibiotic drugs within liposomes results in an increased therapeutic index of these drugs when administered intravenously.

Very interesting results have also been reported by Poste and Fidler on the activation of tumoricidal properties in mouse macrophages by means of liposome-entrapped immunomodulators, such as lymphokines or muramyl dipeptide (MDP).[10]

During the past few years we have studied the interaction of large unilamellar vesicles (LUV) with Kupffer cells in maintenance culture.[11-15] We paid special attention to levels of uptake and rates of intralysosomal degradation of the liposomes. These parameters are highly relevant to the application of liposomes as a drug carrier since the therapeutic effect of a liposome-encapsulated drug may be determined, at least in part, by rate and extent of intracellular release of the drug.

As an extension of these observations we studied the hepatic uptake and intracellular degradation of liposomes on three different levels: (a) in vivo, by injecting [^{3}H]inulin/cholesteryl-[^{14}C]oleate labeled multilamellar vesicles (MLV) into rats. Changes in the ratio of the two liposomal labels after uptake of the vesicles by the liver or spleen will reflect the rate of intracellular degradation in these organs. (b) in vitro, with cultured Kupffer cells, by incubating these with small unilamellar vesicles (SUV) labelled with N-Me- [^{14}C]sphingomyelin. (c) in vitro, in cell-free systems, by incubating [^{14}C]sphingomyelin-labeled SUV at pH 4.8 with crude lysosomal fractions from rat liver.

IN VIVO FATE OF LIPOSOMAL LABELS AFTER UPTAKE BY LIVER AND SPLEEN

After i.v. injection liposomes are cleared from the blood and taken up predominantly by liver and spleen. The rate of elimination is determined, at least in part, by the size of the vesicles: large-size MLV are cleared much more rapidly than the SUV,[16] which is reflected in the uptake by, specifically, the macrophages of the liver (Fig. 1): MLV are taken up to an almost 2-fold higher extent as compared to SUV.

The dominant uptake of MLV by Kupffer cells in addition to the relatively high entrapped aqueous volume (MLV, approx. 3 l/mol lipid vs. SUV: approx. 0.3 l/mol lipid) renders this liposome type an attractive candidate to serve as a drug carrier to liver macrophages. To gain more insight into the intrahepatic degradation of the vesicles we injected MLV (labeled in the aqueous phase with [^{3}H]inulin and in the lipid phase with cholesteryl-[^{14}C]oleate), intravenously into rats. [^{3}H]inulin was used as a parameter of liposome uptake since it is metabolically inert and, if administered in free form, it is not taken up by liver cells but rapidly excreted via the kidneys. Cholesteryl-[^{14}C]oleate, on the other hand, is susceptible to intracellular esterase activity resulting in the liberation of the labeled oleate from the cholesterol moiety. Consequently, intrahepatic degradation of the vesicles will result in an increase in the ^{3}H/^{14}C ratio, if release of oleate from the cells occurs.

Table 1 shows the results of such an experiment: negatively charged MLV were injected intravenously into rats and at various time intervals the ^{3}H/^{14}C ratio was measured in liver and spleen homogenates. The data show that in the liver the ^{3}H/^{14}C ratio increases approx. 18-fold at 24 h after injection indicating extensive intrahepatic degradation of the liposomes and release of ^{14}C-label. In the spleen the increase of the ^{3}H/^{14}C ratio is even more pronounced (approx. 30-fold) suggesting a higher rate of liposomal degradation as compared to the liver. Analysis of labeled lipids revealed that both in liver and spleen liposomal cholesteryl-[^{14}C]-oleate is effectively degraded (Table 2). However, in liver, efficient incorporation of the liberated [^{14}C]oleate is observed into both phospholipids and triglycerides whereas in the spleen oleate incorporation occurs into phospholipids only. Presumably the hepatocytes are responsible for the appearance of oleate in the hepatic triglyceride fraction, a process which occurs only to a limited extent in macrophages as is shown in Table 3 for isolated Kupffer cells in monolayer culture. After uptake by the Kupffer cells of MLV labeled with cholesteryl-[^{14}C]oleate, there is only marginal fatty acid incorporation into triglycerides, the bulk of the incorporation taking place into the phospholipid fraction.

The data in Table 3 also provide evidence that the [^{14}C]oleate is liberated from the cholesterol moiety by a lysosomal esterase, since incubation of the liposomes in the presence of NH$_4$Cl, which is known to raise the intralysosomal pH, results in a substantial inhibition of the degradation of liposomal cholesteryl-[^{14}C]oleate.

In this section we showed that the use of double-labeled liposomes may provide valuable information on the in vivo metabolic fate of intravenously injected vesicles. It should be emphasized, however, that detailed knowl-

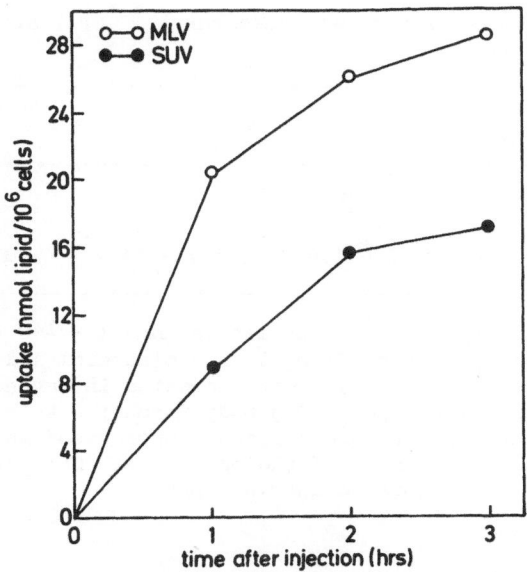

Fig. 1. Uptake of intravenously injected [³H]inulin-labeled
         SUV or MLV by Kupffer cells. SUV consisting of
         sphingomyelin, cholesterol and phosphatidylserine
         (4:5:1 molar ratio) were prepared as described pre-
         viously.[17] MLV of the same lipid composition were
         prepared by extrusion of an unsonicated aqueous
         dispersion of the lipid mixture through a series of
         polycarbonate membrane filters of 1.0, 0.8, 0.6, 0.4
         and 0.2 μm pore-size, respectively. Free, non-
         entrapped inulin was separated from the vesicles by
         gel filtration on Sepharose ®-Cl-2B. The liposome-
         containing fractions were concentrated by ultra-
         filtration through Amicon PM 10 membranes. Rats
         were injected intravenously with the SUV or MLV
         (5 μmol of total lipid per 100 g body weight). At
         times indicated non-parenchymal cells of the liver
         were isolated by pronase perfusion followed by
         Metrizamide gradient centrifugation.[2] Non-parenchymal
         cell fractions were stained for peroxidase activity
         to estimate the relative amount of Kupffer cells (per-
         oxidase-positive). The uptake data, expressed as nmol
         lipid/10⁶ Kupffer cells were corrected for contamin-
         ation by endothelial cells.

edge of the intracellular processing of liposomal labels is mandatory for
accurate interpretation of the experimental data.

UPTAKE AND INTRACELLULAR DEGRADATION OF SUV BY CULTURED KUPFFER CELLS

     As pointed out in the previous section, SUV of certain compositions
display strong affinity for Kupffer cells after intravenous injection
(Fig. 1). On the other hand, we know that these particles can also be
taken up by the hepatocytes.[17] The existence of a fenestrated endothelium
(pore size: 0.1μm) allows the entrance of the vesicles into Disse's space

Table 1. Metabolism of liposomal cholesteryl-[^{14}C]oleate in liver and spleen.

h after injection	1	4	24
^{3}H/^{14}C in liver	10.0 ± 0.9	19.2 ± 0.6	77.7 ± 7.5
^{3}H/^{14}C in spleen	25.2 ± 1.0	65.7 ± 11.2	134.7 ± 36.1

[^{3}H]inulin/cholesteryl-[^{14}C]oleate-labeled MLV (^{3}H/^{14}C ratio = 4.5) composed of phosphatidylcholine, cholesterol and phosphatidylserine (10:4:1 molar ratio) were injected intravenously into rats (0.5 µmol lipid per 100 g body weight). At times indicated liver and spleen were excised, homogenized in water and processed for measurement of radioactivity. Data are given as means ± SD of three determinations.

Table 2. Uptake and metabolism of liposomal cholesteryl-[^{14}C]oleate in liver and spleen in vivo.

Organ	h after injection	apparent uptake (nmol total lipid/g tissue)	% of total lipid label in PL	FFA	TG	CO
liver	1	93.6	25	7	45	20
	4	40.5	42	3	42	10
	24	7.3	43	2	38	7
Spleen	1	216.6	8	22	2	66
	4	29.3	81	5	5	2
	24	11.8	78	4	4	4

Cholesteryl-[^{14}C]oleate-labeled MLV consisting of phosphatidylcholine, cholesterol and phosphatidylserine (molar ratio 10:4:1) were injected intravenously into rats (2 µ mol lipid per 100 g body weight). At times indicated liver and spleen were excised and homogenized in water. Total amount of organ-associated radioactivity was estimated ("apparent uptake") and lipids were extracted and separated by thin-layer chromatography with petroleum ether (40/60) : diethylether : formic acid = 60:40:1.5 as a solvent. Relevant spots were scraped off and analyzed for radioactivity content.

Abbreviations: PL, phospholipids; FFA, free fatty acids; TG, triglycerides; CO, cholesteryl oleate.

upon which they can be internalized by the parenchymal liver cells. Evidence was obtained that uptake of the SUV by this cell type probably also occurs mainly by way of endocytosis.[17,18]

During these in vivo studies we noticed, as others had shown before[19], marked differences in the half life of the SUV in the blood depending on

Table 3. Metabolism of liposomal cholesteryl-[^{14}C]oleate in cultured
          Kupffer cells.

incubation time (min)	apparent uptake (nmol total lipid/mg protein)	% of total lipid label in:			
		PL	FFA	TG	CO
10	12.1	1	13	1	83
30	12.5	14	15	5	63
60	13.2	26	12	6	52
60 (+ NH$_4$Cl)	18.8	3	12	1	82

Kupffer cells in maintenance culture were incubated with
cholesteryl-[^{14}C]oleate-labeled MLV (see Table 2 for lipid
composition) at 37°C in the presence or absence of 10 mM
NH$_4$Cl. 200 nmol vesicle lipid was added to 10^6 Kupffer cells.
At times indicated total cell-associated radioactivity was
estimated ("apparent uptake") and the cells were extracted
with chloroform/methanol; the extracts were chromatographed
on thin-layer plates (see Table 2 for composition of solvent
system). The amounts of label in the relevant spots were
determined and expressed as percent of total chloroform-
soluble radioactivity in the cells.

Abbreviations: See Table 2.

the lipid composition of the vesicles.[18] For example, neutral sphingo-
myelin/cholesterol SUV have an extremely long half life (approx. 12 h)
whereas negatively charged phosphatidylserine vesicles have a relatively
short half life of only 2 h. These differences in half lives of the vesicles are reflected in the rates of uptake by the liver: phosphatidylserine
vesicles are rapidly taken up whereas neutral vesicles are taken up to a
moderate extent.

In vitro experiments with SUV of various lipid composition confirmed
the in vivo observations and suggest that these originate from intrinsic
differences in affinity of the vesicles for the Kupffer cells (Fig. 2).
Neutral vesicles consisting of sphingomyelin/cholesterol are taken up to
a very low extent, while substitution of sphingomyelin by egg phosphatidylcholine results in a slight increase in the uptake. Incorporation of a
positively charged phospholipid like stearylamine causes a significant
further increase in the uptake while negatively charged phosphatidylserine-
containing vesicles are taken up to even much higher extents.

We also modified the SUV by incorporating a mannosyl derivative of
cholesterol into the liposomal membrane in order to obtain an increased
and more specific uptake of the vesicles mediated by the mannose-receptor
localized on the Kupffer cell membrane. However, partial substitution of
the cholesterol by this derivative did not result in an increase in the
uptake (Fig. 3). In contrast, incorporation of a 6-aminomannosyl derivative results in a 15-fold increase in the uptake. These observations are
in line with Baldeschwieler's reports on the increased uptake of the aminomannose modified vesicles by cultured peritoneal macrophages.[20] The
question arises whether the increased uptake resulting from the aminomannose
incorporation is due to a specific interaction with the mannose receptor
rather than to a non-specific electrostatic interaction between the positively charged aminomannose vesicles and the cells. As mentioned above,

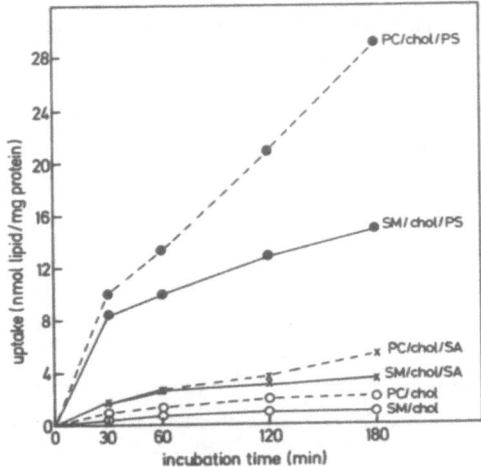

Fig. 2. Effect of lipid composition on the uptake of
cholesteryl-[14C]oleate-labeled SUV by cultured Kupffer cells. SUV (70 nmol lipid) of
various lipid compositions were incubated with
cultured Kupffer cells (1.9 x 10^6 cells) at
37^oC. Liposome compositions were as shown in
Figure with molar ratios of 4:5:1 or 1:1.

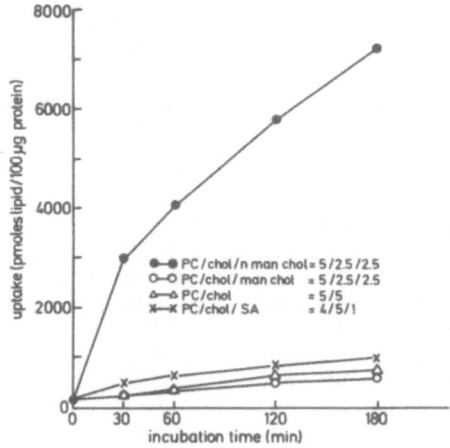

Fig. 3. Uptake of 14C-labeled aminomannose-modified SUV
by cultured Kupffer cells. SUV of different
lipid compositions (40 nmol lipid) were incubated
at 37^oC with Kupffer cells. Liposome uptake is
expressed as nmol of total lipid and calculated
from 14C-label.

Abbreviations: PC, phosphatidylcholine; chol,
cholesterol; n man chol, 6-amino-6-deoxy mannosyl
cholesterol; man chol, mannosyl cholesterol;
SA, sterarylamine.

an increase in uptake is not observed when the vesicles are modified by the mannose derivative of cholesterol. Secondly, we were not able to inhibit the increased uptake of aminomannose vesicles by mannose in a final concentration of up to 25 mM. Thirdly, a recent paper of Rahman and co-workers[21] shows that incorporation of a sufficiently high amount of the positively charged stearylamine also leads to a substantial increase in the uptake of the vesicles by peritoneal macrophages. Finally, Baldeschwieler and coworkers provided convincing evidence that the interaction between the aminomannose vesicles and peritoneal macrophages is of non-specific, electrostatic nature.[22] These data taken together indicate that the increased uptake of the aminomannose vesicles is simply dependent on the presence of large numbers of positively charged amino groups rather than on interaction with a specific receptor.

We also observed prolonged intracellular stability of the aminomannose vesicles as compared to negatively charged phosphatidylserine vesicles. The results are shown in Table 4. Aminomannose-modified and phosphatidyl-serine vesicles were labeled with [14C-Me-] sphingomyelin. After uptake by the cells the liposomal sphingomyelin is subject to degradation by an acid sphingomyelinase, a lysosomal enzyme that catalyzes the hydrolytic formation of phosphorylcholine and ceramide. Appropriate amounts of [14C-] sphingomyelin-labeled aminomannose-modified or phosphatidylserine-containing SUV were incubated with cultured Kupffer cells under conditions such that roughly identical extents of intracellular liposome uptake were obtained for the two vesicle types, so as to achieve equal intralysosomal substrate concentrations. Ammonium chloride was added to the medium to prevent liposomal degradation during the uptake period by inhibiting lysosomal enzyme activity. After changing the liposome-containing medium by a liposome-free medium, the cells were incubated for another 30 min in the presence of NH_4Cl to allow SUV-containing endosomes to fuse with primary lysosomes. The data presented in Table 4 show that the sphingomyelin in the aminomannose vesicles is degraded much more slowly than that in the phosphatidylserine vesicles. This observation could indicate that the aminomannose modified vesicles are only partly internalized and partly remain adsorbed to the cell surface. However, experiments with ethanolamine clearly demonstrated that uptake of the aminomannose vesicles reflects real internalization, as judged by fluorescence microscopy (not shown). This observation seems to confirm Baldeschwieler's report on the excessive stability of the aminomannose vesicles after uptake by macrophages as measured by perturbed angular correlation (PAC) spectroscopy.[20] Apparently, the presence of the positively charged amino groups renders such vesicles poor substrates for sphingomyelinase activity. Comparable observations were published by Wilschut et al.[23] on the decreased susceptibility of dimyristoyl phosphatidylcholine vesicles to phospholipase A_2 activity due to incorporation of positive charge into the vesicles. In addition, the massive accumulation of cationic vesicles may significantly increase the intralysosomal pH thus also contributing to inhibition of lysosomal sphingomyelinase activity.

Decreased susceptibility of the aminomannose vesicles to lysosomal degradation could also be demonstrated by incubating the liposomes, labeled with [14C] sphingomyelin, with lysosomal fractions isolated from rat liver homogenates (Fig. 4): aminomannose vesicles are clearly less sensitive to lysosomal sphingomyelinase than the phsophatidylserine vesicles. Fig. 4 also shows that stearylamine vesicles, which are also positively charged, are degraded at an even lower rate. Fig. 5 shows the rates of hydrolysis of the various liposome preparations at different vesicle lipid concentrations. The data show that the maximal rate (V_{max}) of sphingomyelin hydrolysis in the phosphatidylserine vesicles is at least 2-fold higher than the V_{max} of sphingomyelin degradation in stearylamine and amminomannose vesicles.

Table 4.   Intracellular degradation of aminomannose modified SUV
by Kupffer cells in culture.

Liposome composition	Time of degradation (min)	^{14}C-SM associated with cells (nmol/mg protein)	% of degradation
^{14}C-SM:CHOL:NMANCHOL:SA	0	4.8	0
(4:2.5:2.5:1)	60	4.2	12
	120	3.8	21
^{14}C-SM:CHOL:PS	0	6.2	0
(4:5:1)	60	4.6	26
	120	2.7	57

^{14}C sphingomyelin-labeled aminomannose or phosphatidylserine
SUV were incubated with cultured Kupffer cells (20 nmol and
350 nmol vesicle lipid per 5.2 x 10^6 cells, respectively) in
the presence of 10 mM NH_4Cl at 37°C. After 120 min SUV were
removed from the medium and the cells were incubated for
another 30 min in the presence of NH_4Cl. After 150 min (indicated as zero time) the medium was replaced by an NH_4Cl-
free medium and at 60 and 120 min the percent of label in the
chloroform-soluble cell extract that was associated with the
sphingomyelin spot upon thin-layer chromatography was determined and subtracted from 100 to obtain % of degradation.

Abbreviations: SM, sphingomyelin; CHOL, cholesterol; SA, stearylamine; PS, phosphatidylserine; NMANCHOL, 6-amino-6-deoxy-mannosyl
cholesterol.

The results presented in this section may be relevant to the in vivo
application of liposomes as drug carriers to liver macrophages. By man-
ipulating the lipid composition of the vesicles, rates of uptake and/or
intracellular degradation can be influenced and thereby the rates at which
liposome-encapsulated drugs are released intracellularly and become available to exert their action. In the following section we will present two
applications of liposomes as drug carriers aimed at liver macrophages.

THERAPEUTIC APPLICATION OF LIPOSOMES AS DRUG CARRIERS TO LIVER MACROPHAGES

As pointed out in the introduction, the natural affinity of liposomes
for macrophages has prompted a number of investigators to use the vesicles
as a drug delivery system aimed at this cell type. For example, the intracellular delivery of antimicrobial agents by liposomes to macrophages was
demonstrated in a number of experimental models of intracellular parasitic
and mycotic infections.[6-8] We obtained similar results with the use of
liposome-encapsulated ampicillin in the treatment of Listeria monocytogenes
infections in mice.[9] Clinical experience has indicated that infections
caused by intracellular bacteria such as L. monocytogenes are difficult to
treat especially in patients with underlying malignant disease or in pat-

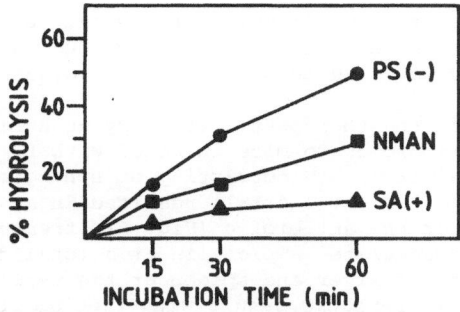

Fig. 4.   Effect of liposomal surface charge on the
degradation of [14C]-sphingomyelin-labeled
SUV at pH 4.8 by lysosomal sphingomyelinase.
Lysosomal fractions were prepared from rat
liver by differential centrifugation. The
fractions were sonicated for 15 min in a
bath sonifier at 10ºC. Lysosomal sphingomyelinase activity was measured at 37ºC in
0.5 ml medium containing 50 mM Na acetate,
pH 4.8, 50 nmol [14C] sphingomyelin-labeled
SUV and 100 µg of lysosomal protein. The
reaction was stopped by adding 3 ml methanol/
chloroform (2:1 v/v) and the lipids were extracted. Sphingomyelinase activity was expressed as the percentage of [14C]choline
released into the water-soluble fraction.

SUBSTRATE-CONCENTRATION CURVES OF
[14C] SPHINGOMYELIN-LABELED SUV
INCUBATED AT pH 4.8 WITH LYSOSOMAL
RAT LIVER FRACTIONS.

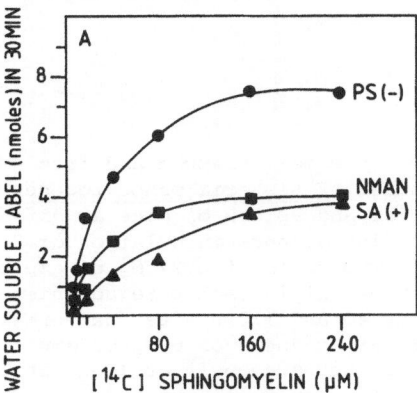

Fig. 5.   Degradation of sphingomyelin in SUV of different compositions by a lysosomal fraction;
dependence on substrate concentration.
See for experimental conditions legend to
Fig. 4.

ients receiving immunosuppressive drugs.[24] The relative inefficiency of
antibiotics in the treatment of listeriosis may be due to their poor penetration into the phagocytic cells in which the microorganisms are lodged.

In order to increase the therapeutic efficacy we administered ampicillin encapsulated in liposomes to mice infected with <u>L. monocytogenes</u>.
Entrapment of ampicillin within MLV consisting of sphingomyelin, cholesterol
and phosphatidylserine (molar ratio 4:5:1) resulted in an 80-fold increase
in therapeutic activity of the antibiotic (Fig. 6); treatment of the
animals with liposome-encapsulated ampicillin at a total dose of 0.54 mg
resulted in sterilization of liver and spleen at the termination of treatment, whereas a similar therapeutic result with free ampicillin could only
be obtained when the mice were treated with a total dose of as much as 48 mg.
Treatment with 0.54 mg of non-entrapped ampicillin mixed with empty liposomes did not result in a therapeutic effect indicating that therapeutic
activity of liposomal ampicillin resulted exclusively from encapsulated
drug. Apparently, encapsulation of ampicillin results in an increased
intracellular availability of the antibiotic. Recently, a clinical study
by Lopez-Berestein and coworkers showed that treatment of fungal infections
with amphotericin-B, encapsulated within liposomes, was effective in eight
out of twelve neutropenic and/or immunocompromised cancer patients.[25]
These patients had previously failed to respond to therapy with the non-
encapsulated drug.

Another example in which the avidity of macrophages for liposomes is
exploited is the application of the vesicles as carriers of immunomodulators to render the cells cytotoxic to tumor cells. In vitro exposure of
monocytes, alveolar and peritoneal macrophages to a variety of immunomod-

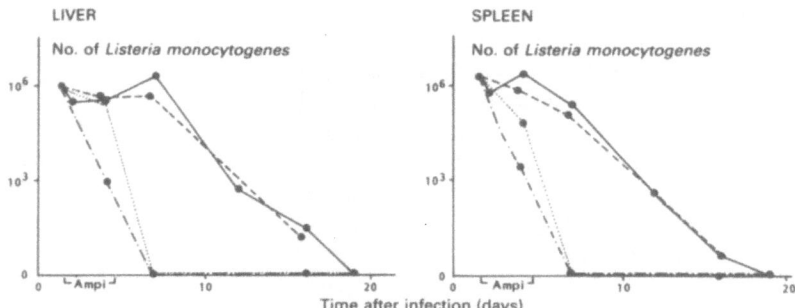

Fig. 6. Effect of liposome-entrapped and free ampicillin
         on the number of <u>Listeria monocytogenes</u> organisms
         in the liver and spleen of mice at different inter-
         vals after intravenous inoculation of 5×10^3
         bacteria. Two doses of 0.27 mg of ampicillin in
         2μ mol vesicle lipid each were administered intra-
         venously at 40 and 112 h after bacterial inoculat-
         ion (...); eight doses of 6 mg of ampicillin were
         administered intravenously at 12-h intervals start-
         ing 40 h after inoculation (-·-); two doses of 2
         μmol of vesicle lipid plus two doses of 0.27 mg
         of ampicillin were administered intravenously at
         40 and 112 h after inoculation (---); untreated
         controls (——). Data are mean values for groups
         of five mice.

lators such as lymphokines[26], γ-interferon[27] and muramyl dipeptide (MDP)[28,29] leads to the activation of these cells to a tumorcytotoxic state. In vivo studies by Fidler and coworkers have shown that administration of lymphokines or MDP, encapsulated in liposomes, results in a significant reduction of experimental lung metastases in mice bearing syngeneic B16-melanoma cells.[30] Similar results were published by Thombre and Deodhar on the inhibition of liver metastases from murine colon adenocarcinoma by intravenous administration of liposome-encapsulated C-reactive protein or crude lymphokines.[31] This therapeutic effect was proposed to be mediated by the activation of hepatic macrophages. We studied the tumoricidal properties of Kupffer cells in vitro following incubation with free and liposome-entrapped MDP (Table 5). The results show that Kupffer cells in monolayer can indeed be activated to a tumorcytotoxic state against xenogeneic murine B16-melanoma cells after exposure to free, non-encapsulated MDP. The percentage of cytotoxicity induced by free MDP reaches a maximum of about 30% after incubation of the Kupffer cells with 2,500-10,000 ng MDP added per well. However, only as little as 10 ng of MDP, encapsulated within 10 nmol of liposomal lipids, is required to obtain a comparable extent of cytotoxicity. From these data it can be concluded that a 250-1000 fold potentiation of the MDP-induced cytotoxicity is achieved by the encapsulation of the drug within liposomes. The data shown in Table 5 also demonstrate that exposure of the cells to liposome-entrapped MDP results in a higher maximal level of cytotoxicity (50-60%) than exposure to free MDP, which produces a maximal cytotoxicity of 30%. Significant levels of Kupffer cell cytotoxicity were also obtained by incubation of the cells with a lipophilic

Table 5. Potentiation of tumoricidal activity of Kupffer cells by muramyl dipeptide (MDP) in liposomes.

Kupffer cell treatment[a]	nmol liposomal lipid per well	ng MDP per well	% cytotoxicity[b]
MDP-liposomes	75	75	54
(1 μg MDP/ mol lipid)	50	50	57
	10	10	38
free MDP	-	10.000	31
	-	5.000	31
	-	2.500	30

[a]Rat Kupffer cells were isolated by pronase perfusion of the liver and purified by centrifugal elutriation. 25×10^4 Kupffer cells were incubated with 200 μl medium (RPMI) containing 20% heat-inactivated fetal calf serum (FCS), free MDP or MDP encapsulated in liposomes (MLV, lipid composition: PC/PS/CHOL, molar ratio 4:1:5). After 4 h ^{3}H-thymidine labeled B16-melanoma cells were added per well; initial ratio of Kupffer cells to tumor cells = 25:1. After a 48-h co-culture period cytotoxicity was assayed by measuring the release of ^{3}H-label from the melanoma cells into the medium.
[b]Percent cytotoxicity was calculated according to the formula: 100 x (A-B)(C-B) , in which A = radioactivity in supernatent of tumor cells co-cultured with treated Kupffer cells, B = radioactivity in supernatant of tumor cells co-cultured with control Kupffer cells, C = radioactivity in the total amount of tumor cells added per well.

derivative of MDP i.e. muramyl tripeptide phosphatidylethanolamine (MTP-PE), which is inserted in the lipid bilayer during liposome preparation (not shown).

From these results we conclude that, like other macrophages, rat Kupffer cells can be rendered tumorcytotoxic. Similar results have recently been published by Fidler and coworkers on the activation of murine Kupffer cells.[32] Therapy with liposome-encapsulated immunomodulators may be developed into a valuable therapeutic modality in the treatment and/or prevention of hepatic micrometastases originating from primary colorectal cancer. We are currently investigating the therapeutic efficacy of this approach in a murine model of a colon adenocarcinoma metastasizing to the liver.

ACKNOWLEDGEMENTS

The authors wish to thank Bert Dontje, Jan Wijbenga, Aletta Veninga and August Lokerse for skillful technical assistance and Rinske Kuperus for typing the manuscript. We thank Merck, Sharpe & Dohme Research Laboratories and Vestar Research Inc. for providing us with the (amino)mannosyl derivatives of cholesterol and CIBA Geigy Ltd. for generous gifts of MDP and MTP-PE. Part of this work was financially supported by the Dutch Cancer Foundation, Koningin Wilhelmina Fonds (KWF).

REFERENCES

1. F.H. Roerdink, E. Wisse, H.W.M. Morselt, J. Van der Meulen and G.L. Scherphof, Cellular distribution of intravenously injected protein-containing liposomes in the rat liver, in: "Kupffer cells and other liver sinusoidal cells," E. Wisse and D. Knook, Eds., Elsevier/North-Holland, Amsterdam (1977).
2. F. Roerdink, J. Dijkstra, G. Hartman, B. Bolscher and G. Scherphof, The involvement of parenchymal, Kupffer and endothelial liver cells in hepatic uptake of intravenously injected liposomes. Effects of lanthanum and gadolinium salts, Biochim. Biophys. Acta 677:79 (1981).
3. G. Poste, C. Bucana, A. Raz, P. Bugelski, R. Kirsh and I.J. Fidler, Analysis of the fate of systemically administered liposomes and implications for their use in drug delivery, Cancer Res. 42:1412 (1982).
4. Y.E. Rahman, E.A. Cerny, K.R. Patel, E.H. Lau and B.J. Wright, Differential uptake of liposomes varying in size and lipid composition by parenchymal and Kupffer cells of mouse liver, Life Sci. 31:2061 (1982).
5. G. Scherphof, F. Roerdink, J. Dijkstra, H. Ellens, R. de Zanger and E. Wisse, Uptake of liposomes by rat and mouse hepatocytes and Kupffer cells, Biol. Cell 47:47 (1981).
6. C.R. Alving, E.A. Steck, W.L. Chapman, Jr., V.B. Waits, L.D. Hendricks, G.M. Swartz, Jr. and W.L. Hanson, Therapy of Leishmaniasis: superior efficacies of liposome-encapsulated drugs, Proc. Natl. Acad. Sci. U.S.A. 75:2959 (1978).
7. G. Lopez-Berestein, R. Mehta, R.L. Hopfer, K. Mills, L. Kasi, K. Mehta, V. Fainstein, M. Luna, E.M. Hersh and R. Juliano, Treatment and prophylaxis of disseminated infection due to Candida albicans in mice with liposome-encapsulated amphotericin B, J. Infect. Dis. 147:939 (1983).
8. G. Lopez-Berestein, R.L. Hopfer, R. Mehta, K. Mehta, E.M. Hersh and R.L. Juliano, Liposome-encapsulated amphotericin B for treatment of disseminated candidiasis in neutropenic mice, J. Infect. Dis. 150:278 (1984).
9. I.A.J.M. Bakker-Woudenberg, A.F. Lokerse, F.H. Roerdink, D. Regts and M.F. Michel, Free vs. liposome-entrapped ampicillin in the treatment

of Listeria monocytogenes infection in normal mice and athymic (nude) mice, J. Infect. Dis. 151:917 (1985).

10. G. Poste, C. Bucana and I.J. Fidler, Stimulation of host response against metastatic tumours by liposome-encapsulated immunomodulators, in: "Targeting of drugs," G. Gregoriadis, J. Senior and A. Trouet, eds., Plenum, New York (1982).

11. J. Dijkstra, W.J.M. van Galen, C.E. Hulstaert, D. Kalicharan, F.H. Roerdink and G.L. Scherphof, Interaction of liposomes with Kupffer cells in vitro, Exp. Cell Res. 150:161 (1984).

12. J. Dijkstra, M. van Galen and G.L. Scherphof, Effects of ammonium chloride and chloroquine on endocytic uptake of liposomes by Kupffer cells in vitro, Biochim. Biophys. Acta 804:58 (1984).

13. J. Dijkstra, M. van Galen, D. Regts and G. Scherphof, Uptake and processing of liposomal phospholipids by Kupffer cells in vitro, Eur. J. Biochem. 148:391 (1985).

14. J. Dijkstra, M. van Galen and G. Scherphof, Influence of liposome charge on the association of liposomes with Kupffer cells in vitro. Effects of divalent cations and competition with latex particles. Biochim. Biophys. Acta 813:287 (1985).

15. J. Dijkstra, M. van Galen and G. Scherphof, Effects of (dihydro)-cytochalasin B, colchicine, monensin and trifluoperazine on uptake and processing of liposomes by Kupffer cells in culture. Biochim. Biophys. Acta 845:34 (1985).

16. R.L. Juliano and D. Stamp, The effects of particle size and charge on the clearance rates of liposomes and liposome encapsulated drugs, Biochem. Biophys. Res. Commun. 63:651 (1975).

17. F. Roerdink, J. Regts, B. van Leeuwen and G. Scherphof, Intrahepatic uptake and processing of intravenously injected small unilamellar phospholipid vesicles in rats, Biochim. Biophys. Acta 770:195 (1984).

18. H.H. Spanjer, Targeting of liposomes to liver cells in vivo, Ph.D. Thesis, State University Groningen, 1985.

19. G. Gregoriadis and J. Senior, The phospholipid component of small unilamellar liposomes controls the rate of clearance of entrapped solutes from the circulation, FEBS Lett. 119:43 (1980).

20. P.-S. Wu, G.W. Tin and J.D. Baldeschwieler, Phagocytosis of carbohydrate-modified phospholipid vesicles by macrophage, Proc. Natl. Acad. Sci. U.S.A. 78:2033 (1981).

21. R.A. Schwendener, P.A. Lagocki and Y.E. Rahman, The effects of charge and size on the interaction of unilamellar liposomes with macrophages, Biochim. Biophys. Acta 772:93 (1984).

22. J.D. Baldeschwieler, Phospholipid vesicle targeting using synthetic glycolipid and other determinants, Ann. N.Y. Acad. Sci. U.S.A. 446:349 (1985).

23. J.C. Wilschut, J. Regts, H. Westenberg and G. Scherphof, Hydrolysis of phosphatidylcholine liposomes by phospholipases A_2, Effects of the local anaesthetic dibucaine. Biochim. Biophys. Acta 433:20 (1976).

24. R.E. Niemann and B. Lorder, Listeriosis in adults: a changing pattern. Report of eight cases and review of the literature, Rev. Inf. Dis. 2:207 (1980).

25. G. Lopez-Berestein, V. Fainstein, R. Hopfer, K. Mehta, M.P. Sullivan, M. Keating, M.G. Rosenblum, R. Mehta, M. Luna, E.M. Hersh, J. Reuben, R.L. Juliano and G.P. Bodey, Liposomal amphotericin B for the treatment of systemic fungal infections in patients with cancer: a preliminary study, J. Infect. Dis. 151:704 (1985).

26. E.S. Kleinerman, A.J. Schroit and W.E. Fogler, Tumoricidal activity of human monocytes activated in vitro by free and liposome-encapsulated human lymphokines. J. Clin. Invest. 72:304 (1983).

27. L. Varesio, E. Blasi, G.B. Thurman, J.E. Talmadge, R.H. Wiltrout and R.B. Herberman, Potent activation of mouse macrophages by recombinant interferon, Cancer Res. 44:4465 (1984).

28. G. Lopez-Berestein, K. Mehta, R. Mehta, R.L. Juliano and E.M. Hersh,

The activation of human monocytes by liposome-encapsulated muramyl dipeptide analogues, J. Immunol. 130:1500 (1983).

29. S. Sone and I.J. Fidler, In vitro activation of tumoricidal properties in rat alveolar macrophages by synthetic muramyl dipeptide encapsulated in liposomes, Cell. Immunol. 57:42 (1981).

30. I.J. Fidler, Z. Barnes, W.E. Fogler, R. Kirsh, P. Bugelski and G. Poste, Involvement of macrophages in the eradication of established metastases following intravenous injection of liposomes containing macrophage activators, Cancer Res. 42:496 (1982).

31. P.S. Thombre and S.D. Deodhar, Inhibition of liver metastases in murine colon adenocarcinoma by liposomes containing human C-reactive protein or crude lymphokine, Cancer Immunol. Immunother. 16:145 (1984).

32. Z.L. Xu, C.D. Bucana and I.J. Fidler, In vitro activation of murine Kupffer cells by lymphokines or endotoxins to lyse syngeneic tumor cells, Am. J. Pathol. 117:372 (1984).

ALTERNATE DELIVERY OF INTERFERONS

Deborah A. Eppstein

Syntex Reserach

Palo Alto, CA 94394

INTRODUCTION

The emergence and mushrooming of recombinant DNA technology and the resultant production of genetically-engineered proteins has, for the first time, made feasible the prospect of utilizing many proteins as therapeutic entities. This is largely due to the potential of producing proteins in greater quantities, while before they were unavailable in sufficient amounts for even sound experimental testing (both in vitro as well as in animal models). Proteins of many classes are included, encompassing lymphokines, monokines, growth promotants and other larger polypeptide hormones, blood coagulation factors, enzymes, etc. In addition, the production of antigens for new vaccines which heretofore were unfeasible is now underway. These new prospects for the widespread use of proteins as therapeutic agents coupled with the necessity of appropriate delivery for such protein thera-peutics places renewed emphasis on development of suitable delivery system for proteins.

The traditional and most widely used method of administration of drugs is by the oral route. However, in the case of proteins, such delivery is not feasible due to the hydrolysis of proteins by digestive enzymes. The methods most commonly used for administration of protein drugs are by repeated injection: intramuscular (IM), subcutaneous (SC) or intravenous (IV) infusion. These methods are acceptable in situations where a very limited number of injections are required, but are undesirable for chronic administration (for example as with insulin therapy). The nature of many of the diseases or disorders being targeted for treatment by these rDNA-produced proteins (or, as in the case of growth promotants, the nature of the desired effect) is <u>chronic</u> rather than acute, thus necessitating fre-quent injections over a prolonged time frame.

The focus of this paper is on delivery of interferons, which were among the first proteins of potential widespread therapeutic interest to be produced in large-scale by rDNA technology. I will first discuss why alternative delivery systems are important for interferons, and raise general considerations and fundamental issues to be taken into account in developing such applications. Then I will describe the approaches we have taken in developing sustained release formulations for various interferons, and the characterization of these systems in vitro and in vivo. Lastly, I will discuss some of the potential medical applications utilizing these controlled-release delivery systems for interferons, and raise some of the fundamental questions that still remain to be answered.

GENERAL CONSIDERATIONS FOR DELIVERY OF INTERFERONS

In order to develop an efficacious delivery system for a therapeutic agent, a basic understanding of the properties of the drug, its pharmaco-kinetics and target organs, mechanism of action (including side effects), and nature of the disease being treated are important.

Protein Nature of Interferons

Interferons (IFN's) fall into three classes: α, β, and γ (see Stewart, 1979 for review of properties of IFN's). All the interferons are proteins in the 20 Kd molecular weight range and thus general considerations for handling of proteins are relevant. The delicate protein nature requires preservation of tertiary structure for maintenance of biological activity, and this obviously affects what formulation techniques can be satisfact-orily applied to IFN's. Furthermore, stability after initial formulation is a prime consideration. This includes not only shelf-life stability, but also stability of the IFN reserves within the animal after administration in a sustained-release formulation.

Nature of Disease Target

The characteristics of the disease being treated will influence the nature of the delivery system of choice for the interferon. Obviously the duration of the controlled-release of interferon must be realistic within the time frame of the target disease (acute vs. chronic), which can vary significantly for different applications (ie, anticancer vs. antiviral). Also, the nature of the disease can dictate whether systemic or localized treatment is most desirable which will influence both the delivery system as well as mode of application.

Mechanism of Action of Interferons

Interferons act on several different types and mediate a multitude of biological effects. Whether the effect is establishment of an antiviral state in a cell, inhibition of proliferation of a cancer cell, or modul-ation of the immune system, (eg, priming/activation of macrophages), the initial requirement is binding of the interferon to its specific cell surface receptor which subsequently triggers the various biological effects. Thus a delivery system for interferon (α, β, or γ) must take this into account. The possibility of down-regulation of receptors for interferon as a result of chronic administration of high levels of interferon must not be ignored. Under some dosing regimes, such loss of cellular binding cap-acity of interferon has been observed (B. Williams, Hospital for Sick Children, Toronto, personal communication; Maxwell et al, 1985). Pulsa-tile rather than continuous administration for high doses, or alternatively continuous administration of low dose levels can be considered.

Knowledge of the pharmacokinetics of the free interferon and the degree of receptor occupancy required for the desired biological effects can be used to help optimize the rate of delivery. With interferons, a low level of receptor occupancy can result in a significant biological effect. It is conceivable that a proper low level of continuous administration would be efficacious.

208

Antibody Formation

The potential immunogenicity of proteins is an important consideration when they are to be used as drugs. In the case of vaccines, it is
of obvious usefulness to maximize the immunogenicity of the specific protein antigen of interest by proper presentation and delivery of the antigen.
Both liposomes (Allison and Gregoriadis, 1974) and more prolonged sustained-
release delivery systems (Amkraut and Martins, 1984) have been shown to act
as adjuvants for certain proteins for increasing the specific immune response. For other therapeutic applications of proteins, however, specific
antibody formation is highly undesirable and can limit the usefulness of
the protein as a drug. Thus, minimizing such antibody formation is a prime
consideration in developing a delivery system for a protein drug.

In the case of interferons, a low degree of formation of neutralizing
antibodies in a small fraction of patients has been observed after repeated
administration of recombinant as well as natural interferons (α and β), but
to date it does not appear to be a significant problem. Recently, utilizing
an ELISA analysis, a much higher incidence of formation of non-neutralizing
antibodies to a recombinant modified β-interferon (rHuIFN-β_{ser17}) was detected in patients after repeated administration, but no apparent clinical
complications arose (Borden et al, 1985). However, with this background,
the importance of minimizing the antigenicity of interferons when alternative delivery systems are developed cannot be overlooked.

Toxicity

Early anticipations that interferons were efficacious without exhibiting undesirable side effects proved to be wrong. The toxicities associated with interferons (α, β, and most recently γ) are reversible, and
include bone-marrow depression and flu-like symptoms (fever, malaise,
myalgia, anorexia, fatigue and the like) and are dose-related. It is conceivable that, as has occurred with other non-protein drugs, a controlled-
release delivery of an interferon could maintain an efficacious drug level
without necessitating the peak concentrations that are associated with the
most severe side effects. Also, in the case of localized disease, a regional administration of the interferon would be a likely choice to minimize
or even prevent systemic side effects. Another consideration in development of a sustained-release system for delivery of interferons is retriev-
ability in the event of development of intolerable side effects.

CONTROLLED RELEASE OF INTERFERONS WITH BIODEGRADABLE DELIVERY SYSTEMS

We have initially taken the approach of developing systems for controlled, sustained release of interferons. Depending on the choice of the
delivery vehicle as well as the type of interferon, the applications can
be either for systemic or local effects, and can encompass administration
by either injection, implantation, or direct application to mucosal membranes including by aerosalization.

Liposomes

Liposomes offer several attractive features for altered delivery of
interferons. They can be made completely biodegradable, and the composition can be varied to result in different release profiles as we showed
previously (Eppstein and Stewart, 1981, 1982; Eppstein, 1982). HuIFN-α,
formulated in either negatively or positively-charged vesicles, was released gradually from the site of IM injection. Inclusion of cholesterol
in the negatively-charged fluid liposomes (egg phosphatidylcholine: brain
phosphatidylserine, 7:3), which would increase the rigidity of the vesicle,

was necessary to obtain increased interferon retention in the vicinity of the local injection site for 24 h. The comparable positively-charged vesicles, however resulted in increased local retention of interferon (>10% remaining after three days) whether or not they contained cholesterol (Eppstein and Stewart, 1982). Size uniformity of the liposomes was obtained by extrusion through 0.2 μ nucleopore filters without reducing the local retention of the interferon (Eppstein, 1982). Similar results were obtained ith natural HuIFN-β (data not shown). Increased retention of interferon (α or β) in the lung, spleen, and liver was obtained after intravenous injection of the liposomal interferon (Eppstein, 1982). To minimize any potential adjuvant effects of the liposomes in enhancing antibody formation, we focused on further optimizing the composition of the negatively charged vesicles to result in a more prolonged release of interferon.

Utilization of a novel type of liposome preparation (lyophilization multilamellar vesicles, LMLV) involving initial lyophilization of the lipids from an organic solvent followed by hydration with aqueous buffer containing interferon resulted in essentially complete incorporation of rHuIFN-β_{ser17} (Fig. 1) as monitored by spiking with ^{125}I-labeled interferon (Felgner et al, manuscript in preparation). The gentle methods involved in preparing LMLV minimize denaturation of the interferon, as evidenced by full retention of its antiviral activity. Furthermore, the ease of preparation should enhance reproducibility from batch to batch. Lipid composition was chosen to increase the rigidity of the liposomes at body temperature. IM or SC injection of "solid" LMLV composed of DAPC:DPPG (7:3) containing rHuIFN-βser17 resulted in significantly increased local retention of interferon, with 50% still remaining near the injection site after three days (Fig. 2).

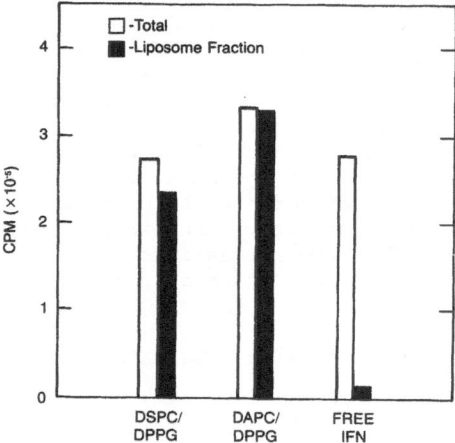

Fig. 1.   Incorporation of β-interferon in liposomes. Liposomes (LMLV) were prepared as described (Eppstein et al., 1985) by lyophilization of lipids from cyclohexane followed by hydration with buffer containing rHuIFN-β_{ser17} (obtained from Cetus/Triton) spiked with ^{125}I-labeled interferon. Liposomes were separated from unincorporated interferon by centrifugation, and the radioactivity in the pellet and supernatant was quantitated. Lipid composition was either distearoyl phosphatidylcholine: dipalmitoyl phosphatidylglycerol (DSPC:DPPG, 7:3) or diarachidoyl phosphatidylcholine: dipalmitoyl phosphatidylglycerol (DAPC:DPPG, 7:3).

Free interferon by contrast was >90% gone from the vicinity of the injection site by one day.

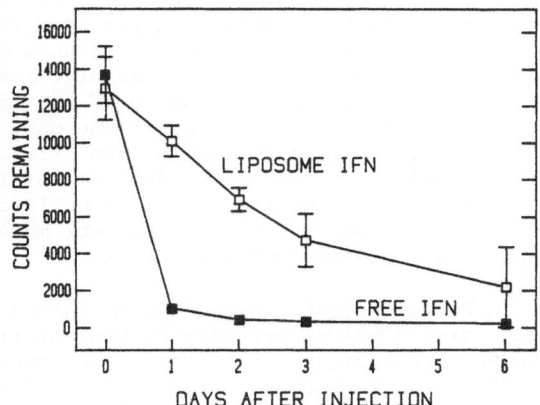

Fig. 2.   Local retention of β-interferon after intramuscular
          injection in mice. Liposomal-rHuIFN-β_{ser17} (DAPC:
          DPPG, 7:3, open squares) or free rHuIFN-β_{ser17} (filled
          squares) spiked with ^{125}I-rHuIFN-β_{ser17}, were injected
          IM into the hind thigh of a mouse (15 µl per injection).
          After 1-6 days, the mice were sacrificed and the radio-
          activity remaining in the injected thigh muscle was
          quantitated in a gamma counter. Each point is the aver-
          age of four injected thighs.

       The characteristics of liposomal-interferon can be summarized as
follows:

1)   Good capture of interferon (50-100%) can be obtained by optimizing
      liposome preparation methods (utilizing LMLV).
2)   Cholesterol increases stability and local retention of interferon
      after IM injection in negatively-charged fluid vesicles. Local
      retention after injection IM or SC is also obtained by increasing
      the rigidity of the vesicles by utilizing lipids of higher thermal
      transition temperatures (>50°).
3)   Serum induces some leakage of interferon, and interaction with cells
      causes still more leakage (Eppstein and Stewart, 1982; Eppstein et al,
      1985).
4)   All biological activity of the interferon is due to interferon
      that has leaked out of the liposome (it is fully neutralizable
      by externally-added antibody) and is subsequently able to bind to
      the sepcific cell-surface receptor; liposome encapsulation does
      not bypass the need for receptor interaction (Eppstein et al, 1985).

5) Liposome formulation of interferon results in controlled-release
    of interferon from one day to two weeks, depending on liposome
    composition.

Polylactic-Polyglycolic Acid Copolymers

Biodegradable polymer matrices formed of copolymers of polylactic acid-
polyglycolic acid [PLGA, poly (d, l-lactide-co-glycolide)] have been used
as a sustained-release delivery system for several therapeutic agents,
including steroids (Anderson et al, 1976; Beck et al, 1979, 1981), chemotherapeutic agents (Yolles and Morton, 1978), and peptides (Kent et al,
1982; Sanders et al, 1984). With the peptides, the release of the drug
from PLGA polymer microspheres appears to be triphasic: first an "initial
burst" of drug appears which results from diffusion of the peptide from
the surface of the microspheres; then follows a slowing down or latent
period; and the release profile culminates with a high release rate apparently caused by bulk (not surface) erosion by hydrolysis of the polymer
(Sanders et al, 1984). The duration of the release profile can be varied
significantly by adjusting the monomer ratio and polymer molecular weight
to result in release profiles ranging from 1-6 months.

Initial attempts to formulate interferon in PLGA polymer matrices were
unsuccessful due to loss of biological activity of the interferon. With
careful procedures, however, techniques were devised to result in 100%
incorporation of the interferon without significant loss of biological
(antiviral) activity (Schryver et al, manuscript in preparation). The
PLGA-interferon preparations were implanted subcutaneously into mice, and
after various time intervals the residual PLGA pellets were removed from
the mice, and the radioactivity of the ^{125}I-rHuIFN-β_{ser17} was determined
(Fig. 3). In addition, the levels of ^{125}I-IFN in the blood were determined

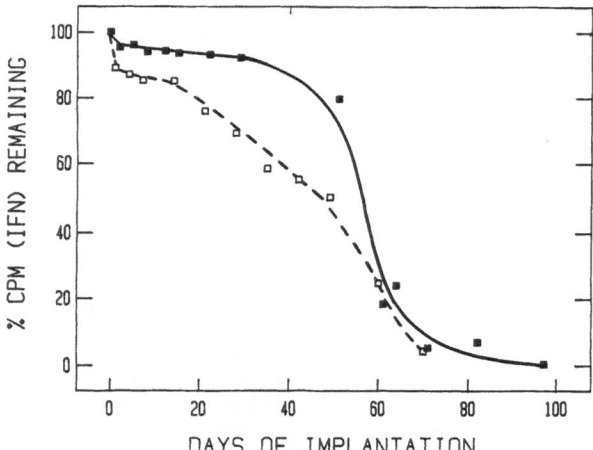

Fig. 3. Local retention of PLGA-formulated β-interferon
         after implantation into mice. rHuIFN-β (spiked
         with ^{125}I-rHuIFN-β) formulated in PLGA polymer
         matrices was implanted SC into mice. After the
         days indicated, the remaining implant was removed
         and the radioactivity in the implant was quantit-
         ated by gamma counting. Results with two differ-
         ent formulations (open and filled symbols) are shown.

(Fig. 4). The pellets were subsequently extracted so that the interferon could be quantitated for biological antiviral activity as well as radioactivity (see Fig. 5 for procedure). From these procedures, it was found that after implantation in vivo, the interferon contained in the hydrated PLGA polymer implant retained its biological activity very well, with only 0.4 $\log_{10}$ loss of activity over a 35 day period. By marked contrast, the activity of identical free interferon in solution at 37°C was reduced by over 100-fold (Fig. 6). It is also noteworthy that the release profile (Fig. 3) of the interferon could be altered from the strikingly triphasic curve (filled symbols) to a much more linear release rate (open symbols) by varying the PLGA preparation. The levels of β-interferon observed in the blood after IM injection of free interferon (peak blood levels ~2% of the

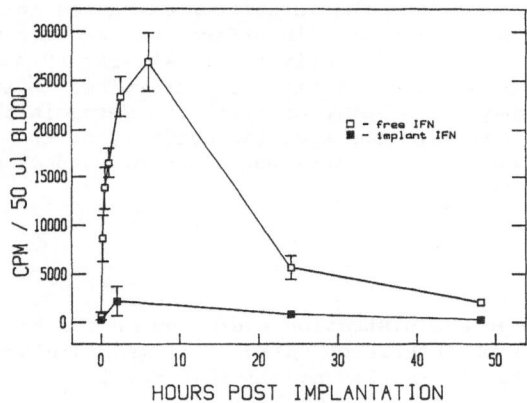

Fig. 4.   Blood levels of β-interferon in mice after injection IM, or after implantation in PLGA formulation. rHuIFN-βser17 spiked with ^{125}I-interferon was injected IM into mice, and at the times indicated 25 µl blood was withdrawn from the retro-orbital sinus and the radioactivity quantitated by gamma counting (open squares, mean $\pm$ S.D. of four mice). PLGA-formulated rHuIFN-βser17 was implanted SC into mice, and blood levels of radioactivity were determined (filled squares, mean $\pm$ S.D. of three mice.

injected dose 2-6 h after injection) are essentially undetectable when the interferon is implanted in the PLGA formulation (Fig. 4). It has been demonstrated in clinical trials that systemic responses can be obtained after IM injection of β-interferon without obtaining significant circulating levels of interferon. This PLGA-interferon formulation results in a slow, controlled-release of interferon over a two month period while avoiding the peak serum levels obtained by injection of free interferon.

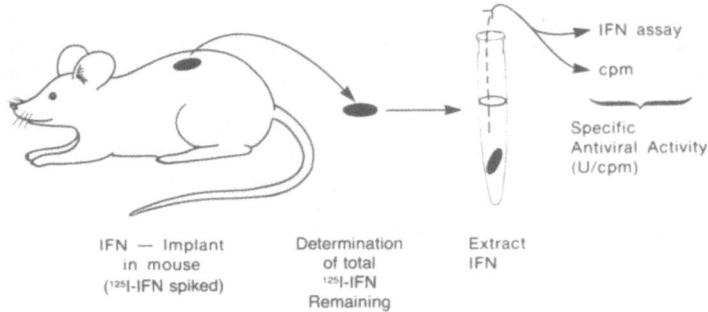

IFN assay

cpm

Specific
Antiviral Activity
(U/cpm)

IFN — Implant   Determination   Extract
in mouse   of total   IFN
(^{125}I-IFN spiked) ^{125}I-IFN
                    Remaining

Fig. 5. Flow diagram describing how the biological specific
         activity of β-interferon remaining in the mouse in the
         PLGA implant was determined. The PLGA implant was re-
         moved from the mouse after varying lengths of time (up
         to five weeks) and the interferon extracted in vitro.
         The amount of extracted interferon was quantitated by
         determining the radioactivity of ^{125}I-rHuIFN-β_{ser17},
         and the biological activity was determined in an anti-
         viral assay. The ratio of these two determinations was
         used to determine the specific activity of the extracted
         interferon. These values were then plotted as described
         in Fig. 6.

It is hoped that such an administration would reduce the toxicities associated with high levels of interferon while still maintaining an efficacious
dose; the answer awaits the results of clinical trials.

TARGETING

     The concept of directed targeting of interferon can be addressed by
either passive or active targeting.

Passive Targeting

     Passive targeting refers to regional localization of a drug at a des-
ired site of action via non-specific physical interaction of the delivery
vehicle with certain body tissues or compartments. One means of enhancing
passive targeting is by localized controlled-release of the drug. This
can be achieved by injection of the controlled-release drug system (such
as PLGA-IFN or liposomal-IFN) in the vicinity of the desired site of action.
For example, treatment for localized viral infections (such as papilloma
or herpes virus infections) can be concentrated in the vicinity of the out-
break by such localized injection. Another possibility is treatment of
solid inoperable tumors.

     Alternately, the drug (with controlled-release system) can be directly
applied to mucosal membranes at the site of desired action to achieve a
local effect while minimizing the systemic action. Potential examples of
such applications include vaginal application for treatment of herpes
simplex virus type 2 (HSV-2) or papilloma virus infections or possibly even
cervical cancer; aerosalization into the lungs (with a liposome vehicle)

214

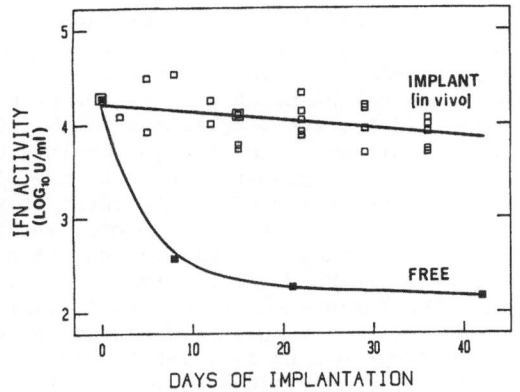

Fig. 6. Stability of β-interferon in PLGA implants in
         mice. rHuIFN-$_{ser17}$, spiked with ^{125}I-interferon,
         was formulated with PLGA and the implanted SC into
         mice. After the times indicated, the remaining
         implants were moved from the mice, the interferon
         was extracted, and the specific activity determined
         as described in Fig. 5 (open squares). Closed
         squares show the biological activity of free β-inter-
         feron at 37°C in comparable buffer. All specific
         activities were normalized to the activity of the
         reference β-interferon sample before plotting.

for treatment of lung infections; intranasal application for short-term
prophylaxis of viral infections of the upper-respiratory tract; or delivery
into the lymphatics. A recent report described a selective delivery of
interferon into the lymphatics via absorption from mixed micelles from the
large intestine of the rat (Yoshikawa et al., 1984). In our initial experience, vaginal application to achieve a localized effect against genital
HSV-2 infections in mice has not been very successful (Fraser-Smith et al,
unpublished results), but further testing with other formulations needs to
be conducted. Intranasal application of interferon has been shown to be
efficacious as a prophylaxis of the common cold, but the side effects associated with the prolonged (3-4 week) treatment regimes and high doses tested
were great enough to initially discourage this application (Farr et al, 1984;
Douglas et al., 1985). Further testing is needed to ascertain if utilization of an appropriate sustained-release delivery system could maintain a
local efficacious dose level without causing the nasal distress associated
with the high doses of free interferon currently needed for prophylactic
efficacy.

     One other type of passive targeting deserves mention here. As inter-
feron-γ is known to be a potent stimulator of macrophages, the question can
be raised whether its ability to activate (prime) macrophages can be enhanced by passive targeting to the macrophages after liposome-encapsulation
as has been shown to occur with the macrophage activator muramyl dipeptide
and analogs (Fidler et al., 1981). γ-Interferon must bind to a specific
cell-surface receptor to subsequently exert its biological effects, includ-

ing those on macrophages (Celada et al., 1984). We have shown that liposome encapsulation cannot bypass this need for binding of γ-interferon to its cell-surface receptor in order to trigger its biological effects, both for establishment of the antiviral state as well as for priming/activation of macrophages to the tumoricidal state (Eppstein et al., 1985). Although liposome-encapsulation cannot directly increase the ability of γ-interferon to prime/activate macrophages, it is possible that the use of a liposome vehicle could alter the pharmacokinetics of γ-interferon in vivo such that its bioavailability, after leaking out of the liposome, was favorably altered. In initial tests of efficacy of rMuIFN-γ against systemic lethal HSV-2 infections of mice, free rMuIFN-γ (Fraser-Smith et al., 1985) showed the same efficacy as did γ-IFN formulated in DOPC:DOPG (7:3) LMLV (injected IP or IV, unpublished results). However, this liposome composition is known to be leaky for γ-IFN (Eppstein et al., 1985); utilization of a more stable formulation potentially could yield altered results.

Active Targeting

Coupling of interferon-α to monoclonal antibodies against osteosarcoma (Pelham et al., 1983) or Epstein-Barr virus (Miescher-Granger et al., 1985) has been reported, with the ultimate goal of utilizing the antibody-interferon conjugates as a targeted treatment for cancer. This approach has the same limitations as does utilization of radiolabeled antibodies for diagnostic imaging: the majority of the antibody conjugate does not reach its target site.

POTENTIAL MEDICAL APPLICATIONS

The beneficial medical applications of free interferon α and β are only now beginning to be understood, and it is reasonable to expect these same diseases to be targeted for treatment with interferon utilizing controlled-release and localized delivery systems.

Localized Controlled Release

1. Papilloma Virus Infections. Genital infections with papilloma virus (condylomata acuminata) are reaching epidemic levels and interferon (α and β) is one of the only therapeutic agents showing promise in treating this disease (Schonfeld et al., 1984; Niimura, 1983; Geffen et al., 1984; Androphy et al., 1984). Although the results from the initial clinical trials are preliminary, the data suggest that multiple repeated injections of interferon (daily or thrice weekly) over a several week period are necessary to achieve ultimate regression of the warts. Conceivably, this may be a good application for a sustained-release delivery system for interferon α or β. Liposome formulations may be sufficient for a 1-2 week sustained release of interferon.
2. Cervical cancer. Cervical dysplasia may respond to treatment with interferon α or β (De Palo et al., 1984, 1985; Vesterinen et al., 1984) and as such a localized sustained-release delivery of the interferon could be quite appropriate for this application.
3. Tumors at the time of surgery. Although surgery can be used to remove many types of tumors, the danger of recurrence of the tumor at the site of surgical removal in addition to metastatic occurrence still exists. For cancers that have a positive response profile to interferon, implantation of a PLGA-interferon pellet at the site of surgery might be appropriate in helping curtail recurrences. If unacceptable side effects occur at some time after the implantation, the remaining interferon-containing implant can be removed.

Systemic Controlled Release

1. <u>Chronic treatment of certain cancers.</u> Patients with cancers that are known to respond favorably to prolonged treatment with interferon-α, such as hairy cell leukemia (Quesada et al., 1984) and renal cell carcinoma (Quesada et al., 1985), may benefit from treatment with a sustained-release formulation such as PLGA-interferon. For example, in the case of hairy cell leukemia, daily IM injections of interferon-α were given for six or more months in order to achieve complete remission in several patients. It would be of obvious advantage to reduce these daily injections to once every two or three months.

2. <u>Combination Therapies.</u> Interferons show promise for use in combination therapies against both viral infections (Fraser-Smith et al., 1984a, b,; 1985; Eppstein et al, 1984) as well as in cancers. For such combination treatments, it may be desirable to use a sustained-release delivery of the interferon, adding the second drug treatment by either oral or parenteral routes as appropriate.

CONCLUSIONS

Alternative delivery systems for interferons can take the form of sustained-release formulations which can be administered for systemic localized therapy. Liposome formulations were developed which resulted in 50-100% capture of the various interferons, and which could give a controlled-release of the interferon over a several day to two week period, depending on composition. Such formulations may be applicable to localized viral infections where one to two week treatment regimes are desirable.

More prolonged controlled-release preparations of interferons could be achieved by formulation in matrices composed of copolymers of polylactic-polyglycolic acid. These preparations can be implanted subcutaneously to result in gradual release of the interferon over a two to three month period. Potential applications include systemic treatment of cancers known to respond to prolonged repeated injections with interferon.

REFERENCES

Allison, A.C. and Gregoriadis, G., 1974, Liposomes as immunological adjuvants, <u>Nature,</u> 252:252.

Amkraut, A.A. and Martins, A.B., 1984, Method for administering immuno-potentiator, U.S. Patent 4, 484:923.

Anderson, L.C., Wise, D.L. and Howes, J.F., 1976, An injectable sustained release fertility control system, <u>Contraception,</u> 13:375.

Androphy, E.J., Dvoretzky, I., Maluish, A.E., Wallace, H.J. and Lowey, D.R., 1984, Response of warts in epidermodysplasia verruciformis to treatment with systemic and intralesional alpha interferon. <u>J. Am. Acad. Dermatol.,</u> 11: 197.

Beck, L.R., Cowsar, D.L., Lewis, D.H., Gibson, J.W. and Flowers, C.E., 1979, New long-acting injectable microcapsule contraceptive system, <u>Am. J. Obstet. Gynecol.,</u> 135:419.

Beck, L.R., Ramos, R.A., Flowers, C.E., Lopez, G.Z., Lewis, D.H. and Cowsar, D.R., 1981, Clinical evaluation of injectable biodegradable contraceptive system, <u>Am. J. Obstet. Gynecol,</u> 140:799.

Borden, E., Hawkins, M., Anderson, S., Schiessel, J., Sielaff, K., DeMets, D., Davis, T., Horning, S., Rosno, S., Merigan, T. and Konrad, M., 1985, Phase I evaluation of a synthetic mutant of Interferon β. <u>Cancer Res.</u> (in press).

Celada, A., Gray, P.W., Rinderknecht, E. and Schreiber, R.D., 1984, Evidence for a gamma-interferon receptor that regulates macrophage tumoricidal activity, J. Exp. Med., 160:55

De Palo, G., Stefanon, B., Rilke, R., Pilotti, S. and Ghione, M., 1984, Human fibroblast interferon in cervical and vulvar intraepithelial neoplasia associated with papilloma virus infections. Int. J. Tiss. Reac., 6:523.

De Palo, G., Stefanon, B., Rilke, F., Pilotti, S, and Ghione, M., 1985, Human fibroblast interferon in cervical and vulvar intraepithelial neoplasia associated with viral cytopathic effects, J. Reproduct. Med., 30:404.

Douglas, R.M., Albrecht, J.K., Miles, H.B., Moore, B.W., Read, R., Worswick, D.A., and Woodward, A.J., 1985, Intranasal interferon-α_2 prophylaxis of natural respiratory virus infection, J. Infect. Dis., 151:731.

Eppstein, D.A., 1982, Altered pharmacologic properties of liposome-assoc-iated human interferon-alpha, J. Interferon Res., 2:117.

Eppstein, D.A., Marsh, Y.V., Fraser-Smith, E.B. and Matthews, T.R., 1985, Potent synergistic inhibition of herpes simplex virus-2 by combin-ation of α, β or γ interferons with DHPG (9-[1,3-dihydroxy-2-propoxy)-methyl]guanine): differences in vitro vs. in vivo, in: "The Biology of the Interferon System 1984", H. Kirchner and H. Schellekens, eds., Elsevier Science Publishers, Inc., New York.

Eppstein, D.A. Marsh, Y.V., Van der Pas, M.A., Felgner, P.F. and Schreiber, A.B., 1985, Biological activity of liposome-encapsulated murine interferon γ is mediated by a cell membrane receptor, Proc. Natl. Acad. Sci. USA, 82:3688.

Eppstein, D.A. and Stewart II, W.E., 1982, Altered pharmacological prop-erties of liposome-associated human interferon alpha, J. Virol. 41:575.

Farr, B.M., Gwaltney Jr., J.M., Adams, K.F., and Hayden, F.G., 1984, Intra-nasal interferon-α_2 for prevention of natural rhinovirus colds, Antimicrob. Agents Chemother., 26:31.

Fidler, I.J., Sone, S., Fogler, W.E. and Barnes, Z.L., 1981, Eradication of spontaneous metastases and activation of alveolar macrophages by intravenous injection of liposomes containing muramyl dipeptide, Proc. Natl. Acad. Sci. USA, 78:1680.

Fraser-Smith, E.B., Eppstein, D.A., Marsh, Y.V. and Matthews, T.R., 1984a, Enhanced efficacy of the acyclic nucleoside 9-(1,2-dihydroxy-2-propoxymethyl)-guanine in combination with beta-interferon against herpes simplex virus type 2 in mice, Antimicrob. Agents Chemotherap., 25:563.

Fraser-Smith, E.B., Eppstein, D.A., Marsh, Y.V., and Matthews, T.R., 1984b, Enhanced efficacy of the acyclic nucleoside 9-(1,3-dihydroxy-2-propoxymethyl)-guanine in combination with alpha-interferon against herpes simplex virus type 2 in mice, Antimicrob. Agents Chemotherap., 26:937.

Fraser-Smith, E.B., Eppstein, D.A., Marsh, Y.V. and Matthews, T.R., 1985 Enhanced efficacy of the acyclic nucleoside 9-(1,3-dihydroxy-2-propoxymethyl)-guanine in combination with gamma interferon against herpes simplex virus type 2 in mice, Antiviral Research, 5:137.

Geffen, J.R., Klein, R.J., and Friedman-Kien, A.E., 1984, Intralesional administration of large doses of human leukocyte interferon for the treatment of condylomata acuminata, J. Infect. Dis., 150:612.

Kent, J.S., Sanders, L.M., McRae, G.I., Vickery, B.H., Tice, T.R. and Lewis, D.H., 1982, In-vivo controlled release of an LHRH analog from in-jected polymeric microcapsules, Contracept. Deliv. syst., 3:58.

Maxwell, B.L., Talpaz, M., and Gutterman, J.U., 1985, Down-regulation of peripheral blood cell interferon receptors in chronic myelogenous leukemia patients undergoing human interferon (HuIFN-α) therapy, Int. J. Cancer,, 36:23.

Miescher-Granger, S., Hochkeppel, H.K., Braun, D.G. and Alkan, S.S., 1985, Biological activities of human recombinant interferon α/β targeted by anti-Epstein-Barr virus monoclonal antibodies, FEBS Lett. 179:29.

Niimura, M., 1983, Intralesional human fibroblast interferon in common warts, J. Dermatolog., 10:217.

Pelham, J.M., Flannery, G.R., Gray, J.D., Pimm, M.V., and Baldwin, R.W., 1983, Interferon conjugation to human osteogenic sarcoma monoclonal antibody, Cancer Immunol. Immunother., 15:210.

Queseda, J.R., Reuben, J., Manning, J.T. and Gutterman, J.U., 1984, Alpha interferon for induction of remission in hairy-cell leukemia, New Engl. J. Med., 310:15.

Quesada, J.R., Swanson, D.A., and Gutterman, J.U., 1985, Phase II study of interferon-alpha in metastatic renal carcinoma: a progress report, J. Clin. Oncol., 3:1086.

Sanders, L.M., Kent, J.S., McRae, G.I., Vickery, B.H., Tice, T.R., and Lewis, D.H., 1984, Controlled release of a luteinizing hormone-releasing hormone analogue from poly (d, l-lactide-co-glycolide)-microspheres, J. Pharmaceut. Sci., 73:1294.

Schonfeld, A., Schattner, A., Crespi, M., Levavi, H., Shoham, J., Nitke, S., Wallach, D., Hahn, T., Yarden, O., Doerner, T., and Revel, M., 1984. Intramuscular human interferon-β injections in treatment of condylomata acuminata, Lancet, 1:1038.

Stewart II, W.E., 1979, "The Interferon System," Springer-Verlag, New York.

Vesterinen, E., Meyer, B., Purola, E., and Cantell, K., 1984, Treatment of vaginal flat condyloma with interferon cream, Lancet, 1:157.

Yolles, S. and Morton, D.F., 1978, Timed-release depot for anti-cancer agents, Acta Pharm. Svec., 15:382.

Yoshikawa, H., Takada, K., Muranishi, S., Yu-ichiro, S. and Naruse, N., 1984, A method to potentiate enteral absorption of interferon and selective delivery into lymphatics, J. Pharm. Dyn., 7:59.

LIPOSOMES IN ANTIMICROBIAL THERAPY

Gabriel Lopez-Berestein,[1] Rudolph L. Juliano,[3] Kapil Mehta,[1]
Reeta Mehta,[3] Teresa McQueen,[1] and Roy L. Hopfer[2]

Departments of Clinical Immunology and Biological Therapy[1]
and Laboratory Medicine,[2] The University of Texas M.D.
Anderson Hospital and Tumor Institute at Houston, and the
Department of Pharmacology,[3] The University of Texas Medical
School at Houston, Houston, Texas

INTRODUCTION

The development of new therapeutic modalities involves the synthesis
of new drugs or drug analogues, or a modification of the therapeutic index
of established active drugs. The latter can be accomplished by the use of
drug carriers, such as liposomes, which can modify the pharmacokinetics and
distribution of drugs.

The successful use of liposomes as drug carriers depends on several
considerations, including: 1) the exploitation of their natural localization;
2) the appropriate selection of drugs to be entrapped; and 3) the proper
selection of the type of liposome and its lipid composition. The localization of liposomes in the mononuclear phagocyte system (MPS) prompted
Gregoriadis et al.[1] to suggest that liposomes could be used for the treatment of parasitic diseases which invade the MPS. This concept was later
supported by data from several laboratories.[2-4] Earlier work in experimental parasitic diseases demonstrated that liposomal encapsulation of the
highly toxic antimonial compounds markedly enhanced the therapeutic index
by decreasing the toxicity and enhancing the therapeutic activity.[2-4] The
success of these early experiments triggered a major interest in the use of
liposomes as carriers of antimicrobials. This is not intended to be an
exhaustive review of the literature, but to highlight some applications of
liposomal-carriers in each of the major areas of antimicrobial therapy.

LIPOSOMES IN THE TREATMENT OF BACTERIAL DISEASES

The facultative intracellular bacteria (i.e. Salmonella, Listeria,
Brucella, etc.) preferentially infect the MPS. Earlier in-vitro studies
demonstrated that liposome-entrapped dihydrostreptomycin enhanced the intracellular killing of Staphylococcus aureus.[1] Desiderio and Campbell,[5,6]
initially demonstrated that liposomal cephalothin was superior to the free
drug in the eradication of the intracellular organism in vitro. These investigators successfully used liposome-encapsulated cephalothin for the
treatment of experimental murine salmonellosis. The higher intracellular
concentrations of cephalothin that were achieved by using the liposomal

form probably accounted for the enhanced activity. When administered intravenously to mice, liposomal cephalothin was rapidly cleared from the circulation (4 min compared with 30 min for the free drug), while achieving higher levels of drug in the liver and spleen. There was a significant decrease in the number of bacteria recovered from the spleen of mice treated with liposomal cephalothin as compared with spleens from mice treated with the free drug or with untreated mice.

LIPOSOMES IN THE TREATMENT OF PARASITIC DISEASES

Leishmania is a parasite transmitted by the bite of an insect vector, the phlebotomine fly, that infects more than 100 million people worldwide.[7] The parasite affects mainly the MPS in any of its three clinical forms.[8] The drugs of choice are the pentavalent antimonial compounds, but their use is limited due to their kidney, liver, and heart toxicity. This combination of factors led to the use of liposomal-antimonials in experimental leishmaniasis (for a review, see ref. 7). Liposome-encapsulated antimonials were shown to be more effective than the free antimonials in experimental visceral leishmaniasis in hamsters[9] and dogs.[10] Although the liposomal drug was more toxic than the free drug (possibly by achieving higher levels in the liver), the therapeutic index of the liposomal form was 35-40 times higher than the free drug.[7] Amphotericin B (AmpB) entrapped in liposomes has been successfully used for the treatment of experimental visceral and cutaneous leishmaniasis.[8,11,12] In addition, liposome-encapsulated lymphokines were also shown to be effective in the treatment of experimental visceral leishmaniasis.[13]

Malaria, caused by <u>Plasmodium</u>, an obligate intracellular protozoan, has a worldwide prevalence of more than 100 million cases. Primaquine, the most effective antimonial compound, is associated with severe cardiotoxicity; when entrapped in liposomes it was shown to be less toxic and as effective as free primaquine in the treatment of murine malaria.[14] As shown for other liposome-entrapped drugs, the liver and spleen were the primary sites for the accumulation of liposomal primaquine.[15,16]

LIPOSOMES IN THE TREATMENT OF FUNGAL INFECTIONS

Animal studies. Systemic fungal infections are a frequent complication of chronic debilitating diseases and in immunocompromised patients. Systemic fungal infections are often fatal in patients undergoing antineoplastic chemotherapy for hematologic malignancies.[17,18] The most effective antifungal drug, AmpB, is highly nephrotoxic. Graybill et al.[19,20] used liposomal-AmpB (L-AmpB) in the treatment of experimental murine cryptococcosis and histoplasmosis. They were able to demonstrate that the efficacy of L-AmpB was maintained while its toxicity was reduced (17 times less toxic than free AmpB). Graybill et al.[19,20] found less antifungal activity when L-AmpB was given at the same dose as free AmpB. This was in contrast with the observations of Alving et al.[7] in leishmaniasis. Their results[7] suggested that liposomal delivery would allow for a reduction in the total dose of antimonials for effective treatment of leishmaniasis.

L-AmpB was also shown to be effective in the treatment of disseminated murine candidiasis in non-neutropenic mice.[21-23] When neutropenia, a condition known to facilitate fungal infection in humans, was induced in mice, L-AmpB was found to be even more effective than free AmpB at the same cumulative dose.[24] This could have been due to the administration of higher single doses in liposomal form. Although in vitro antifungal activity was observed with liposomes containing ergosterol, the activity was lower than that of liposomes containing phospholipids and cholesterol or phospholipids

Table 1. Treatment of disseminated fungal infections in patients with cancer[a]

Patient no.	Age/Sex	Underlying[b] disease/Status[c]	Fungal disease Diagnosis	Main Site	Previous Amp B dosage (mg)	No. of doses/concentration per dose (mg/kg)	L-AmpB Treatment Cumulative dose AmpB (mg)	Lipid (g)	Duration[d] (days) response
1	13/F	PDL(CR)	Aspergillus	Liver	6900	15/0.4-0.8	381	6.0	84/CR
2	22/M	ALL(R)	Aspergillus	Lung	3500	4/0.4-0.8	189	2.4	9/PR
3	22/F	APL(CR)	Candida	Liver	5000	15/0.4-0.8	501	10.1	45/CR
4	32/M	CML(R)	Aspergillus	Sinus	300	6/0.4-0.8	288	4.7	13/CR
5	48/M	HCL(AD)	Mucor	Liver	286	7/0.4-1.0	420	5.5	12/NR
6	35/M	CML(R)	Candida	Blood	661	6/0.4-0.8	200	2.8	12/PR
7	16/M	ALL(R)	Candida	Blood	201	4/0.4-0.8	145	1.47	8/NR
8	17/M	ALL(R)	Aspergillus	Sinus	280	16/0.4-1.0	816	6.77	52/PR
9	48/F	AML(R)	Aspergillus	Lungs	250	4/0.8-1.0	190	3.5	4/NR
10	23/M	ALL(R)	Aspergillus	Sinus	1680	43/0.8-1.5	3,100	66.5	52/PR
11	49/M	KS(AD)	Histoplasma	Lung	8900	35/0.8-1.0	2,136	57.3	35/PR
12	34/M	AML(R)	Aspergillus	Lung	700	6/0.8-1.2	509	8.6	7/NR

[a] Adapted from Reference 36.

[b] PDL, poorly differentiated lymphoma; ALL, acute lymphocytic leukemia; APL, acute progranulocytic leukemia; CML, chronic myelogenous leukemia; HCL, hairy cell leukemia; AML, acute myelogenous leukemia; KS, Kaposi sarcoma

[c] CR, complete remission; R, relapse; AD, active disease

[d] CR, complete remission; resolution of all signs and symptoms of infection for a minimum follow-up period of eight weeks; PR, partial remission; complete or partial resolution of symptoms, but with disease recurring within two months of onset of therapy; NR, no response

[e] Modified from Reference 36.

alone.[25] The presence of sterols, however, did not affect the antifungal efficacy of L-AmpB in murine candidiasis.[26] The enhanced therapeutic index observed with L-AmpB is probably related to maintained or increased antifungal activity while there was a decreased mammalian cell toxicity.[27]

Systemic fungal infections[28] affect the kidney, lung, spleen, liver, and bone marrow[29-31] which are also a site of preferential uptake of liposomes.[32,33] Thus, liposomal encapsulation promotes AmpB delivery to the organs[34] most frequently infected by these fungi. This may result in an enhanced interaction of AmpB with the yeast and pseudohyphal cells, particularly in the presence of damaged endothelium.[28]

In as much as there is a high frequency of fungal infections in patients with hematologic malignancies, particularly among those who are receiving chemotherapy, the use of L-AmpB as a prophylactic agent would likely prove beneficial. L-AmpB was shown to be effective in the prophylaxis against Candida albicans infections in non-neutropenic mice.[35] In these mice, one dose of L-AmpB (4 mg/kg) given five days prior to injection of C. albicans was effective against the development of a systemic infection (Fig. 1). Mice pretreated with L-AmpB were free of infection. This was not true for those pretreated with free AmpB.

Human trials. Twelve patients with hematologic malignancies complicated by systemic fungal infections were treated with L-AmpB.[36] All patients had biopsy- or culture-proven progression of the fungal infection while they were receiving conventional AmpB. Three patients were cured, five had partial resolution of signs or symptoms of the infection, and four had no response (Table 1). Two of the cured patients had hepatosplenic fungal infections; the third had maxillary sinus aspergillosis. Two additional partial responses were observed in patients with maxillary sinus aspergillosis. In the cases with hepatosplenic involvement, the higher concentrations of drug achieved with the L-AmpB in these organs may account for its successful use. In other situations such as the effective treatment of sinusitis (which was usually associated with severe facial swelling and erythema) other mechanisms may be involved. Capillary damage secondary to endothelial invasion may give rise to a "leaky" capillary, allowing better penetration of L-AmpB. It is worthwhile to point out that in all

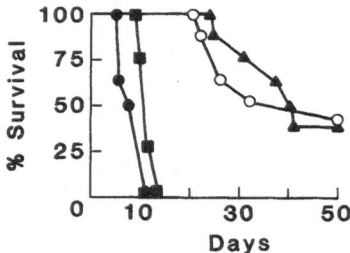

Fig. 1. Prophylactic effect of free and L-AmpB against
Candida albicans infection in mice. Groups of
eight to ten Hale-Stoner mice, six- to eight-
weeks old were inoculated intravenously with 7 x
10^5 cfu of C. albicans strain 336 previously
isolated from a patient with candidiasis. Untreated controls (-●-); free AmpB 0.8 mg/kg (-■-);
free AmpB 0.8 mg/kg daily x five days (-o-); L-AmpB
4 mg/kg (-▲-) (From Reference 24).

cases in which response was observed, improvement occured within 48-72 h following initiation of treatment.

L-AmpB also lacked the acute and chronic toxicity commonly seen with free AmpB. Temperature increases of greater than 2°C occurred in 10 occasions out of 160 doses administered. No problems with liver or kidney functions were observed. L-AmpB was administered in a dosage of 20-50 ml 0.9% NaCl solution over 15-30 min. This is an additional advantage of the liposomal preparation, since free AmpB usually requires volumes of 500-1000 ml and the concomittant administration of meperidine or dyphendramine to reduce side effects. This preliminary study is encouraging and shows the feasibility of using a liposomal carrier in clinical trials.

LIPOSOME-ENCAPSULATED MACROPHAGE ACTIVATORS IN ANTIMICROBIAL THERAPY

Liposomes are an efficient way of delivering macrophage activators to the MPS.[37-39] These observations have been extensively exploited by several investigators in order to treat a wide variety of experimental metastatic tumors.[40] The same approach has been successfully used in the prevention of experimental viral, fungal, and bacterial infections.[41-45] Using combinations of L-AmpB and a liposome-encapsulated muramyl dipeptide analogue we have demonstrated (submitted for publication) that lower doses of L-AmpB could be used to achieve a protection from candida infection in mice (Table 2). The survival of mice injected with lower doses of L-AmpB and low doses

Table 2. Effect of administration of L-6-0-S [abu^1] MDP and L-AmpB in the prophylaxis of C. albicans infection in mice[a]

L-AmpB (mg/kg)	L-6-0-S [abu^1] MDP[b] (mg/kg)	Survivors/ Total	Median survival (days)	Animals free of infection/ Alive at day 50
No treatment		0/8	3	-
"empty" liposomes		0/8	4	-
0	1.2	0/8	6	-
0.8	0	0/7	10	-
2.0	0	4/7	28	0/4
3.0	0	3/7	> 50	1/3
4.0	0	6/8	> 50	1/6
2.0	1.2	4/6	30	0/4
3.0	1.2	3/6	29	2/3
4.0	1.2	8/8	> 50	7/8

[a]Mice were inoculated with 7 x 10^5 cfu of C. albicans strain 335. Treatment was administreed two days prior to the yeast inoculum. Total lipid in the liposomal preparation given was 400 mg/kg body weight. Surviving mice were killed on day 50 and the organs were assessed microbiologically for Candida growth using Saboraud Dextrose agar.

[b]6-0-stearoyl-N-acetylmuramyl-L-α-aminobutyryl-D-isoglutamine 6-0-S [abu^1] MDP.

of the liposomal MDP analogue was as good as those treated with L-AmpB alone. These observations are encouraging since a synergistic effect of lower doses of both agents in liposomes will make this an attractive approach to the prevention of fungal infections in susceptible patients.

CONCLUSIONS AND PROSPECTS

L-AmpB seems to be superior to free AmpB in the treatment of systemic fungal infections. L-AmpB was effective in patients with fungal infections refractory to AmpB, was devoid of the side effects of AmpB, and was easier to administer. Further clinical trials are necessary to establish L-AmpB as a new drug in the antifungal armamentarium. It was, however, very encouraging to observe antifungal activity in the presence of reduced toxicity. This finding alone leads to a definite improvement of the therapeutic index. The evidence accumulated in the prophylactic use of the L-AmpB in experimental fungal infection indicates that it may provide a unique tool for the prevention of opportunistic infections in highly susceptible individuals.

It is evident from the experimental and clinical information now available that liposomes may be a safe and effective say of transporting antimicrobials. Clinical efficacy will likely be achieved in intracellular infections. Since higher doses of antimicrobials can be administered safely and effectively, the use of liposomal delivery systems becomes a very attractive means for the treatment of a variety of infectious diseases.

ACKNOWLEDGEMENTS

This work was funded in part by NIH-BRSG 5511 and a Scholar grant from the Leukemia Society of American to Dr. Lopez-Berestein.

REFERENCES

1. P.F. Bonventre and G. Gregoriadis, Killing of intraphagocytic Staphylococcus aureus by dihydrostreptomycin entrapped within liposomes, Antimicrob. Agents Chemother. 13:1049 (1978).
2. R.R.C. New, M.L. Chance, S.C. Thomas and W. Peters, Antileishmanial activity of antimonials entrapped in liposomes, Nature 272:55 (1978).
3. C.R. Alving, E.A. Steck, W.L. Hanson, P.S. Loizeauz, W.L. Chapman, Jr. and V.B. Waits, Improved therapy of experimental leishmaniasis by use of a liposome-encapsulated antimonial drug, Life Sci. 22:1021 (1978).
4. R.R.C. New and M.L. Chance, Treatment of experimental cutaneous leishmaniasis by liposome-entrapped Pentostam, Acta. Trop. 37:253 (1980).
5. J.V. Desiderio and S.G. Campbell, Intraphagocytic killing of salmonella typhimurium by liposome-encapsulated cephalothin, J. Infect. Dis. 148:563 (1983).
6. J.V. Desiderio and S.G. Campbell, Liposome-encapsulated cephalothin in the treatment of experimental murine salmonellosis, J. Reticuloendothel. Soc. 34:279 (1983).
7. C.R. Alving, Delivery of liposome-encapsulated drugs to macrophages, Pharmacol. Ther. 22:407 (1983).
8. J.S. Weldon, J.F. Munnell, W.L. Hanson and C.R. Alving, Liposomal chemotherapy in visceral leishmaniasis: An ultrastructural study of an intracellular pathway, Z. Parasitenkd. 69:415 (1983).
9. C.R. Alving, G.M. Swartz, Jr., L.D. Hendricks, W.L. Chapman, Jr., V.B. Waits and W.L. Hanson, Liposomes in leishmaniasis: Effects of

parasite virulence on treatment of experimental leishmaniasis in hamsters, Ann. Trop. Med. Parasitol. 78:279 (1984).

10. W.L. Chapman, Jr., W.L. Hanson, C.R. Alving and L.D. Hendricks, Antileishmanial activity of liposome-encapsulated meglumine antimonate in the dog, Am. J. Vet. Res. 45:1028 (1984).

11. R.R.C. New, M.L. Chance and S. Heath, Antileishmanial activity of amphotericin and other antifungal agents entrapped in liposomes, J. Antimicrob. Chemother. 8:371 (1981).

12. C.B. Panosian, M. Barza, F. Szoka and D.J. Wyler, Treatment of experimental cutaneous leishmaniasis with liposome-intercalated amphotericin B, Antimicrob. Agents Chemother. 25:655 (1984).

13. S.G. Reed, M. Barral-Netto and J.A. Inverso, Treatment of experimental visceral leishmaniasis with lymphokine encapsulated in liposomes, J. Immunol. 132:3116 (1984).

14. P. Pirson, R.F. Steiger, A. Trouet, J. Gillet and F. Herman, Primaquine liposomes in the chemotherapy of experimental murine malaria, Ann. Trop. Med. Parasitol. 74:383 (1980).

15. J.E. Smith and R.E. Sinden, Studies on the kinetics of uptake and distribution of free and liposome-entrapped primaquine and of sporozoites by isolated perfused rat liver, Ann. Trop. Med. Parasitol. 77:379 (1983).

16. P. Pirson, R. Steiger and A. Trouet, The disposition of free and liposomally encapsulated antimalarial primaquine in mice, Biochem. Pharmacol. 31:3501 (1982).

17. G.P. Bodey, V. Rodriguez, H.-Y. Chang and G. Narboni, Fever and infection in leukemic patients. A study of 494 consecutive patients, Cancer 41:1610 (1978).

18. L.S. Young, Nosocomial infections in the immunocompromised adult, Am. J. Med. 70:398 (1981).

19. R.L. Taylor, D.M. Williams, P.C. Draven, J.R. Graybill, D.J. Drutz and W.E. Magee, Amphotericin B in liposomes: A novel therapy for histoplasmosis, Am. Rev. Respir. Dis. 125:610 (1981).

20. J.R. Graybill, P.C. Carven, R.L. Taylor, D.M. Williams and W.E. Magee, Treatment of murine cryptococcosis with liposome-associated amphotericin B, J. Infect. Dis. 145:748 (1982).

21. J. Ahrens, J.R. Graybill, P.C. Carven and R.L. Taylor, Treatment of experimental murine candidiasis with liposome-associated amphotericin B, Sabouraudia: J. Med. Vet. Mycol. 22:163 (1984).

22. C. Tremblay, M. Barza, C. Fiore and F. Szoka, Efficacy of liposome-intercalated amphotericin B in the treatment of systemic candidiasis in mice, Antimicrob. Agents Chemother. 26:170 (1984).

23. G. Lopez-Berestein, T. McQueen and K. Mehta, Protective effect of liposomal-amphotericin B against C. albicans in mice, Cancer Drug Delivery, In press.

24. G. Lopez-Berestein, R.L. Hopfer, R. Mehta, E.M. Hersh and R.L. Juliano, Liposome-encapsulated amphotericin B for treatment of disseminated candidiasis in neutropenic mice, J. Infect. Dis. 150:278 (1984).

25. R.L. Hopfer, K. Mills, R. Mehta, G. Lopez-Berestein, V. Fainstein and R. L. Juliano, In-vitro antifungal activities of amphotericin and liposome-encapsulated amphotericin B, Antimicrob. Agents Chemother. 25:387 (1984).

26. G. Lopez-Berestein, R. Mehta, R. Hopfer, K. Mehta, E.M. Hersh and R. Juliano, Effects of sterols on the therapeutic efficacy of liposomal amphotericin B in murine candidiasis, Cancer Drug Deliv. 1:37 (1983).

27. R. Mehta, G. Lopez-Berestein, R. Hopfer, K. Mills and R.L. Juliano, Liposomal amphotericin B is toxic to fungal cells but not to mammalian cells, Biochim. Biophys. Acta 770:230 (1984).

28. B.D. Fisher, D. Armstrong, B. Yu and J.W.M. Gold, Invasive aspergillosis. Progress in early diagnosis and treatment, Am. J. Med. 71:571 (1981).

29. M.W. Degregorio, W.M. Lee, C.A. Linker, R.J. Jacobs and C.A. Ries,

Fungal infections in patients with acute leukemia, <u>Am. J. Med.</u> 73: 543 (1982).

30. H.S. Mirsky and J. Cuttner, Fungal infection in acute leukemia, <u>Cancer</u> 30:348 (1972).

31. R.L. Myerowitz, G.J. Pazin and C.M. Allen, Disseminated candidiasis. Changes in incidence, underlying disease, and pathology, <u>Am. J. Clin. Pathol.</u> 68:29 (1977).

32. L.P. Kasi, G. Lopez-Berestein, K. Mehta, M. Rosenblum, H.J. Glenn, T.P. Haynie, G. Mavligit and E.M. Hersh, Distribution and pharmacology of intravenous ^{99m}Tc-labeled multilamellar liposomes in rats and mice, <u>Int. J. Nucl. Med. Biol.</u> 2:35 (1984).

33. G. Lopez-Berestein, L. Kasi, M.G. Rosenblum, T. Haynie, M. Jahns, H. Glenn, R. Mehta, G.M. Mavligit and E.M. Hersh, Clinical pharmacology of ^{99m}Tc-labeled liposomes in patients with cancer, <u>Cancer Res.</u> 44:375 (1984).

34. G. Lopez-Berestein, M.G. Rosenblum and R. Mehta, Altered tissue distribution of amphotericin B by liposomal encapsulation: Comparison of normal mice to mice infected with <u>Candida albicans</u>. <u>Cancer Drug Deliv.</u> 1:199 (1984).

35. G. Lopez-Berestein, R. Mehta, R.L. Hopfer, K. Mills, L. Kasi, K. Mehta, V. Fainstein, M. Luna, E.M. Hersh and R. Juliano, Treatment and prophylaxis of disseminated infection due to <u>Candida albicans</u> in mice with liposome-encapsulated amphotericin B, <u>J. Infect. Dis.</u> 147:939 (1983).

36. G. Lopez-Berestein, V. Fainstein, R. Hopfer, K. Mehta, M.P. Sullivan, M. Keating, M.G. Rosenblum, R. Mehta, M. Luna, E.M. Hersh, J. Reuben, R.L. Juliano and G.P. Bodey, Liposomal amphotericin B for the treatment of systemic fungal infections in patients with cancer: A preliminary study, <u>J. Infect. Dis.</u> 151:704 (1985).

37. G. Lopez-Berestein, K. Mehta, R. Mehta, R.L. Juliano and E.M. Hersh, The activation of human monocytes by liposome-encapsulated muramyl dipeptide analogues, <u>J. Immunol.</u> 130:1500 (1983).

38. K. Mehta, G. Lopez-Berestein, E.M. Hersh and R.L. Juliano, Uptake of liposomes and liposome-encapsulated muramyl dipeptide by human peripheral blood monocytes, <u>J. Reticuloendothel. Soc.</u> 32:155 (1982).

39. K. Mehta, R.L. Juliano and G. Lopez-Berestein, Stimulation of macrophage protease secretion via liposomal delivery of muramyl dipeptide derivatives to intracellular sites, <u>Immunol.</u> 51:517 (1984).

40. G. Lopez-Berestein, L. Milas, N. Hunter, K. Mehta, E.M. Hersh, C.G. Kurahara, M. VanderPas and D.A. Eppstein, Prophylasis and treatment of experimental lung metastases in mice after treatment with liposome-encapsulated 6-0-stearoyl-N-acetylmuramyl-L-alpha-amino-butyryl-D-isoglutamine, <u>Clin. Expl. Metastasis</u> 2:127 (1984).

41. W.C. Koff, I.J. Fidler, S.D. Showalter, M.K. Chakrabarty, B. Hampar, L.M. Ceccorulli and E.S. Kleinerman, Human monocytes activated by immunomodulators in liposomes lyse herpesvirus-infected but not normal cells, <u>Science</u> 224:1007 (1984).

42. LaBonnardiere C. Donnees preliminaires sur l'effet protecteur de l'interferon couple a des liposomes dans le modele souris-virus de l'hepatite murine, <u>Ann. Microbiol.</u> 129:397 (1978).

43. W.C. Koff, Protection of mice against fatal herpes simplex type 2 infection by liposomes containing muramyl tripeptide, <u>Science</u> 228: 495 (1985).

44. W.C. Koff, S.D. Showalter, D.A. Seniff and B. Hampar, Lysis of herpes-virus-infected cells by macrophages activated with free or liposome-encapsulated lymphokine produced by a murine T cell hybridoma, <u>Infect. Immun.</u> 42:1067 (1983).

45. E.B. Fraser-Smith, D.A. Eppstein, M.A. Larsen and T.R. Matthews, Protective effect of a muramyl dipeptide analog encapsulated in or mixed with liposomes against <u>Candida albicans</u> infection, <u>Infect. Immun.</u> 39:172 (1983).

DESIGN, CHARACTERIZATION AND ANTI-TUMOR ACTIVITY OF ADRIAMYCIN-CONTAINING

PHOSPHOLIPID VESICLES

Alberto Gabizon,* Dorit Goren, Avner Ramu, and Yechezkel
Barenholz**

Hadassah University Hospital and Hebrew University-Hadassah
Medical School**, Jerusalem, Israel
*Correspondence: Cancer Research Institute, 1282-M, School
of Medicine, University of California, San Francisco, CA
94143, USA

SUMMARY

In this report we describe the design of adriamycin (ADM)-containing
liposome preparations aiming at optimization of various pre-established
parameters. Regarding liposome composition phosphatidylserine (PS) and
phosphatidylglycerol (PG) appear to be suitable negatively charged phospholipids which combined with phosphatidylcholine (PC) and cholesterol (CHOL),
confer to liposomes high loading capacity for ADM and reasonable stability
in plasma. Intravenous administration of these negatively-charged liposomes resulted in a favorable tissue distribution of ADM in both normal and
tumor-bearing mice, characterized by decreased cardiac uptake of drug, and
increased and sustained drug levels in the liver. Moreover, enhanced accumulation of drug also occurred in metastatic tumor cells isolated from
the liver when ADM was injected in the liposome-associated form. This
passive drug targeting resulted in an improved therapeutic efficiency of
liposome-associated ADM in a tumor model of liver metastases. Liposome
delivery of ADM was also shown to increase significantly its cytoreductive
effect on spleen-infiltrating leukemia cells and to maintain the same cytoreductive efficiency on bone marrow residing leukemia cells with an overall
favorable effect on survival in the BCL1 leukemia model. The reduction of
ADM toxicity by liposome association together with the anti-tumor results
indicate that liposomes represent a useful drug-delivery system for the
treatment of major neoplastic conditions.

INTRODUCTION

Administration of liposome-encapsulated drugs results in important
changes in the tissue distribution and pharmacokinetics when compared to
the pattern of the respective free drugs (Rahman and Wright, 1975; Kimelberg et al., 1976: Juliano and Stamp, 1978). These changes may have profound implications with regard to the tissue-specific toxicity of some
cytotoxic drugs. Examples of tissue-specific toxicities are the pulmonary
fibrosis caused by bleomycin, (Blum et al, 1973) the vincristine-induced
neuropathy (Rosenthal and Kaufman, 1974) and the cardiotoxicity of anthracyclines (Minow et al., 1975; Von Hoff et al., 1979). Among these, the

mechanism of the cardiomyopathy induced by anthracyclines and ways of preventing it have been the subject of extensive research since the first reports on the cardiotoxic effect of these drugs. The myocardial damage induced by the anthracyclines is cumulative and irreversible and severely restricts the clinical use of these drugs. One of the strategies aimed at overcoming the ADM induced cardiotoxicity may involve the delivery of ADM by a carrier capable of preventing the drug from reaching the heart muscle. Liposomes could adequately fulfill this task given their relative inability to cross continuous capillaries and the lack of phagocytic reticuloendothelial cells in the myocardial tissue (Poste, 1983). Moreover, liposomes offer the opportunity of targeting their contents to the liver, a common site of metastatic disease (Segal et al., 1974; Poste et al., 1982).

In the present study, we summarize the results of a systematic approach designed at engineering an optimal liposome composition for ADM delivery. The successive steps involved in this approach are aimed at selecting one or various liposome compositions that would fulfil the following criteria: high efficiency of drug capture; preservation of the full biological activity of the drug after liposome encapsulation as shown by in-vitro cytotoxicity tests; stability in the presence of plasma as indicated by retention of > 80% of the liposome-associated drug after 1 hour of incubation; favorable tissue distribution pattern with reduced uptake of liposome-associated ADM in the heart muscle and increased level in tumor cells infiltrating potential target organs accessible to liposomes such as the liver; decreased toxicity including cardiotoxicity; increased therapeutic index in a variety of relevant tumor models. Our results indicate that properly designed ADM-liposome preparations can significantly improve the therapeutic index of ADM in selected neoplastic processes.

MATERIAL AND METHODS

BALB/c and C57BL mice from the Animal Breeding Center of the Hebrew University (Jerusalem, Israel) were used throughout this study. The tumor cell lines used in these experiments included J6456 lymphoma, BCL1 leukemia, and P388 leukemia cells. Either multilamellar vesicles (MLV) or probe-sonicated liposomes were used in this study. Detailed descriptions of the preparations have been reported previously (Gabizon et al., 1982 and 1985). Extraction and fluorometric quantitation of ADM was done as described previously (Gabizon et al., 1982). Fractionation of metastatic J-6456 cells from the liver was done on Percoll gradients as described by Gabizon et al. (1983).

RESULTS

1. Effect of liposome composition on drug capture. The results obtained with different lipid compositions of sonicated and unsonicated preparations are presented in Table 1. Clearly, the entrapment of ADM was much higher when negatively-charged phospholipids such as cardiolipin(diphosphatidylglycerol) (DPG), PS or PG were used, as compared to neutral liposomes. A four-to-five-fold raise in the amount of captured drug could be achieved by substituting 20 to 30% of the PC by a negatively-charged phospholipid. The addition of CHOL did not change significantly the amount of entrapped ADM per mol phospholipid. Interestingly semisynthetic phospholipids with saturated fatty acid chains such as DPPG and DPPC were significantly less efficient than natural phospholipids in the capture of ADM. Finally SUV showed a reduced ability for drug entrapment as compared with other types of liposomes of the same composition.

Table 1. Efficiency of ADM Entrapment in Liposomes[a]

Liposome Composition	Initial Molar Ratio	ADM-Entrapment	
		% Total	mmol ADM/mol phospholipid
1) Non-sonicated (MLV)[b]			
PC:CHOL:ADM	4:4:1	14	35
PC:ADM	4:1	10	26
DPG:PC:CHOL:ADM	1:4:5:1	64	128
DPG:PC:ADM	1:4:1	58	116
PS:PC:CHOL:ADM	3:7:10:2.2	59	129
2) Sonicated[c]			
PC:CHOL:ADM	5:5:0.5	15	16
DPG:PC:CHOL:ADM	1:4:2:1	47	94
DPG:PC:CHOL:ADM	1:4:5:1	45	90
DPG:CHOL:ADM	5:2.5:1	90	180
PS:PC:CHOL:ADM	3:7:4:2	50	100
PG:PC:CHOL:ADM	3:7:4:2	61	128
DPPG:PC:CHOL:ADM	3:7:4:2	39	78
DPPG:DPPC:CHOL:ADM	3:7:4:2	28	56
3) Fractionated SUV[d]			
PS:PC:CHOL:ADM	3:7:4:2:	22	49

(a) Phospholipid, 20-40 μ mol/ml; ADM, 2-5 mg/ml.
(b) Size range, 0.5 - 5 μ m.
(c) Relatively heterogeneous population, containing oligo- and unilamellar vesicles, size range 20-200 nm.
(d) Homogeneous population of vesicles (SUV) obtained from the supernatant of sonicated vesicles after ultracentrifugation. Size < 50 nm.

2. **Effect of liposome composition on the stability of liposome-entrapped drug in the presence of plasma.** An important requirement for the use of liposomes as in-vivo carriers is their ability to retain the entrapped drug in the presence of whole blood or its constituents. Table 2 presents the result obtained with a variety of liposome preparations incubated for 1 h at 37°C in 50% plasma. Cardiolipin-containing liposomes were very unstable releasing more than 60% of the encapsulated drug whether or not CHOL was present. Other types of liposomes tested were much more stable especially if CHOL was present. This effect of CHOL in diminishing leakage of ADM from the vesicles was less marked in PS- and PG-containing liposomes.

3. **In vitro cytotoxicity of liposome-encapsulated drug.** Negatively charged and neutral liposomes were found to be non-toxic to in-vitro cultures of J-6456 and P388 tumor cells at lipid concentrations required to achieve a reasonable dose range of ADM (10^{-8} to 10^{-6}M) (data not shown). When ADM-loaded liposomes were used, the IC_{50} of the encapsulated drug was not significantly different from that of free drug. An example is presented in Fig. 1. We concluded that the cytotoxic activity of ADM is not hindered by association with liposomes, nor do liposomal lipids contribute significantly.

Table 2. Release of Adriamycin (ADM) in the Presence of Plasma[a]

Liposome Composition	Molar Ratio	% Liposome-Retained ADM
DPG:PC:CHOL (ADM)	1:4:5	39
DPG:PC:CHOL (ADM)	1:4:0	28
PS:PC:CHOL (ADM)	3:7:10	98
PS:PC:CHOL (ADM)	3:7:0	62
PC:CHOL (ADM)	4:4	72
PC:CHOL (ADM)	4:2	62
PC:CHOL (ADM)	4:1	76
PC:CHOL (ADM)	4:0	29
PG:PC:CHOL (ADM)	3:7:5	99
PG:PC:CHOL (ADM)	3:7:4	96
PG:PC:CHOL (ADM)	3:7:2.5	81
PG:PC:CHOL (ADM)	3:7:0	70

(a) Liposomes incubated for 1 h at 37°C in the presence of 50% human
plasma. Liposome-retained drug separated from released drug either
by sephadex G-50 filtration or by ultracentrifugation.

Table 3. Biodistribution of L-ADM and F-ADM in normal mice[a]

Liposomes	μ g ADM-equivalents/g wet tissue weight		
	Liver	Spleen	Heart
F-ADM	2.6	1.3	1.14
L-ADM	13.8	29.4	0.28

(a) BALB/c mice injected i.v. with 4 mg/kg ADM in either free (F-ADM) or
liposome-associated form (L-ADM). Non sonicated PS:PC:CHOL:ADM vesicles were used in this experiment. ADM was extracted from tissues
3 h after injection and measured fluorometrically as described by
Gabizon et al. (1982).

4. <u>Tissue Distribution Studies in Normal Mice</u>. Given the low entrapment
efficiency of neutral PC liposomes and the poor stability of DPG-containing
liposomes, vesicles containing other negatively-charged phospholipids were
selected for further studies. The results shown in Table 3 indicate that
the use of PS-containing liposomes brings about a reduced uptake of ADM by
the heart muscle. Similar changes in tissue distribution were observed
with PG-containing liposomes (data not shown).

Table 4. Biodistribution of F-ADM and L-ADM in tumor-bearing mice[a]

Liposomes	μg ADM-equivalents[b]/g wet tissue weight							
	1h				24h			
	Liver	Spleen	Heart	Tumor[c]	Liver	Spleen	Heart	Tumor[c]
F-ADM	13.72	6.39	15.59	30	3.57	5.54	6.10	16
L-ADM	43.68	52.62	3.93	204	26.62	18.73	0.82	116

(a) BALB/c mice inoculated 14 days previously with 10^6 J-6456 lymphoma cells, received i.v. 10 mg/kg of ADM in either free or liposome-associated form. Sonicated PG:PC:CHOL vesicles were used in this experiment.

(b) ADM and its metabolites were extracted from tissues and measured fluorometrically as described in Gabizon et al., 1982.

(c) Metastatic J-6456 cells infiltrating the liver were isolated as described by Gabizon et al. (1983). Results expressed in ng ADM-equivalents per 10^7 tumor cells.

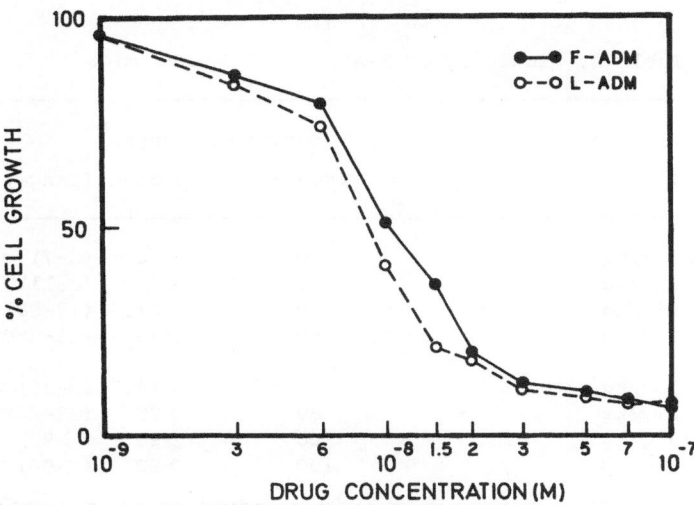

Fig. 1. Cytotoxicity of F-ADM and L-ADM on P388 leukemia cells after 72 h of continuous exposure. Results expressed as percentage of control cell counts. F-ADM, IC50 = 1.6 x 10^{-8}M; L-ADM, IC50 = 1.1 x 10^{-8}M. Sonicated PG:PC:CHOL:ADM vesicles were used in this experiment.

5. Tissue Distribution Studies in Tumor-bearing Mice. We next investigated whether the same pattern of tissue distribution is observed in tumor-
bearing mice. For this purpose mice were inoculated with 10^6 J-6456
lymphoma cells and after reaching a tumor burden of 10^8 cells, equal doses
of ADM in free or liposome-encapsulated form were injected. In addition
to fluorometric quantitation of ADM in various organs, cell fractionation
was performed after enzymatic digestion of the liver using Percoll density
gradients (Gabizon et al., 1983). A cell fraction composed of more than
95% J-6456 tumor cells as demonstrated by fluorescence staining with monoclonal anti-Thy 1.2 and by morphological appearance in May-Gruenwald-
Giemsa stained smears, was obtained. An example of one of such experiment
using PG-containing liposomes is presented in Table 4. As found in normal
mice, liver and spleen drug levels were increased while heart muscle drug
levels were decreased using the liposome-encapsulated form. The ADM levels
were also significantly increased in the intrahepatic metastatic tumor cells
indicating that delivery via liposomes could achieve a drug-concentration
effect on liver metastases.

6. Toxicity of the liposome-associated drug. Obviously, one of the most
important goals to achieve with a drug delivery system in cancer therapy
is the reduction of toxicity. Indeed, the systemic toxicity of ADM can be
significantly reduced using liposomes as carriers. As shown in Table 5,
ADM delivered by PG:PC:CHOL liposomes was markedly less toxic than soluble
ADM by a factor of twofold approximately. Two phases of toxicity were
recognized: an acute toxic death occurring within two weeks from injection,
and a delayed toxic death generally observed between 1 to 3 months after
injection. Similar types of liposomes have also been shown to protect
against ADM-induced cardiotoxicity (Olson et al., 1982; Gabizon et al.,
submitted for publication).

Thus, liposome-associated ADM, appears to be a useful system to achieve
reduction of the acute toxicity and to circumvent the problem of cardiotoxicity, a major manifestation of chronic toxicity.

Table 5. Toxicity of F-ADM and L-ADM in mice[a]

Type of Treatment	Survival (days)[b]	
	Percent	Median (range)
F-ADM, 25mg/kg	0	6 (6-7)
F-ADM, 20mg/kg	0	8 (8-11)
F-ADM, 15mg/kg	0	27.4 (17-86)
F-ADM, 10mg/kg	80	>90 (42->90)
L-ADM, 25mg/kg	0	14.5 (9-30)
L-ADM, 20mg/kg	60	>90 (23->90)
L-ADM, 15mg/kg	100	>90 (>90)
L-ADM, 10mg/kg	100	>90 (>90)

(a) BALB/c mice injected i.v. with ADM in either free or liposome-assoc-
    iated form. Sonicated PG:PC:CHOL vesicles were used in this experi-
    ment.
(b) Follow-up, 90 days. Differences between F-ADM and L-ADM treatments
    were statistically significant at the $p < 0.01$ level (Wilcoxon test)
    for the 10, 15 and 20 mg/kg doses.

Table 6. Superior Therapeutic Activity of L-ADM in the J-6456
          Metastatic Tumor Model[a]

| Experiment No. | Survival time (days) | | | P[b] |
	Median	(range)	Mean ± S.E.	
1. Untreated	22	(21-28)	23.3 ± 0.8	
F-ADM	35.2	(31-60)	38.6 ± 2.2	< 0.01
F-ADM+PS:PC:CHOL	34	(30-55)	39.0 ± 2.9	n.s.
PS:PC:CHOL:ADM	48	(36-70)	49.2 ± 2.9	< 0.01
2. Untreated	18.3	(18-23)	19.4 ± 0.7	
F-ADM	31.4	(29.35)	32.4 ± 0.6	< 0.01
PS:PC:CHOL (25%):ADM	42	(37-59)	44.1 ± 1.9	< 0.01
PS:PC:CHOL (100%):ADM	45	(24-64)	47.8 ± 3.9	n.s.
PG:PC:CHOL (25%):ADM	44.5	(38-55)	46.0 ± 1.8	n.s.
PG:PC:CHOL (100%):ADM	46.8	(35-51)	46.3 ± 2.1	n.s.

(a) BALB/c mice inoculated with 10^6 J-6456 cells and treated i.v. 3 days
     later with 8mg/kg ADM in either free or liposome-associated form.
     Treatment schedules: Expt. 1: day 3, 10, 17, 36, 46; Expt. 2: day 3,
     10, 17.
(b) Wilcoxon test; n.s. = not significant.

7. <u>In-vivo anti-tumor efficacy.</u> The therapeutic activity of liposome
associated ADM was examined using the J-6456 tumor model with predominant
dissemination to the liver. The results of a representative experiment
are shown in Table 6. The administration of L-ADM resulted in a significant prolongation of survival when compared to the effect of free ADM.
Association of ADM was required for the enhancement of the anti-tumor effect
to occur since free drug mixed with plain liposomes was not more effective
than free drug alone. There was no significant difference in the anti-
tumor activity using either PS or PG liposomes and decreasing the CHOL:
phospholipid ratio from 100% to 25%. Another tumor model in which ADM
liposomes were tested is the BCL1 leukemia. After i.v. inoculation of
BCL1 cells, the main anatomic compartments becoming infiltrated are the
spleen and the bone marrow. Adoptive transfer assays of spleen cells and
bone marrow cells from F-ADM and L-ADM treated BCL1-inoculated mice indicated that spleen-residing tumor cells were killed more effectively with
L-ADM, while bone-marrow-residing tumor cells were killed to the same
degree by L-ADM and F-ADM (Fig. 2). When survival of BCL1-inoculated mice
was examined, L-ADM was significantly superior to F-ADM (Table 7). Sonicated vesicles were used in the anti-tumor experiments for two main reasons:
the penetration of small vesicles into the hepatic parenchyma is more
efficient than that of MLV (Poste et al, 1982) a critical factor for the
delivery of drug to liver metastatic tumors; and sonicated vesicles could
be easily sterilized by filtration through 0.2 μm polycarbonate membranes.

DISCUSSION

     As shown by several groups of investigators (reviewed in Mayhew and
Papahadjopoulos, 1983), liposomes represent a potentially useful delivery

Table 7. Superior Therapeutic Activity of L-ADM in the BCL1 Leukemia Model[a]

Treatment	Survival time (days)		P[b]
	Median (range)	Mean ± S.E.	
Untreated	21.8 (20-40)	24.9 ± 2.0	
F-ADM	63.4 (22-75)	60.0 ± 5.8	< 0.01
L-ADM	81.0 (68-144)	85.1 ± 5.9	< 0.01

(a) BALB/c mice inoculated with 10^6 BCL1 leukemia cells and treated 4 days later with 8 mg/kg ADM in either free or liposome-associated form. Sonicated PS:PC:CHOL:ADM vesicles were used in this experiment. Treatment schedule: days, 4, 11, 18.
(b) Wilcoxon test.

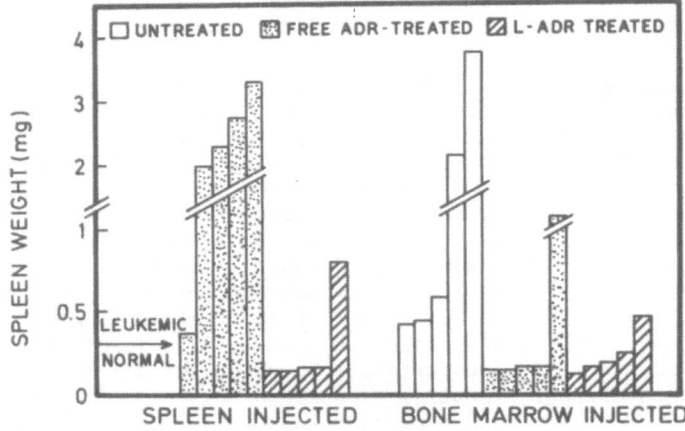

Fig. 2. Comparative bioassay of the anti-tumor activities of L-ADM and F-ADM on spleen and bone-marrow residing BCL1 leukemia cells. Mice inoculated with BCL1 cells were treated with either F-ADM or L-ADM, 10 mg/kg, i.v. Forty-eight hours later, spleen and bone-marrow cells from treated and untreated mice were transferred i.v. into syngeneic recipients. Spleen weight as indicator of leukemia was recorded 7 weeks after transfer. The presence of leukemia in all spleens with weights above the normal/leukemic cut-off line was confirmed histopathologically. All mice injected with spleen cells from untreated animals were dead by the time of examination and therefore are not displayed. Sonicated PS:PC:CHOL:ADM were used in this experiment.

system for ADM. This is mainly derived from the ability of liposomes to concentrate their contents in accessible tissues, while bypassing drug-sensitive tissues such as the heart muscle in the case of ADM. In the present study, we have followed a systematic approach to achieve optimally engineered vesicles for the delivery of ADM. Stable vesicles with a high efficiency of drug capture, favorable tissue distribution reduced toxicity and enhanced anti-tumor effect in proper in-vivo anti-tumor models were designed. Reduced cardiotoxicity of L-ADM has been shown by several groups (Rahman et al., 1980; Forssen and Tokes, 1981; Olson et al., 1982; Van Hoesel et al., 1984) and is probably the result of a diminished drug uptake. Attenuation of the systemic and cardiac toxicity of ADM is by itself most important and exploitable in cancer therapy. This is in contrast to strictly phase-specific drugs such as ara-C and methotrexate in which liposomes tend to increase toxicity (Mayhew and Papahadjopoulos, 1983).

Obviously, if liposomes are to be useful therapeutically, the anti-tumor activity of ADM must be preserved to the extent that an improved therapeutic index can be achieved. Because of the particular aspects determining the tissue distribution of liposome-associated drugs, one can anticipate that tumor responses will vary according to site. In this and other reports, (Gabizon et al., 1983) we have shown that L-ADM is more effective than F-ADM at equal doses on liver and spleen residing tumor cells. However, it has also been found (Mayhew et al., 1983; Gabizon et al., 1985) that s.c. implanted tumors were more sensitive to free than liposomal drug, thus illustrating the site-sepcific anti-tumor responses. The difficulty in predicting the anti-tumor efficacy of the liposome-interaction of ADM with membranes (Goormaghtigh and Ruysschaert, 1984) and its consequences on drug transfer, make prohibitive any generalization of these results. However, within the context of selected neoplastic conditions, such as liver metastases, the possibility of a passive target-ing of ADM liposomes represents an attractive area for initial clinical testing.

ACKNOWLEDGEMENTS

Supported by the Israel National Council for Research and Development and by the Robert Szold Institute.

REFERENCES

Blum, R.H., Carter, S.K., and Agree, K., 1973, A clinical review of bleo-mycin: a new antineoplastic agent, Cancer, 31:903.
Forssen, E.A., and Tokes, Z.A., 1981, Use of anionic liposomes for the reduction of chronic doxorubicin-induced cardiotoxicity, Proc. Natl. Acad. Sci. U.S.A., 78:1973.
Gabizon, A., Dagan, A., Goren, D., Barenholz, Y., and Fuks, Z., 1982, Lipo-somes as in-vivo carriers of Adriamycin: reduced cardiac uptake and preserved anti-tumor activity in mice, Cancer Res., 42:4734.
Gabizon, A., Goren, D., Fuks, Z., Barenholz, Y., Dagan, A., and Meshorer, A., 1983, Enhancement of adriamycin delivery to liver metastatic cells with increased tumoricidal effect using liposomes as drug carriers, Cancer Res., 43:4730.
Gabizon, A., Goren, D., Fuks, Z., Meshorer, A., and Barenholz, Y., 1985, Superior therapeutic activity of liposome-associated adriamycin in a murine metastatic tumor model, Br. J. Cancer, 51:681.
Goormaghtigh, E. and Ruysschaert, J.M., 1984, Anthracycline glycoside membrane interactions, Biochim. Biophys. Acta, 779:271.
Juliano, R.L. and Stamp, D., 1978, Pharmacokinetics of liposome-encapsulated anti-tumor drugs, Biochem. Pharmacol., 27:21.

Kimelberg, H.K., Tracy, T.F., Biddlecome, S.M., and Bourke, R.S., 1976, The effect of entrapment in liposomes on the in vivo distribution of ^{3}H-methotrexate in a primate, <u>Cancer Res.</u>, 36:2949.

Mayhew, E. and Papahadjopoulos, D., 1983, Therapeutic applications of liposomes, <u>in</u>: "Liposomes," M.J. Ostro, ed., Marcel Dekker, New York.

Mayhew, E., Rustum, Y., and Vail, W.J., 1983, Inhibition of liver metastases of M5076 tumor by liposome-entrapped adriamycin, <u>Cancer Drug. Deliv.</u> 1:43.

Minow, R.A., Benjamin, R.S., Gottlieb, J.A., 1975, Adriamycin (NSC-123127)-cardiomyopathy: an overview with determination of risk factors, <u>Cancer Chemother</u>. Rep., 6:195.

Olson, F., Mayhew, E., Maslow, D., Rustum, Y., and Szoka, F., 1982, Characterization, toxicity and therapeutic efficacy of adriamycin encapsulated in liposomes, <u>Eur. J. Cancer Clin. Oncol.</u>, 18:167.

Poste, G., Bucana, C., Raz, A., Bugelski, P., Kirsh, R., and Fidler, I.J., 1982, Analysis of the fate of systemically administered liposomes and implications for their use in drug delivery, <u>Cancer Res.</u>, 42:1412.

Poste, G., 1983, Liposome targeting in vivo: problems and opportunities, <u>Biol. Cell.</u>, 47:19.

Rahman, Y., Wright, B.J., 1975, Liposomes containing chelating agents: Cellular penetration and a possible mechanism of metal removal, <u>J. Cell.Biol.</u>, 65:112.

Rahman, A., Kessler, A., More, N., Sikic, B., Rowden, G., Woolley, P., and Schein, P.S., 1980, Liposomal protection of adriamycin-induced cardiotoxicity in mice, <u>Cancer Res.</u>, 40:1532.

Rosenthal, S. and Kaufman, S., 1974, Vincristine neurotoxicity, <u>Ann. Intern. Med.</u>, 80:733.

Segal, A.W., Wills, E.J., Richmond, J.E., Slavin, G., Black, C.D.V., and Gregoriadis, G., 1974, Morphological observations on the cellular and subcellular destination of intravenously administered liposomes, <u>Br. J. Exp. Pathol.</u>, 55:320.

Van Hoesel, Q.G.C.M., Steerenberg, P.A., Crommelin, D.J.A., Van Dijk, A., Van Oort, W., Klein, S., Douze, J.M.C., deWildt, D.J., and Hillen, F.C., 1984, Reduced cardiotoxicity and nephrotoxicity with preservation of antitumor activity of doxorubicin entrapped in stable liposomes in the LOU/M Wsl rat, <u>Cancer Res.</u>, 44:3698.

Von Hoff, D.D., Layard, M.W., Basa, P., Davis, H.L., Von Hoff, A.L., Rozencweig, M., and Muggia, F.M., 1979, Risk factors for doxorubicin-induced congestive heart failure, <u>Ann. Intern. Med.</u>, 91:710.

PARTICLE CHARGE AND SURFACE HYDROPHOBICITY OF COLLOIDAL DRUG CARRIERS

R.H. Muller,[1] S.S. Davis,[1] L. Illum,[2] and E. Mak[1]

[1]Pharmacy Department, University of Nottingham
 Nottingham NG7 2RD
[2]Pharmaceutics Department, Royal Danish School Pharmacy
 2 Universitetsparken, Denmark

INTRODUCTION

The fate of colloidal particles in the body is determined by different factors such as particle size and shape,[1-8] particle charge[9-22] and surface hydrophobicity.[23-32] Among them, surface hydrophobicity seems to play an important role in particle phagocytosis. Particles with different surface charges will show, after incubation with serum, a similar negative charge[9,33,34] of about -11 mV to -18 mV (calculated as the Zeta potential). This is due to the adsorption of serum components which shield the original charge and create a new one. Therefore, for subsequent interactions with macrophages, i.e. clearance by the reticuloendothelial system (RES), the colloidal particles may have similar surface charges but not similar surface hydrophobicity. The repulsive forces between negatively charged macrophages and the negatively charged particles will thus be normally the same, but due to different van der Waals (hydrophobic) interactions the attractive forces will be of different magnitudes. The size of these attractive forces is proportional to the hydrophobicity of the particle surface layer of blood components (opsonisation).

The composition of the adsorbed layer itself will be influenced by two factors: (a) the initial surface hydrophobicity and surface polarity of the particle (the higher the hydrophobicity and the lower the polarity the higher will be the amount of more lipophilic blood components adsorbed;[35] (b) the sign and magnitude of particle charge (opsonic components with the same sign of charge will be adsorbed less, oppositely charged ones will be adsorbed more). Thus, differently charged particles can have adsorption layers with different compositions. However, it should be pointed out that the opsonisation process and the composition of the adsorbed layer is mainly determined by surface hydrophobicity and polarity. Hydrophilic particles will demonstrate less opsonisation and less clearance by the RES than more hydrophobic ones.[36]

Various papers have reported on the dependance of particle phagocytosis by macrophages on the particle charge.[9-22] For liposome systems the variations in charge can be achieved by changing the composition of the phospholipid layer while for solid carriers the addition of different charged surface groups can be undertaken. However, such procedures will not only change the surface charge but also the surface hydrophobicity and polarity.

The question is then how much of the measured effect is due to the change in charge and how much is due to the simultaneous alteration (but not measured) in hydrophobicity and polarity? Changes in hydrophobicity and polarity may be an explanation for contradictory results concerning the influence of phagocytosis by particle charge.

To elucidate the dependance of phagocytosis on particle properties it is necessary to characterise particles in terms of (a) particle size, size distribution (and shape); (b) particle charge; (c) surface hydrophobicity and polarity. Monodisperse latex particles (polymer microspheres) with different properties can be used as model systems for studying the different methods of characterisation.

PARTICLE CHARGE

For the interactions between particles and blood components or macrophages, it is not the charge on the particle surface but the charge on the effective surface which is the relevant parameter. This charge is influenced by the charge on the surface and the concentration and valency of the ions in the surrounding medium. Most particles in dispersion (solid particles or emulsion droplets) possess a negative surface charge due to ionization of surface groups or to the dispersion process itself. In salt solutions the particles possess an adsorbed monolayer of ions on their surfaces (Fig. 1). This layer consists of fixed, dehydrated and (in most cases) negative ions (even if the surface itself is negatively charged) and is called the inner Helmholtz plane.[37-45] The negative ions are preferentially adsorbed because of their lower degree of hydration in comparison to the positive ions, leading to stronger van der Waals attractive forces between particle and negative ions. During the adsorption process they lose their water of hydration. Due to the adsorption of negative ions the negative surface potential ψ_0 will be increased. The next monolayer of adsorbed ions (the outer Helmholtz plane) consists of fixed, hydrated, positive ions which decrease the negative potential of the inner Helmholtz plane to ψ_δ. The inner and outer Helmholtz plane together are called the Stern plane,[42] the associated potential is the Stern potential ψ_δ whereby δ is the thickness of the Stern plane. The adsorbed positive ion layer cannot level out the charge of the negative layer because of the large volume of the hydrated positive ions. An exponential decay of the potential to zero occurs in the so-called diffuse electrical double layer which contains positive ions and an increasing number of negative ions as the distance from the particle surface increases. The ions in this layer are not fixed and are hydrated. The thickness of the diffuse layer is $1/\varkappa$ whereby $\varkappa$ is the Debye Huckel parameter. When a particle moves in a dispersion medium a part of the diffuse layer will be stripped off, and the effective surface of the particle is the surface of shear. The effective potential for interactions with other particles is the potential at the shear plane (the so called Zeta potential, ZP[44,45]). The higher the Zeta potential, the higher are the repulsive forces towards particles with the same sign of charge. Higher electrolyte concentrations or a higher valency of the ions will cause a faster decay of the potential in the diffuse layer and thereby reduce the zeta potential and the repulsive forces.

There are two different forces of interaction between particles: attractive van der Waals and repulsive electrostatic forces (same sign of charge). The resulting total potential energy of interaction depends on the distance from the particle surface and can be calculated according to the DLVO theory (Derjaguin, Landau, Verwey and Overbeek). The shape of the potential energy curve determines whether the particles can approach each other and aggregate, e.g. whether a negatively charged carrier can adhere to a negatively charged macrophage (not considering the hydrophobic

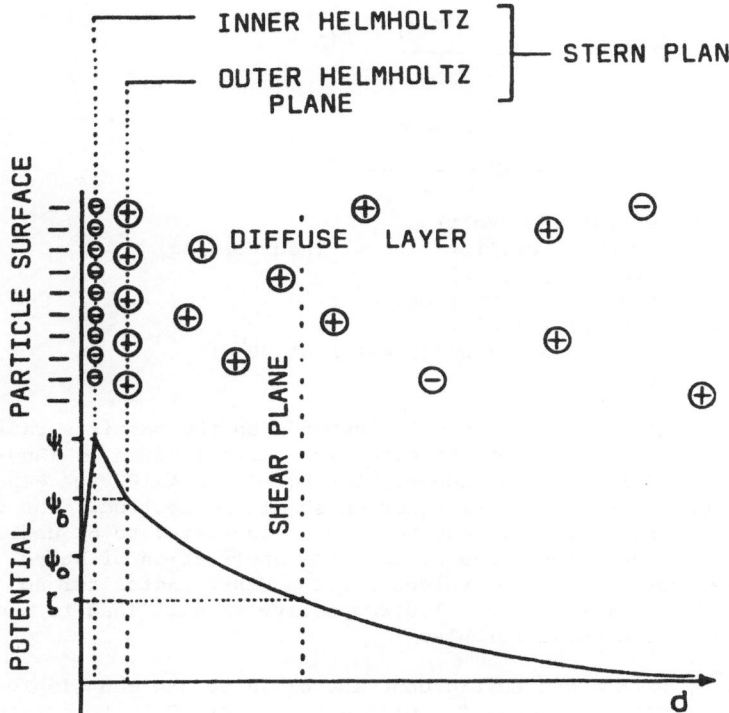

Fig. 1. Composition of the ion layers on a negative
         particle surface and the dependance of the
         potential curve on the distance from the
         particle.

interactions). The repulsive forces create an energy barrier which must be
overcome before particles can aggregate or cells and particles can adhere.
The height of this energy barrier is proportional to the magnitude of the
Zeta potential, which itself depends on the charge on the particle surface
and the concentration and the type of ions in the surrounding medium. In
physiological salt concentrations the Zeta potential is very low or even
zero.

Laser Doppler Anemometry (LDA)

     The charge carried by colloidal particles can be determined by meas-
uring the particle velocity in an electrical field, then using the
Smoluchowski, Henry or Debye-Hückel equations a Zeta potential can be calculated. The particle velocity v is usually expressed in relation to the
applied field strength E as electrophoretic mobility μ :

$$\mu = v / E \quad (\, (\mu\,m/s) \, / \, (V/cm)$$

The zeta potential is obtained from the electrophoretic mobility by one of
the following equations:

Smoluchowski:

$$ZP = \mu \; \frac{4 \, \pi \, \eta}{\epsilon}$$

Henry:

$$ZP = \mu \; \frac{4 \; \pi \; \eta}{\epsilon} \; f(a\varkappa)$$

Debye-Huckel:

$$ZP = \mu \; \frac{6 \; \pi \; \eta}{\epsilon}, \; \text{where}$$

ZP: zeta potential (esu Volt)
μ : electrophoretic mobility ((cm/s)/(esu Volt/cm))
v : particle velocity (cm/s)
E : field strength (esu Volt/cm)
η : dynamic viscosity (Poise)
ϵ : dielectric constant (80 for water at 20°C)
f(a$\varkappa$): Henry factor

The application of the different formulae depends on the particle radius a (Angstrom) and the thickness of the electric double layer 1/$\varkappa$ (Angstrom). For a$\varkappa$ < 0.1 the Debye-Huckel, for 0.1 < a$\varkappa$ < 100 the Henry and for a$\varkappa$ > 100 the Smoluchowski equation should be applied. The formulae are valid only for esu units and have to be converted to SI units. The Henry equation covers the range between the application of Debye-Huckel and Smoluchowski. Setting values for the Henry factor for a$\varkappa$ = 0.1 and a$\varkappa$ = 100 (equal to 1.5 and 1.0 respectively) will lead to the Debye-Huckel and Smoluchowski formulae.

For water at 25°C, with a dielectric constant of 78.5, one can simply multiply the electrophoretic mobility with a factor of 12.8, to obtain the Smoluchowski Zeta potential in mV:

$$ZP \; (mV) = 12.8 \; x \; \mu \; ((\mu m/s)/(V/cm)$$

   Measurement of the particle velocity in an electrical field can be performed with a Tiselius or a cylindrical cell apparatus;[46] the latter one determines the particle velocity using a microscope. These methods are not suitable for particles smaller than 200 nm, especially not for those particles in the size range from 30 up to 100 nm. As an alternative, laser light scattering techniques can be used. Laser Doppler Anemometry (LDA)[47-62] is a very fast and objective method with high resolution for the determination of particle velocities. The optical set up (Fig. 2) consists of a laser, a beam splitter which divides the beam into two beams of the same intensity, a lens focussing the beams into the measuring volume and another lens collecting the light scattered in the forward direction and projecting it onto a photo detector. The beam crossover forms an ellipsoid measuring volume (Fig. 3a) (extent 0.05 mm). In this region a pattern of interference fringes is formed due to the coherence of the two laser beams. It contains alternate light and dark zones (Fig. 3b). Particles moving through the fringe system will scatter light with a frequency f_{sc} which is different from the incident beam due to the Doppler effect. This frequency shift (the Doppler frequency f_d) is correlated with the particle velocity by the equation:

$$f_d = \frac{2 \sin (\frac{\Theta}{2})}{\lambda} \; v, \; \text{where}$$

f_d: Doppler frequency of the scattered signal
θ : detection angle
λ : laser wavelength
v: particle velocity

For measuring the sign of velocity and potential one of the laser beams can be modulated, e.g. with 250 Hz as in the Malvern Zetasizer.[62] This causes

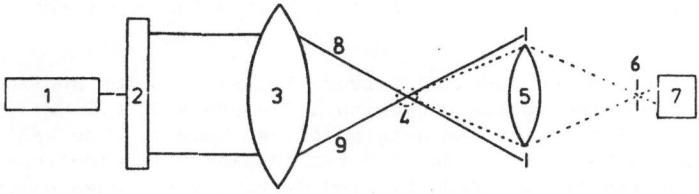

Fig. 2. Optical set up for Laser Doppler Anemometry (LDA)
1, Laser; 2, beam splitter; 3,5, lenses; 4, measuring
volume; 6, pinhole; 7, photo detector; 8,9, laser
beams.

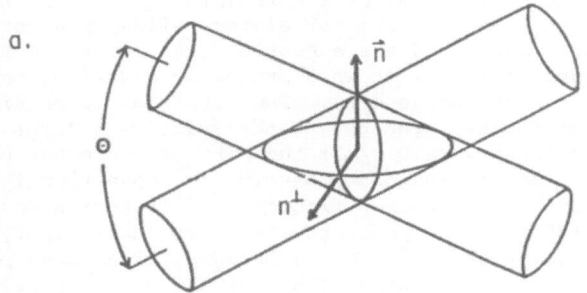

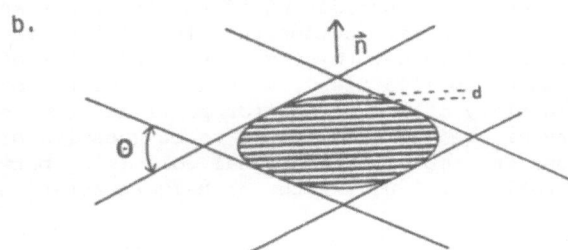

Fig. 3a,b. Ellipsoid measuring volume formed by the crossing laser beams with interference fringe pattern
(light and dark zones).

a drift of the interference fringe pattern parallel to the direction of the particle motion. If particles have no charge they are stationary. Because of the moving fringe system they will scatter light with a Doppler frequency equal to the modulation frequency, e.g. 250 Hz. For particles moving along with the fringe system, the Doppler frequency will be lower, for particles moving against the fringe system, the Doppler frequency will be higher than 250 Hz. Synchronisation of modulating frequency and the applied polarity changing field will give information about the sign of the charge.

In the LDA system produced by Malvern Instruments the applied voltage is a polarity reversing square wave with a frequency of 0.5 Hz. This reduces electrode polarisation and gassing of the sample. The cell contains two platinised platinum electrodes and two sensing gold electrodes for the determination of the field strength. The driving electrodes are separated from the measurement capillary by two semipermeable membranes. The electrode chamber is filled with pure dispersion medium without particles. The two available measurement capillaries have cross sectional diameters of 4.0 mm and 0.7 mm. The smaller one is used for liquids with high conductivity, e.g. physiological salt solution. The larger one has the advantage that particle size measurements can be performed by Photon Correlation Spectroscopy (PCS) to obtain information about changes in the particle size characteristics of the sample due to the applied electrical field (e.g. coagulation, dissaggregation).

The Doppler frequency of the scattering signal can be determined using either a frequency analyser or a correlator. The Malvern Zetasizer is based on the latter method. The correlator builds up a correlation function[62] which is transferred via a Fourier transformer to the frequency spectrum. The diagram (Fig. 4) shows 3 frequency distributions of particles with different size and density. The first one is obtained from a titanium dioxide suspension. Because of the relatively large size of 300 nm and the high density (4.0 g/ml) the particles are not strongly influenced by Brownian motion and disturbancies by convection in the sample. As a result they give a very narrow frequency distribution corresponding to a potential of +40 mV. The peak obtained with 1 μm polystyrene latex particles with a density of 1.05 g/ml is broader. The latex particles were dispersed in 0.01 M phosphate buffer (pH 7.0) and showed a potential of -73 mV. Under the same conditions the peak for 60 nm polystyrene latex shows a strong diffusion broadening. Because of their small size they have a high diffusion motion. This broadening of the frequency spectrum is the limiting factor in Laser Doppler Anemometry. In spite of the broad peak there were no difficulties in determining a zeta potential of -64 mV and also zero values without applied field, with a high degree of reproducability. The applicability of the method for particles smaller than 50 nm, especially with very diluted samples, seems to be questionable. For example, determinations on samples of colloidal gold (size between 15 and 20 nm) in a concentration of 0.01 % with the Malvern Zetasizer were not possible.

Latex particles with different surface properties are being used as model systems (Table 1) to test the applicability of the LDA method for the characterisation of drug carriers. Three polystyrene particles with sizes of 60, 170 and 1000 nm have been chosen to study the influence of size and to enable the results to be compared with previous studies using the same system. A range of polystyrene particles with different added surface groups has also been investigated. Fluorescent labelled polystyrene and fluorescent carboxylated polystyrene particles were chosen to test the effects of such labelling on surface properties. Fig. 5 shows the pH-mobility profiles of some model particles in phosphate buffer where the electrophoretic mobility is plotted on ordinate axis rather than Zeta

Table 1. Standard particles with different surface properties.

PARTICLES	SIZE (μm)	
POLYSTYRENE (PS)	0.06	PS-0.06
	0.17	PS-0.17
	1.00	PS-1.00
PS-FLUORESCENT (PSF)	0.19	PSF-0.19
POLYSTYRENE – COOH	0.19	PS-COOH-0.19
PSF-COOH	0.28	PSF-COOH-0.28
POLYSTYRENE – NH$_2$ (aromatic)	0.18	PS-AR-NH$_2$-0.18
POLYSTYRENE – NH$_2$ (aliphatic)	0.19	PS-AL-NH$_2$-0.19
POLYSTYRENE – OH	0.25	PS-OH-0.25

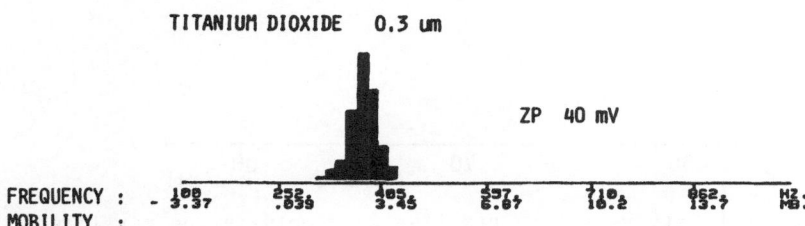

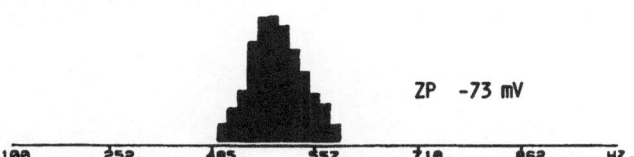

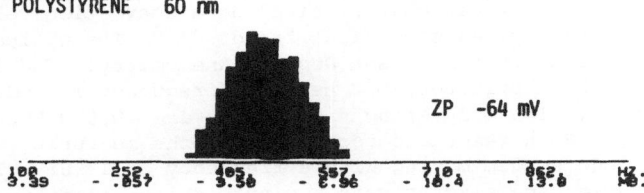

Fig. 4. Frequency versus mobility spectra from titanium dioxide (0.3 μm) and latex particles (1.0 μm and 60 nm) (Malvern Zetasizer).

potential. It is clear that there is a distinct difference in the mobility and therefore in the Zeta potential for the 60 nm polystyrene particles as compared with the 170 and 1000 nm particles. That is in spite of being made from the same material. It can also be noted that the particles carrying amino groups are slightly more negative than the carboxylated ones.

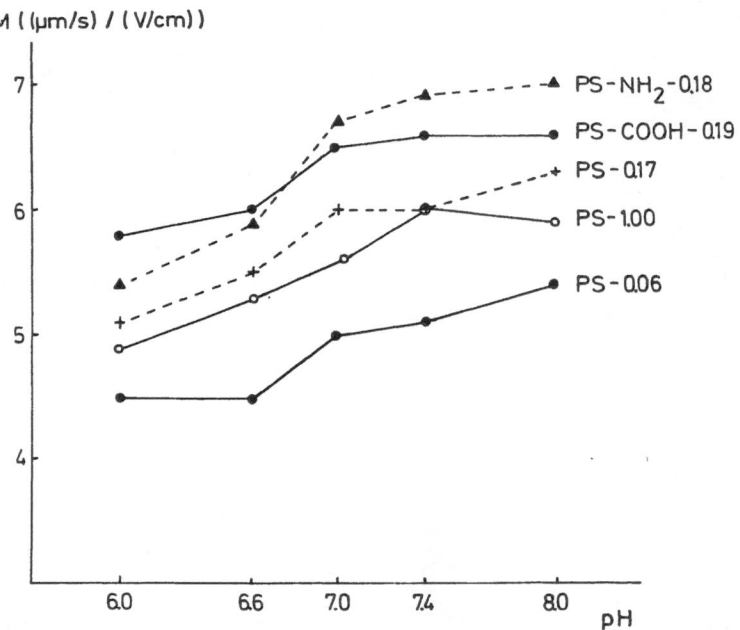

Fig. 5. Mobility versus pH profiles from polystyrene particles with different sizes (PS-0.06, PS-0.17 and PS-1.00) and polystyrene particles with carboxyl (PS-COOH-0.19) and aromatic amino groups (PS-NH -0.18) in 0.01 M phosphate buffer.

Amplitude Weighted Phase Structuration (AWPS)

AWPS is a new signal processing method which enables the simultaneous determination of particle charge, size and a convection velocity in the direction of the applied electrical field.[63-66] The optical set up (Fig. 6) used is similar to that in Laser Doppler Anemometry. The laser beam is split up by a beam splitter, but here the frequency is modulated by a pair of Bragg cells. The modulating frequencies are 40,000 kHz and 40,001 kHz respectively. Both beams are focussed onto the measuring volume where they create a fringe system moving with a frequency of 1 kHz. To reduce electrode polarisation and electrolysis a sinusoidal oscillating field is applied. Frequencies from 30 to 100 Hz can be used. The scattered light is detected in the forward direction by an avalanche diode. The analogue photo signal (Fig. 7) is amplified, high and low pass filtered, and fed to a phase and an amplitude detector. The amplitude detector integrates the signal and digitizes it using a 16 bit A/D converter. The phase is measured relative to the optical shift frequency. The Bragg cell drive system allows the

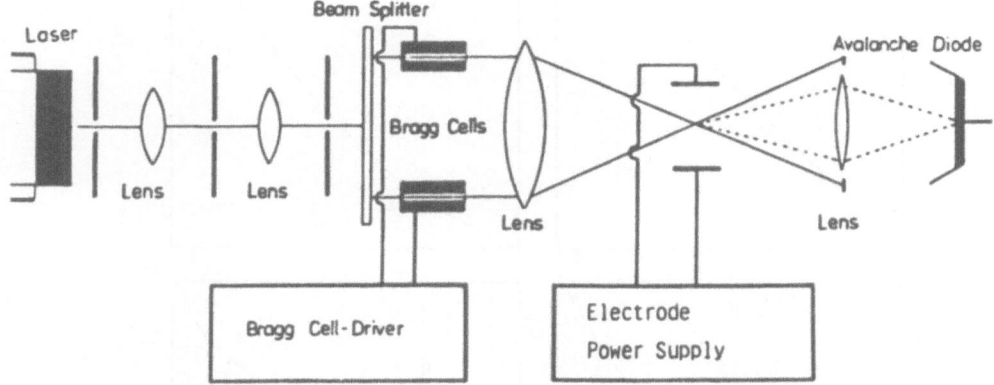

Fig. 6.   Optical set up for Amplitude Weighted Phase
          Structuration (AWPS).

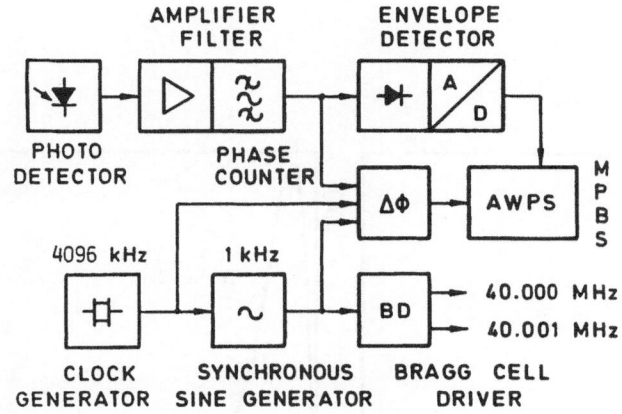

Fig. 7.   Electronic set up for Amplitude Weighted Phase
          Structuration (AWPS).

          MPBS, Micro Processor Based Structurator

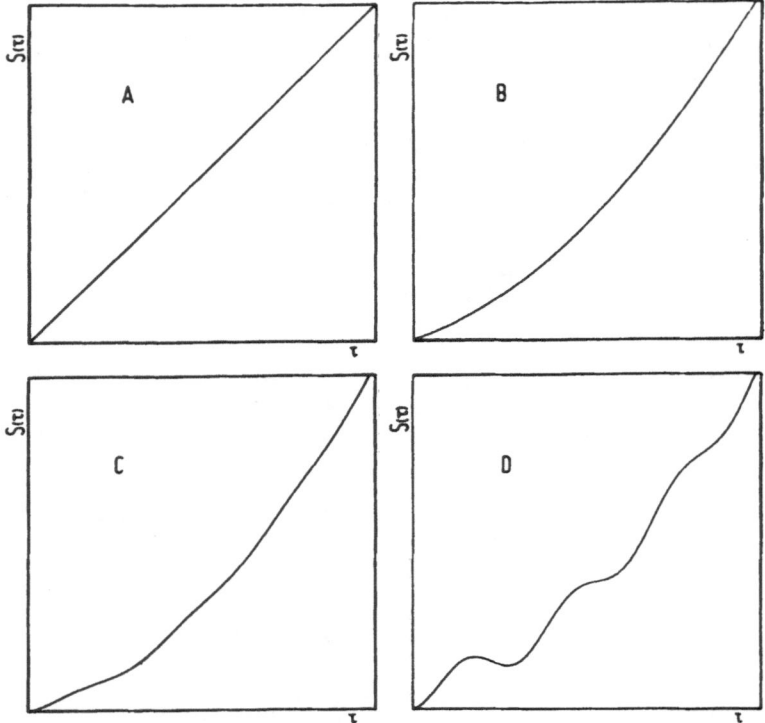

Fig. 8. Structure functions for diffusion without convection. (A), diffusion with present convection; (B), diffusion with simultaneous convection and a field velocity of 4 μm/s (C) and 90
μm/s (D).

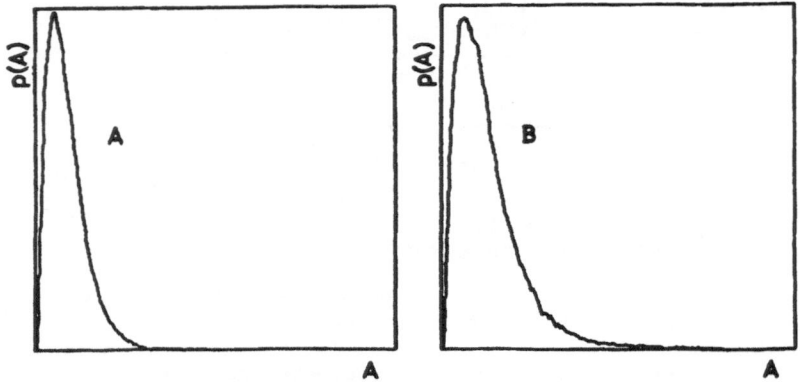

Fig. 9. AWPS amplitude histograms showing a Rayleigh
distribution (A) and a Rayleigh distribution
with distortions due to aggregates (B).

direct input of the 1 kHz shift frequency. For measuring the phase with a resolution of $2\pi/4096$ this frequency is derived from a 4096 kHz clock and both frequencies are used in the phase counter. Digitized phase and amplitude are fed to a 68,000 computer system for processing a 64 channel structure function $S(\tau)$. The amplitude weighted phase difference:

$$Q(\tau) = \int_{0}^{T} A(\tau)d\phi t$$

is squared and averaged which results in an amplitude weighted structure function equal to:

$$S_{(\tau)} = \ \ <Q(\tau)^2> = 9^2\,\frac{\pi}{4}<A^2>\ \left[v_c\,\tau^2 + (\mu E/\omega_E)^2(1-\cos\omega_E)\right] + 9^2<A^2>2D\tau$$

whereby A is the amplitude and ϕ the phase of the scattered light. The structure function $S(\tau)$ comprises:

q: $2\pi/s$
s: fringe distance
v_c: a constant collective velocity, e.g. the thermal convection component parallel to the direction of the electrical field
τ: relaxation time
μ: the electrophoretic mobility, defined as field velocity v over field strength E
E: field strength
ω_E: oscillation frequency of the electrical field
D: diffusion constant

The measured structure function is normalised with the averaged squared amplitude $<A^2>$ and fitted to the model function

$$y = a\,\tau^2 + b\,(1-\cos\omega_E\tau) + c\tau$$

Using the three fit parameters a, b and c the convection velocity, the electrophoretic mobility and the particle size can be calculated simultaneously. From the fit function it can be seen that without convection and field mobility the resulting structure function is a straight line (Fig. 8, A). In the presence of convection a parabolic part is added (Fig. 8, B). If there is a field mobility from the particles, the structure function will include a sinusoidal part which increases quadraticly with the field velocity (Fig. 8, C and D). Function C arises from a velocity of 4 µ.m/s and function D from a velocity of 90 µm/s. Furthermore the histogram of the intensity amplitude of the scattering signal (fig. 9A) can be used to check the validity of the data and purity of the sample. For many-particle scattering the intensity amplitude A = |I| follows a Rayleigh distribution. For small amplitudes the density is proportional to A. Dust and coagulated particles produce distortions of the Rayleigh type distribution (Fig. 9B). The same is true for very polydisperse samples. The following data were obtained from different systems used as drug carriers:

Poly(butylcyanoacrylate) Particles.

The structure function in Fig. 10 was obtained from poly(butylcyanoacrylate) (PBCA) particles with a size of 190 nm. Because the convection was extremely low and the particles were uncharged the function obtained is almost a straight line. The Rayleigh distribution shows no distortions due to the low polydispersity and purity of the sample. Using 35 nm particles, a structure function was obtained with a parabolic part due to convection in the cell and the higher sensitivity of the smaller particles to convection. Because of the relatively high salt concentration (0.1 M phosphate buffer, pH 7.4) the Zeta potential of the particles is zero.

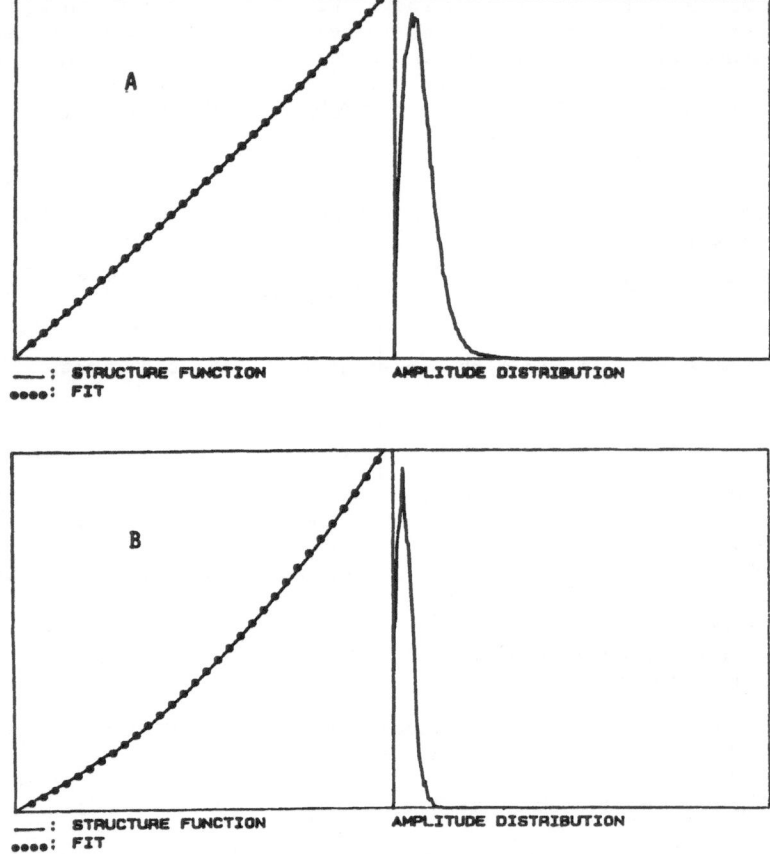

Fig. 10. Structure functions and amplitude histograms of
poly(butylcyanoacrylate) particles with a Zeta
potential of 0 mV in 0.1 M phosphate buffer.
A, 190 nm particles, very low convection;
B, 35 nm particles with superimposed convection.

Albumin Particles

Fig. 11 shows the function obtained using albumin particles with a
mean size of 1.8 μm. The system was very polydisperse and included particles down to 80 nm but also particles larger than 1.8 μm. In 0.01 M phosphate buffer the measured potential was -24 mV (Smoluchowski). The structure function is now superimposed by a weak sinusoidal part. In this case
the Rayleigh distribution shows distortions due to the high polydispersity
and the very low concentration of the sample. The strong amplification of
the signal necessary for data analysis also amplified the effect of dust
and coagulated particles. A ten fold dilution of the sample with distilled
water made this effect more obvious. A second broad peak appeared behind
the first peak of the Rayleigh distribution. The dilution increased the
Zeta potential to -38 mV and gave the function a distinct sinusoidal shape.
The particles were fractionated into different sizes and remeasuring the
Zeta potential of the 100 nm fraction in 0.01 M phosphate buffer, provided
a value of -23 mV. This indicates the independance of the potential on
the particle size.

250

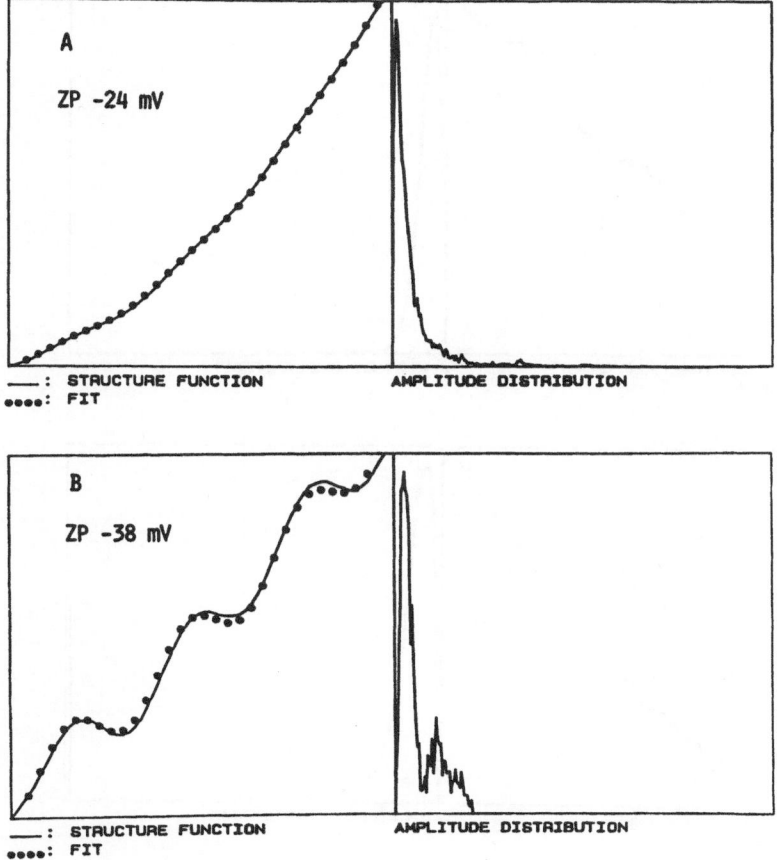

Fig. 11. Structure functions and amplitude histograms of
albumin particles with a mean size of 1.8 μm in
0.01 M (A) and 0.001 M (B) phosphate buffer.

Polystyrene Particles

The function and amplitude distribution shown in Fig. 12 were obtained
from monodisperse latex particles with a size of 190 nm. The particles
suspended in 0.01 M phosphate-citrate buffer at pH 2.2 had a Zeta potential
of -53 mV (Smoluchowski). The intensity amplitude follows a Rayleigh
distribution. In contrast, the second graph is the distribution obtained
using 170 nm latex with macroscopically visible aggregates (phosphate-
citrate buffer 0.01 M, pH 5.0 (Smoluchowski; Zeta potential -42 mV). An
amplitude signal is obtained over the whole of the abscissa range. The
curve does not reach the base line and shows a narrow peak on the right
side, indicating an amplitude overflow due to the presence of very large
particles.

HYDROPHOBICITY

Adsorption Measurements

The surface hydrophobicity of model particles can be characterized by
adsorption measurements using Rose Bengal.[67-72] This hydrophobic dye shows

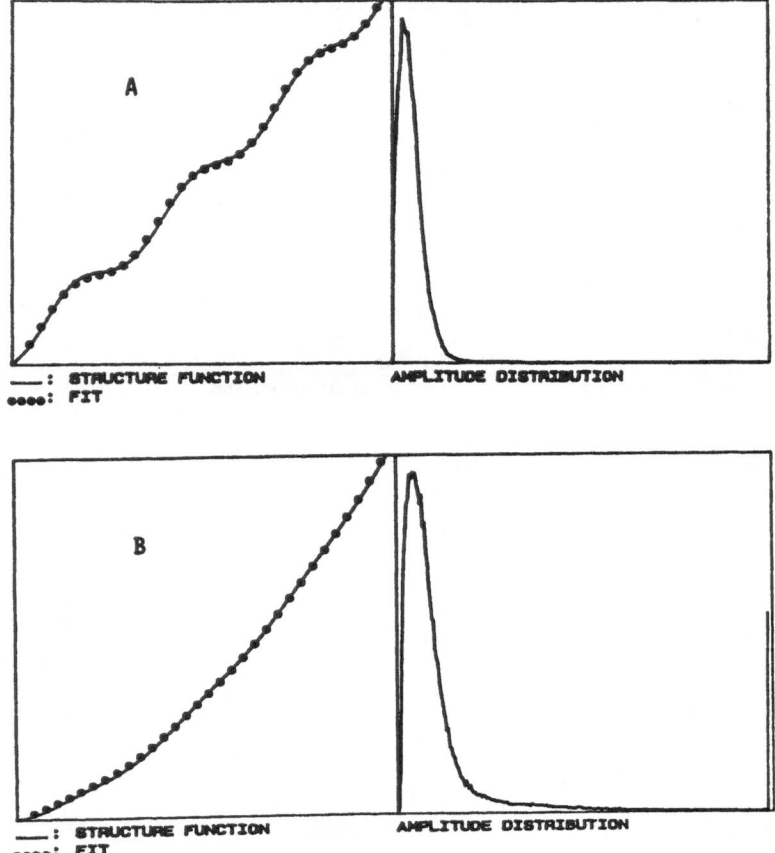

Fig. 12. Structure functions and amplitude histograms of
latex particles in 0.01 M phosphate-citrate buffer.
A, 190 nm, pH 2.2, monodisperse; B, 170 nm, pH 5.0,
with macroscopically visible aggregates.

different degrees of affinity for the particle surface depending on the
surface hydrophobicity. Experiments were performed in 0.1 M phosphate
buffer at pH 7.4. A salt concentration of 0.1 M was chosen to suppress
the charge effect and to increase hydrophobic interactions. After an incubation time of 3 h the particles were centrifuged and the concentration
of free Rose Bengal was determined in the supernatant by absorbance measurements (542.7 nm). Individual adsorption isotherms were determined for the
range of model particles described above. The increase in the amount of
Rose Bengal adsorbed with concentration was similar for the polystyrene
particles of sizes 170 and 1000 nm (binding constants 0.289 and 0.279 ml/µg
respectively), but the increase was more pronounced for the particles of
size 60 nm (binding constant 0.40 ml/µg), indicating a higher affinity of
Rose Bengal to the surface. Clearly distinct differences in the surface
hydrophobicity of particles can exist even when made from the same polymer
but with different sizes. The particles carrying amino groups show a very
low affinity with a corresponding binding constant of 0.135 (ml/µg).

The adsorption isotherms for the particles reached different plateau
levels. This could be due to differences in the surface polarity of the
particles. In this regard Kronberg and Stenius have published data for the

adsorption of non-ionic surfactants (containing polyethyleneoxide chains) onto microspheres with different polarities.[73-76] The plateau of the adsorption isotherms decreased with increasing polarity of the surface. Therefore, they suggested determination of the surface polarity by the

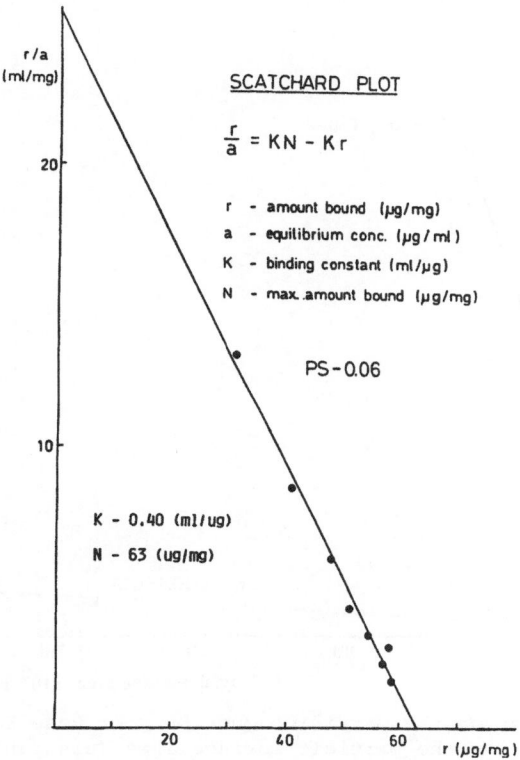

Fig. 13. Scatchard Plot of the adsorption isotherm data from polystyrene particles 60 nm (PS-0.06).

adsorption of the same known surfactant onto the unknown surface. The binding constants given above for Rose Bengal were determined using a Scatchard plot approach[77] (Fig. 13) and the equation:

$$r/a = KN - Kr$$

where r is the amount bound (µg/mg), a the equilibrium concentration (µg/ml), K the binding constant (ml/µg) and N the maximum amount bound (µg/mg). The

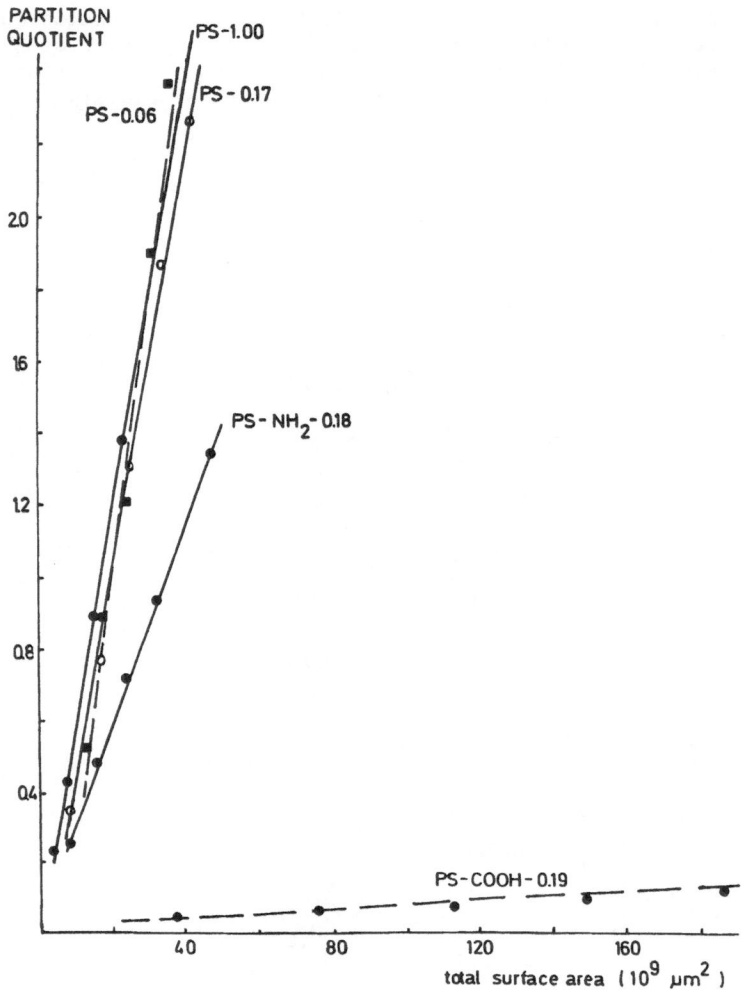

Fig. 14. Plot of the partition quotients of Rose Bengal
versus the particle surface area (constant Rose
Bengal concentration, increasing particle concentration).

equation, which is valid for isotherms of the Langmuir type, yielded
straight lines for all the model systems. The data in Fig. 13 was obtained
with 60 nm polystyrene particles. The binding constants K can be employed
for estimation of the free energy of binding[78] using the equation:

$$\Delta G_O = - RT \ln K$$

Further, the partition quotients of Rose Bengal between the particle
surface as one phase and the dispersion medium as the other phase were
determined. Constant Rose Bengal concentrations and increasing particle
concentrations were used. The partition quotient was calculated as the
ratio of the amount bound onto the surface to the amount unbound. Plotting
the quotients against the surface area (Fig. 14) gave straight line relationships. The slope of these lines can be taken as a measure of hydro-

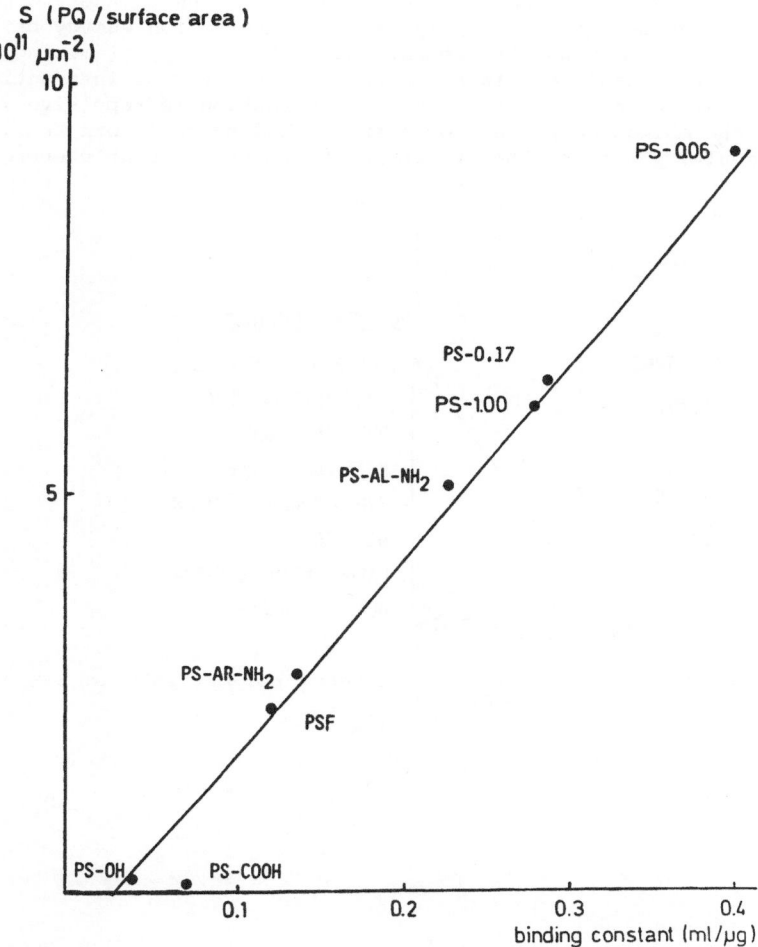

S (PQ / surface area)
$(10^{11} \mu m^{-2})$

Fig. 15. Plot of the ratio of partition quotients to
          surface area (from the partitioning experiments)
          against the binding constants obtained from the
          adsorption isotherms (using a Scatchard Plot).

phobicity. There are distinct differences between the slopes obtained for
the simple polystyrene particles and those for the microspheres carrying
amino and carboxyl groups. For the polystyrene particles the lines for
170 nm and 1000 nm spheres are parallel, indicating that the surfaces are
similar. In contrast the 60 nm particles show a steeper slope indicating
a more hydrophobic surface.

Plotting the partition quotients against the binding constants obtained
from the adsorption isotherms (Fig. 15) shows a good correlation between
the two methods. Such methods allow the particles to be placed in an order
of increasing hydrophobicity. It is clear that the 60 nm polystyrene
spheres are the most hydrophobic, while the 170 nm and the 1000 nm have
similar surface characteristics.

Two-Phase Partitioning

The mixing of dextran and polyethylene glycol (PEG) solutions will result in a two phase system above certain concentrations of the two polymers.[79] This is because for large polymers the interaction energy between the molecules themselves tends to dominate over the entropy of mixing per mole. Therefore the result of mixing depends on the type of interaction between the two different polymers. If the interaction is repulsive in character and the molecules prefer to be surrounded by their own kind, two phases will result. That is, the system has the most favourable energetic

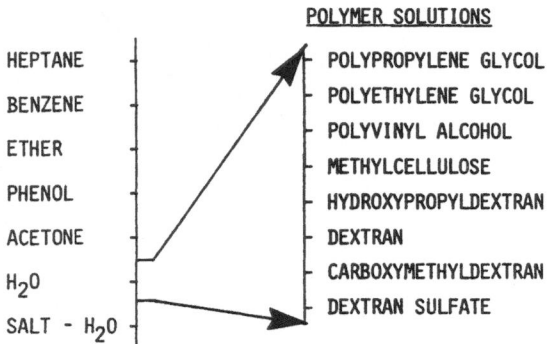

Fig. 16. Hydrophobic ladder of aqueous polymer solutions (after ref. 79).

state when the two polymers are separated. The PEG/dextran system will separate in a PEG-rich upper phase and a dextran-rich lower phase. There is no principal difference between the phase separation mechanism of polymer solutions and the classic water/octanol partition system. The polymer solutions can be ranked in a hydrophobic ladder (Fig. 16). This ladder falls within a very narrow part of the conventional solvent spectrum which lists the solvents according to their hydrophilic-hydrophobic character.[79] Partitioning between two phases is an established method for the separation of macromolecules, bacteria, cells, etc.[80-91] In relation to the surface properties of the distributing material it will have a greater affinity for one of the two phases.

The forces which determine the rate and extent of distribution of the particles are the Brownian motion and the interfacial tension.[79] Brownian motion will distribute the particles randomly throughout the whole system. The interfacial force will move them to the phase where they have the lowest interfacial free energy. There are three possible positions for a particle in a two phase system. It can be located in the lower phase, in the interface or in the upper phase. If it takes a position in the interface, a certain area of the liquid/liquid interface will disappear and with it a decrease in the free surface energy of the whole system. Depending on the interfacial tensions between the particle and the upper and lower phase, γ_{P1} and γ_{P2}, and the interfacial tension between the two phases, γ_{12}, the particle can take the following positions:

(a) if (γ_{P1} - γ_{P2})/ γ_{12} is $>$ or = 1, the particle will transfer to the lower phase for $\gamma_{P1} > \gamma_{P2}$, and will transfer to the upper phase for $\gamma_{P1} < \gamma_{P2}$.

(b) if (γ_{P1} - γ_{P2})/ γ_{12} is $<$ 1, the particle will locate in the interface. The exact position within the interface will depend on the magnitude of γ_{P1} and γ_{P2}.

By the addition of salts to the system a potential difference can be created between the two phases due to the uneven distribution of ions. A system containing 0.11 M sodium phosphate buffer will result in a positive upper phase with a potential difference of about 3 mV to the lower phase.[92] Such systems can be used for the determination of charge effects in addition to hydrophobic interactions. Data published by Reitherman[92] show that 90% of negatively charged human red blood cells (HRBC) will transfer to the upper phase in a system with a positive potential. In a neutral or slightly negative system (0.01 M phosphate buffer, 0.13/0.15 M NaCl) almost 100% of the particles were found at the interface.

Only a few papers have described partitioning of latex particles, expecially polystyrene in two phase polymer systems.[93] In the reported systems the particles moved only to one phase. In our own work we have tested particle systems with dextran 20, 70 and 500 and PEG with molecular weights from 2,000 up to 35,000. With dextran 500 and PEG 2,000 one found no distribution. 100% of the particles went to the upper (PEG) phase. Increasing the molecular weight of a polymer will move particles away from the polymer phase, while lowering the molecular weight of the polymer will increase the affinity for the polymer phase. Therefore the molecular weight of the dextran was lowered to 20×10^3 and for PEG increased stepwise to 35,000. Data from this preliminary study (Table 2) show that a distribution between the upper phase and interface could be achieved; no particles were found in the bottom phase. The particles behave similarly to the HRBC. The system with dextran 20 and PEG 35,000 appears to be the most suitable but it has still to be optimised by varying salt concentrations. In addition, a part of the PEG will be replaced by PEG fatty acid esters. The PEG esters will partition in the same way as the unsubstituted polymer.[82] By adding these hydrophobic materials the PEG phase will be more apolar. If the particles are more hydrophobic they will increase their concentration in the PEG phase due to hydrophobic interactions. If they have a more hydrophilic character they will be pushed away from the now more apolar PEG phase. The difference in partitioning between the two phase systems for unsubstituted (partition coefficient K_1) and esterified PEG (K_2), calculated as $\Delta \log K$ value ($\log K_2 - \log K_1$), can be taken as a measure of hydrophobic interactions and hydrophobicity of the particles. $\Delta \log K$ is positive if the hydrophobic part of their surface dominates. It is negative if the hydrophilic interaction is higher, and zero if both effects are equal.

Other Methods

In order to complete the characterisation of particles in terms of their surface hydrophobicity, other methods can be used such as hydrophobic interaction chromatography (HIC), the adsorption of fatty acids onto the surface and contact angle measurements.

In hydrophobic interaction chromatography[94-100] Sepharose linked with different hydrocarbon chains[94,101-103] (from ethyl to octyl) is a suitable matrix. The particles will stick onto the gel due to hydrophobic interactions. By changing the properties of the eluent (lowering the polarity, the addition of surfactants (e.g. Triton X-100 or Berol) with a gradient

Table 2. Partitioning of polystyrene (PS-0.17) and polystyrene
particles with added carboxyl (PS-COOH-0.19) and hydroxy
groups (PS-OH-0.25) in a Dextran 20/PEG 35,000 system.

Particles	Top phase (%)	Interface (%)	Bottom phase (%)
PS-0.17	89.5	10.5	nil
	91.5	9.5	nil
	91.7	9.3	nil
PS-COOH-0.19	78.2	21.8	nil
	76.6	23.4	nil
	74.1	25.9	nil
PS-OH-0.25	85.8	14.2	nil
	83.8	16.2	nil
	85.9	14.1	nil

mixer the particles will leave the column according to the strength of their
hydrophobic interactions with the matrix. The retention volume can be taken
as a measure of hydrophobicity. Another possible method for characterisation is the determination of the adsorption of radiolabelled fatty acids
(C 14 - C 18) it should be possible to calculate the free energy of adsorption for a methylene group onto different particles. Finally, contact
angle measurements[106,107] on polymer slabs or compressed discs of particulate material can be performed in order to correlate the results obtained
by the other methods described above with literature data.

REFERENCES

1. D.B. Zilversmit, G.A. Boyd and M. Brucer, The effect of particle size
   on blood clearance and tissue distribution of radioactive gold colloids, J. Lab. Clin. Med., 40:255 (1952).
2. R.L. Juliano and D. Stamp, Biochem. Biophys. Res. Commun., The effect
   of particle size and charge on the clearance rates of liposomes and
   liposome-encapsulated drugs, 63:651 (1975).
3. H.G. Schroeder, G.H. Simmons and P.P. De Luca, Distribution of sub-
   visible microspheres after intravenous administration to Beagle dogs,
   J. Pharm. Sci., 67:504 (1978).
4. M. Kanke, G.H. Simmons, D.L. Weiss, B.A. Bivans and P.P. De Luca,
   Clearance of [141]Ce-labelled microspheres from blood and distribution
   in specific organs following intravenous and intraarterial administration in Beagle dogs, J. Pharm. Sci., 69:775 (1980).
5. M. Frier, in: "Progress in Radiopharmacology", P.H. Cox ed., Vol. 2,
   Elsevier/North Holland Biomedical Press, Amsterdam (1981).
6. L. Illum, S.S. Davis, C.G. Wilson, M. Frier, J.G. Hardy and N.W. Thomas,
   Blood clearance and organ deposition of intravenously administered
   colloidal particles: effects of particle size, nature and shape, Int.
   J. Pharmaceut., 12:135 (1982).
7. L. Illum and S.S. Davis, The targeting of drugs parenterally using
   microspheres, J. Parent. Sci. Tech., 36:242 (1982).
8. E. Tomlinson and J.G. McVie, New directions in cancer chemotherapy. 2.
   Targeting with microspheres, Pharm. Int., 4:281 (1983).

9. D.J. Wilkins and P.A. Myers, Studies on the relationship between the electrophoretic properties of colloids and their blood clearance and organ distribution in the rat, Brit. J. Exp. Path., 47:568 (1966).
10. D.J. Wilkins and P.A. Myers, Anomalous electrophoretic behaviour of certain adsorbed protein derivatives, Proc. Soc. Exp. Biol. Med., 133:255 (1970).
11. D.J. Wilkins and P.A. Myers, in: "Surface chemistry of biological systems", M. Blank, ed., Plenum Press, New York (1970).
12. T.P. Stossel, R.J. Mason, J. Hartwig and M. Vaughan, Quantitative studies of phagocytosis by polymorphonuclear leukocytes: use of emulsions to measure the initial rate of phagocytosis, J. Clin. Invest., 51:615 (1972).
13. T.P. Stossel, Quantitative studies of phagocytosis: Kinetic effects of cations and heat-labile opsonin, J. Cell. Biol., 58:346 (1973).
14. W.E. Magee, C.W. Gott, J. Schoknecht, M.D. Smith and K. Cherian, Interaction of cationic liposomes containing entrapped horseradish-peroxidase with cells in culture, J. Cell. Biol., 63:492 (1974).
15. V.K. Jansons, P. Weiss, T. Chen and W.R. Redwood, In-vitro interaction of L1210 cells with phospholipid vesicles, Cancer Res., 38:531 (1978).
16. S.S. Davis, Colloids as drug delivery systems, Pharmaceutical Technology, 71: (1981).
17. G. Capo, F. Garrouste, A.M. Benoliel, P. Bongrand and R. Depieds, Non-specific binding by macrophages: Evaluation of the influence of medium-range electrostatic repulsion and short-range hydrophobic interaction, Immunol. Commun., 10:35 (1981).
18. C. Capo, P. Bongrand, A.M. Benoliel, A. Tyter and R. Depieds, Particle macrophage interaction: Role of surface charges, Annales D'Immunologie D132, No. 2-3:165 (1981).
19. A. Raz, C. Bucana, W.E. Fogler, G. Poste and I.J. Fidler, Biochemical, morphological and ultrastructural studies on the uptake of liposomes by murine macrophages, Cancer Res., 41:487 (1981).
20. M.J. Hsu and R.L. Juliano, Interaction of liposomes with the reticuloendothelial system. II. Non-specific and receptor-mediated uptake of liposomes by mouse peritoneal macrophages, Biochim. Biophys. Acta, 720:411 (1982).
21. R.C. Roozemond and D.C. Urli, Peculiar behaviour of rabbit thymocytes in interaction with liposomes of different compositions shown by fluorescence polarisation studies, lipid analysis, and uptake of vesicle-entrapped carboxyfluorescein, Biochim. Biophys. Acta, 689:499 (1982).
22. R.A. Schwendener, P.A. Lagocki and Y.E. Rahman, The effect of charge and size on the interaction of unilamellar liposomes with macrophages, Biochim. Biophys. Acta, 772:93 (1984).
23. C.J. Van Oss and M.W. Stinson, Immunoglobulins as specific opsonins. I. The influence of polyclonal and monoclonal immunoglobulins on the in-vitro phagocytosis of latex particles and staphylococci by human neutrophils, J. Reticuloendoth. Soc., 8:397 (1970).
24. C.J. Van Oss and C.F. Gillman, Phagocytosis as a surface phenomenon. II. Contact angles and phagocytosis of encapsulated bacteria before and after opsonisation by specific antiserum and complement, J. Reticuloendoth. Soc., 12:497 (1972).
25. C.J. Van Oss and C.F. Gillman, Phagocytosis as a surface phenomenon. III. Influence of C1423 on the contact angle and on the phagocytosis of sensitised encapsulated bacteria, Immunol. Commun., 2:415 (1973).
26. C.J. Van Oss, C.F. Gillman and A.W. Neumann, in: "Phagocytic Engulfment and Cell Adhesiveness", Marcel Dekker, New York (1975).
27. C.J. Van Oss and C.F. Gillman, Phagocytosis and immunity, Int. Convoc. Immunol., 4:505 (1975).
28. I. Stjernestrom, K.-E. Magnusson, O. Stendahl and C. Tagesson, Hydrophobic and charge interaction of smooth Salmonella typhimurium 395-MS sensitized with anti-MS immunoglobulin G and complement, Infect. Immun., 18:261 (1977).

29. C.J. Van Oss, Phagocytosis as a surface phenomenon, Ann. Rev. Micro-
    biol., 32:19 (1978).
30. K-E. Magnusson, O. Stendahl, I. Stjernestrom and L. Edebo, Reduction
    of phagocytosis, surface hydrophobicity and charge of Salmonella
    typhimurium 395 MR 10 by reaction with secretory IgA (SIgA),
    Immunology, 36:439 (1979).
31. T.A. Horbett and P.K. Weathersby, Adsorption of proteins from plasma
    to a series of hydrophilic-hydrophobic copolymers. I. Analysis with
    the in-situ radioiodination technique, J. Biomedical Materials Res.,
    15:403 (1981).
32. T.A. Horbett, Adsorption of proteins from plasma to a series of hydro-
    philic-hydrophobic copolymers. II. Compositional analysis with the
    prelabelled protein technique, J. Biomedical Materials Res., 15:673
    (1981).
33. D.J. Wilkins, The biological recognition of foreign from native
    particles as a problem in surface chemistry, J. Coll. Interf. Sci.,
    25:84 (1967).
34. C.J. Van Oss and C.F. Gillman, Phagocytosis as a surface phenomenon.
    I. Contact angles and phagocytosis of non-opsonised bacteria,
    J. Reticuloendoth. Soc., 12:283 (1972).
35. D.R. Absolam, C.J. Van Oss, W. Zingg and A.W. Neumann, Phagocytosis
    as a surface phenomenon: Opsonisation by aspecific adsorption of
    IgG as a function of bacterial hydrophobicity, J. Reticuloendoth.
    Soc., 31:59 (1982).
36. L. Illum and S.S. Davis, The organ uptake of intravenously administ-
    ered colloidal particles can be altered using non-ionic surfactant
    (Poloxamer 388), FEBS Lett., 167:79 (1984).
37. O. Stern, Zur Theorie der elektrischen Doppelschicht, Z. Elektrochemie.,
    30:508 (1924).
38. J.Th.G. Overbeek and L. Lyklema, Electric potentials in colloidal
    systems, in: "Electrophoresis," Vol. I., Bier, M. ed., Academic
    Press, New York (1959).
39. M.T. Riddick, "Control of colloid stability through zeta potential,"
    Vol. 1, Livingstone Publ. Co., Wynnewood (1968).
40. H. Sontag and K. Strenge, "Koagulation und Stabilitat disperser
    Systeme," VEB Deutscher Verlag der Wissenschaften, Berlin (1970).
41. P. Ney, "Zetapotentiale und Flotierbarkeit von Mineralien," Springer
    Wien-New York (1973).
42. H. Van Olphen, "An introduction to colloid and surface chemistry,"
    John Wiley and Sons, New York (1977).
43. A.M. James, Electrophoresis of particles in suspensions, in: "Surface
    and Colloid Science," Vol. 11, R.J. Good and R.R. Stromberg, eds.,
    Plenum Press, New York (1979).
44. R.J. Hunter, "Zeta Potential in Colloid Science: principles and app-
    lications," Academic Press, London (1981).
45. G. Lagaly, Energetische Wechselwirkungen in Dispersionen und Emuls-
    ionen," H. Asche, D. Essig and P.C. Schmidt, eds., Wissenschaftliche
    Verlagsgesellschaft, Stuttgart (1984).
46. A. Tiselius, A new apparatus for electrophoretic analysis of colloidal
    mixtures, Trans. Faraday Soc., 33:524 (1937).
47. J.C. Earnshaw and M.W. Steer, eds., "The application of laser light
    scattering to the study of biological motion," Plenum Press, New
    York (1983).
48. B.R. Ware and W.H. Flygare, The simultaneous measurement of the elec-
    trophoretic mobility and diffusion coefficient in bovine serum
    albumin solutions by light scattering, Chem. Phys. Lett., 12:81
    (1971).
49. B.R. Ware and W.H. Flygare, Light scattering in mixtures of BSA, BSA
    dimers and fibrinogen under the influence of electric fields,
    J. Colloid Interface Sci., 39:670 (1972).

50. B.R. Ware, Electrophoretic light scattering, Advan. Colloid. Interface Sci., 4:1 (1974).
51. E.E. Uzgiris, Electrophoresis of particles and biological cells measured by the Doppler shift of scattered laser light, Opt. Commun., 6:55 (1972).
52. A.J. Bennet and E.E. Uzgiris, Laser doppler spectroscopy in an oscillating electric field, Phys. Rev., A8:2662 (1973).
53. E.E. Uzgiris, Laser doppler spectrometer for study of electrokinetic phenomena, Rev. Sci. Instrum., 45:74 (1974).
54. E.E. Uzgiris and J.H. Kaplan, Laser doppler spectroscopic studies of the electrokinetic properties of human blood cells in dilute salt solution, J. Colloid Interface Sci., 55:148 (1976).
55. E.E. Uzgiris and D.C. Golibersuch, Excess scattered light intensity fluctuation from hemoglobin, Phys. Rev. Lett., 32:37 (1974).
56. D.D. Haas and B.R. Ware, Design and construction of a new electrophoretic light scattering chamber and applications to solutions of hemoglobin, Anal. Biochem., 74:175 (1976).
57. H.Z. Cummins and E.R. Pike, eds., "Photon Correlation and Light Beating Spectroscopy," Plenum Press, New York (1973).
58. H.S. Cummins and E.R. Pike, eds., "Photon Correlations Spectroscopy and Velocimetry," Plenum, New York (1977).
59. T.S. Durrani and C.A. Greated, "Laser Systems in Flow Measurement," Plenum, New York (1977).
60. B. Eliasson and R. Dandliker, A theoretical analysis of laser doppler flow meters, Optica Acta, 21:119 (1974).
61. K. Schatzel, Advances in cross-beam rate correlation, Optica Acta, 27:45 (1980).
62. Malvern Application Report MAR301, Malvern Instruments, Malvern (GB) (1984).
63. K. Schatzel and J. Merz, Measurement of small electrophoretic mobilities by light scattering and analysis of the amplitude weighted phase structure function, J. Chem. Phys., 81:2482 (1984).
64. R.H. Muller and J. Merz, Moderne Messmethoden zur Bestimmung der Teilchenladung II, Pharmazeutische Verfahrenstechnik, 1:141 (1985).
65. B.W. Muller, J. Merz and R.H. Muller, Simultane Bestimmung von elektrophoretischer Beweglichkeit und Teilchengrosse bei uberlagerter Konvektion an Suspensionen und Emulsionen, Colloid Polymer Sci., 263:342 (1985).
66. J. Merz, Simultane Messung von Diffusionskonstante und elektrophoretischer Beweglichkeit hochdisperser Systeme mit Hilfe der amplitudengewichtetan Phasenstrukturfunktion, Ph.D. Thesis, Department of Pharmacy, University of Kiel (FRG) (1985).
67. L. Brand, J.R. Gohlke and D.S. Rao, Evidence for binding of Rose Bengal and Anilinonaphthalene-sulphonates at the active regions of liver alcohol dehydrogenase, Biochemistry, 6:3510 (1967).
68. K.K. Rohatgi and A.K. Mukhopadhyay, Isolation of unique dimer spectra of dyes from the composite spectra of aggregated solutions, Photochem. Photobiol., 14:551 (1971).
69. I.M. Issa, R.M. Issa and M.M. Ghoneim, Spectrophotometric studies on fluorescein derivatives in aqueous solutions, Z. Physik. Chem., (Leipzig), 250:161 (1972).
70. C-W. Wu and F.Y-H. Wu, Rose Bengal: a spectroscopic probe for ribonucleic acid polymerase, Biochemistry, 12:4349 (1973).
71. F. Y-H. Wu and C-W. Wu, Rose Bengal: inhibitor of ribonucleic acid chain elongation, Biochemistry, 12:4343 (1973).
72. J.J.M. Lamberts and D.C. Neckers, Rose Bengal and non-polar derivatives, J. Am. Chem. Soc., 105:7465 (1983).
73. B. Kronberg, L. Kall and P. Stenius, Adsorption of non-ionic surfactants on latexes, J. Dispersion Sci. and Techn., 2:215 (1981).
74. B. Kronberg, Thermodynamics of adsorption of non-ionic surfactants on latexes, J. Colloid Interf. Sci., 96:55 (1983).

75. B. Kronberg and P. Stenius, The effect of surface polarity on the
    adsorption of non-ionic surfactants. I. Thermodynamic considerat-
    ions, J. Colloid Interf. Sci., 102:410 (1984).
76. B. Kronberg, P. Stenius and G. Igeborn, The effect of surface polarity
    on the adsorption of non-ionic surfactants. II. Adsorption to poly-
    (methyl-methacrylate) latex, J. Colloid Interf. Sci., 102:418 (1984).
77. G. Scatchard, The attractions of proteins for small molecules and
    ions, Ann. N.Y. Acad. Sci., 51:660 (1949).
78. G. Derzelic, N. Derzelic and Z. Telisman, The binding of human serum
    albumin by monodisperse polystyrene latex particles, Eur. J.
    Biochem., 23:575 (1971).
79. P-A. Albertson, "Partition of Cell Particles and Macromolecules,"
    John Wiley and Sons, New York (1971).
80. H. Walter and F.W. Selby, Effects of DEAE-dextran on the partition of
    red blood cells in aqueous dextran-polyethylene glycol two-phase
    systems, Biochim. Biophys. Acta, 148:517 (1967).
81. H. Walter, R. Garza and R.P. Coyle, Partition of DEAE-dextran in
    aqueous dextran-polyethylene glycol phases and its effect on the
    partition of cells in such systems, Biochim. Biophys. Acta, 156:
    409 (1968).
82. V.P. Shanbag and C-G. Axelsson, Hydrophobic interaction determined by
    partition in aqueous two-phase systems. Partition of proteins in
    systems containing fatty acid esters of poly(ethylene glycol), Eur.
    J. Biochem., 60:17 (1975).
83. E. Eriksson, P-A. Albertsson and G. Johansson, Hydrophobic surface
    properties of erythrocytes studied by affinity partition in aqueous
    two-phase systems, Mol. and Cell. Biochem., 10:123 (1976).
84. H. Walter and J. Krob, Hydrophobic affinity partition in aqueous two-
    phase systems containing poly(ethylene glycol)-palmitate of right-
    side-out and inside-out vesicles from human erythrocyte membranes,
    FEBS Lett., 61:290 (1976).
85. K-E. Magnusson and G. Johansson, Probing the surface of Salmonella
    typhimurium SR and R bacteria by aqueous biphasic partitioning in
    systems containing hydrophobic and charged polymers, FEMS Lett.,
    2:225 (1977).
86. K-E. Magnusson, O. Stendahl, C. Tagesson, L. Edebo and G. Johansson,
    The tendency of smooth and rough Salmonella typhimurium bacteria
    and lipopolysaccharide to hydrophobic and ionic interaction, as
    studied in aqueous polymer two-phase systems, Acta Path. Microbiol.
    Scand., Sect. B 85:212 (1977).
87. P-A. Albertsson, Partition between polymer phases, J. Chromatogr.,
    159:111 (1978).
88. E. Eriksson and P-A. Albertsson, The effect of the lipid composition
    on the partition of liposomes in aqueous two-phase systems, Biochim.
    Biophys. Acta, 507:425 (1978).
89. H. Miorner, E. Myhre, L. Bjorck and G. Kronvall, Effect of specific
    binding of human albumin, fibrinogen and immunoglobulin G on surface
    characteristics of bacterial strains as revealed by partition experi-
    ments in polymer phase systems, Infect. Immun., 29:879 (1980).
90. H. Miorner, P-A. Albertsson and G. Kronvall, Isoelectric points and
    surface hydrophobicity of gram-positive cocci as determined by cross-
    partition and hydrophobic affinity partition in aqueous two-phase
    systems, Infect. Immun., 36:227 (1982).
91. P.T. Sharpe and G.S. Warren, The incorporation of glycolipids with
    defined carbohydrate sequence into liposomes and the effects on
    partition in aqueous two-phase systems, Biochim. Biophys. Acta,
    772:176 (1984).
92. R. Reitherman, S.D. Flanagan and S.H. Barondes, Electromotive phenom-
    ena in partition of erythrocytes in aqueous polymer two phase syst-
    ems, Biochim. Biophys. Acta, 297:193 (1973).
93. P-A. Albertsson, Particle fractionation in liquid two-phase systems:

The composition of some phase systems and the behaviour of some
model particles in them. Application to the isolation of cell
walls from microorganisms, Biochim. Biophys. Acta, 27:378 (1958).

94. S. Hjerten, J. Rosengren and S. Pahlman, Hydrophobic interaction
chromatography: The synthesis and the use of some alkyl and aryl
derivatives of agarose, J. Chromatogr., 101:281 (1974).

95. G. Halperin and S. Shaltiel, Homologous series of alkyl agarose dis-
criminate between erythrocytes from different species, Biochem.
Biophys. Res. Commun., 72:1497 (1976).

96. S. Hjerten, Hydrophobic interaction chromatography of proteins on
neutral adsorbents, in: "Methods of Protein Separation"
N. Calsimpoolas, ed., Vol. 2, Plenum, New York (1976).

97. B.H.J. Hofstee and N.F. Otillio, Non-ionic adsorption chromatography
of proteins on neutral adsorbents, J. Chromatogr., 159:57 (1978).

98. C.J. Smyth, P. Jonsson, E. Olsson, O. Soderlind, J. Rosengren,
S. Hjerten and T. Wadstrom, Differences in hydrophobic surface
characteristics of porcine enteropathogenic Escherichia coli with
or without K88 antigen as revealed by hydrophobic interaction
chromatography, Infect. Immun., 22:462 (1978).

99. S. Tylewska, S. Hjerten and T. Wadstrom, Contribution of M protein to
the hydrophobic surface properties of Streptococcus pyogenes, FEMS
Lett., 6:249 (1979).

100. A. Faris, T. Wadstrom and J.H. Freer, Hydrophobic adsorptive and
hemagglutinating properties of Escherichia coli possessing colon-
isation factor antigens (CFA-I or CFA-II), Type I pili or other
pili, Curr. Microbiol., 5:67 (1981).

101. S. Shaltiel, Hydrophobic chromatography, Methods Enzymol., 34:126
(1974).

102. J. Rosengren, S. Pahlman, M. Glad and S. Hjerten, Hydrophobic interaction chromatography on non-charged sepharose derivative: Binding
of a model protein, related to ionic strength, hydrophobicity of
the substituent, and degree of substitution, Biochim. Biophys. Acta,
412:51 (1975).

103. J-I. Ochoa, Hydrophobic (interaction) chromatography, Biochimie,
60:1 (1978).

104. T. Malmqvist, Bacterial hydrophobicity measured as partition of
palmitic acid between the two immiscible phases of cell surface and
buffer, Acta Path. Microbiol. Scand., Sect. B 91:69 (1983).

105. S. Kjelleberg, C. Lagercrantz and Th. Larsson, Quantitative analysis
of bacterial hydrophobicity studied by the binding of dodecanoic
acid, FEMS Lett., 7:41 (1980).

106. D.A. Rawlins and J.B. Kayes, Pharmaceutical suspension studies. I.
A comparison of adsorption of polyvinylalcohol at the diloxanide
furoate BP and polystyrene latex water interface, Int. J. Pharm.,
13:145 (1983).

107. J. Kreuter, Physicochemical characterisation of polyacrylic nanoparticles, Int. J. Pharm., 14:43 (1983).

PARTICLE SIZE ANALYSIS OF COLLOIDAL SYSTEMS BY PHOTON CORRELATION SPECTROSCOPY

S.J. Douglas and S.S. Davis

University of Nottingham, Department of Pharmacy
University Park, Nottingham NG7 2RD

INTRODUCTION

Photon correlation spectroscopy (PCS), also known as quasielastic light scattering, offers many advantages over more classical techniques; such as electron microscopy, for the particle size analysis of colloidal systems. Particularly advantageous are the rapid analysis time (approximately 1 minute), minimal sample preparation, wide applicable size range (5-3000nm) and no prior requisite knowledge of sample concentraion.[1] In addition, PCS has many biological and medical applications[2] and may be used to study various colloidal properties such as the nature of particle surfaces,[3] kinetics of colloid formation[4] and antigen-antibody induced particle aggregation.[5]

Obviously, it is not possible in the space of this article to cover all aspects of PCS since previous NATO advanced study institutes have been devoted solely to this subject.[6,7] Therefore, it is hoped to provide an overview based on experience gained in the use of PCS during our work with particulate drug targeting systems.

PCS THEORY

Only a brief outline of PCS theory will be given here since full treatments exist elsewhere[8,9] and need not be repeated herein.

When a colloidal suspension is illuminated by a laser light source the relative phases of the scattered light waves from different particles vary as the particles undergo Brownian motion, causing the intensity of scattered light at the detector to fluctuate with time. Although fluctuations for individual particles occur randomly, there is a well defined lifetime for the build up and decay of the total scattered light as shown in Fig. 1. The technique of PCS makes use of the fact that the time dependence of these intensity fluctuations can be related to the translational diffusion coefficient of the scatterers in the illuminated medium.

The particle size of the colloid is found through the technique of photocount autocorrelation.[10] To understand this process, consider the intensity curve of Fig. 1 being divided into time intervals length T, which in practice is the correlator sample time. During any given sample

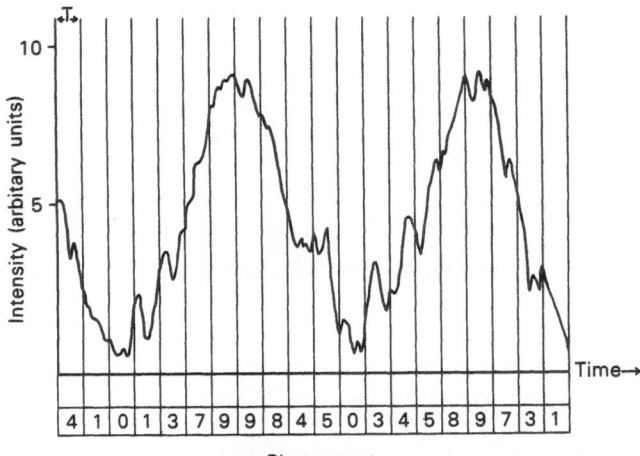

Fig. 1. Typical trace of fluctuating scattered light intensity from a colloidal dispersion and the corresponding distribution of photocounts.[10] Assigning the photocount (n) in the first time interval (t_1) as nt_1 and the total number of intervals as N, then the channel contents for a 64 channel autocorrelator would contain the following sums of products;

Channel 1 = $(nt_1 \cdot nt_2) + (nt_2 \cdot nt_3) + (nt_3 \cdot nt_4) + \ldots$

$\ldots (nt_{N-64} \cdot nt_{N-63})$

Channel 2 = $(nt_1 nt_3) + (nt_2 \cdot nt_4) + (nt_3 \cdot nt_5) + \ldots$

$\ldots (nt_{N-64} \cdot nt_{M-62})$

Channel 64 = $(nt_1 \cdot nt_{65}) + (nt_2 \cdot nt_{66}) + (nt_3 \cdot nt_{67})$

$+ \ldots (nt_{N-64} \cdot nt_N)$.

time the scattering intensity is proportional to the number of detected photons (n). The photocount in sampling interval i, centred on time t_1, is defined as $n(t_i)$. The autocorrelator then constructs the sum of products

$$\sum_{i=1}^{i=(N-M)} n(t_i) n(t_{i+m})$$ [1]

for a range of delay times mT (also given as τ), where m = 1,2,3,...., M, and M is the number of channels built into the autocorrelator. N is the number of sampling intervals of length T in the experimental run. As the experiment proceeds the second order correlation function, $G_2(\tau)$, accumulates in the correlator channels until time NT. The resulting correlation function is normalised to give the normalised second order autocorrelation function, $g_2(\tau)$, which is an exponential curve (Fig. 2, upper) described by the equation

$$g_2(\tau) = 1 + \exp(-2DK^2\tau)$$ [2]

where D is the translational diffusion coefficient of the colloid and K is the magnitude of the scattering vector given as

$$K = \frac{4\pi n \sin(\theta/2)}{\lambda}$$ [3]

in which n is the refractive index of the scattering medium, θ is the scattering angle and λ is laser wavelength.

Equation [2] may be written as

$$\ln[g_2(\tau)-1] = -2DK^2\tau$$ [4]

so that a plot of $\ln[g_2(\tau)-1]$ vs τ gives a straight line of slope $-2DK^2$ (Fig. 2, lower). Since K is essentially a constant, D is readily obtained and used to calculate the spherical hydrodynamic diameter, d_h, by recourse to the Stokes-Einstein equation

$$d_h = \frac{kT}{3\eta D\pi}$$

where k is Boltzmann's constant, T the absolute temperature and η the solvent viscocity.

The above treatment is only applicable for a monodisperse system of ideal spherical noninteracting particles that are small compared to the incident light wavelength. For a polydisperse system $g_2(\tau)$ consists of a sum of exponentials and the log-plot deviates from linearality (Fig. 2, lower). However, the correlation function may be adequately described by a polynomial of order up to three[11,12] from which the z-average particle diameter, d_z, is obtained,[13] together with the z-average normalised variance of the distribution known as the polydispersity index or quality parameter, Q.[10] For a monodisperse system Q should be zero, theoretically, although this is difficult to achieve in practice (for example, we found for a monodisperse latex $Q \simeq 0.03$). Generally, for a relatively narrow distribution, Q is less than 0.1.[12]

PCS data can also be analysed by a number of mathematical techniques[14,15] to yield a particle size distribution. However, such treatments are highly involved and require very precise data which are susceptible to experimental error. Alternatively, a theoretical size distribution can be generated from d_z and Q according to the relatively simple method developed by Pearce.[16] Although this theory is only applicable to log-normal unimodal size distributions it does allow the calculation of the number average particle diameter, d_n, which may be directly compared to results obtained by electron microscopy.[17]

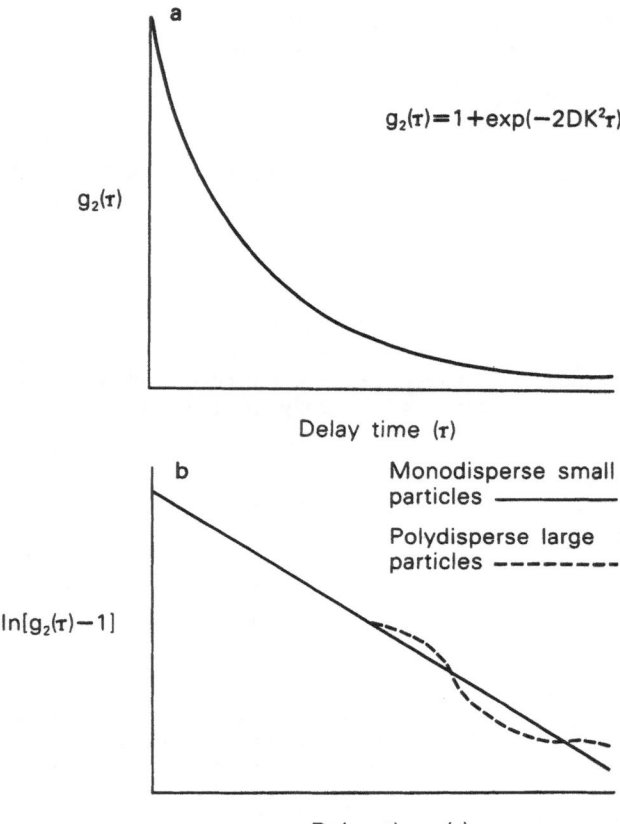

$$g_2(\tau) = 1 + \exp(-2DK^2\tau)$$

Fig. 2. Exponential decay of the normalised second order
correlation function (upper); log-plot of the
correlation function for monodisperse small particles (-) and large polydisperse particles (---)
(lower).

PCS INSTRUMENTATION

The PCS apparatus is represented diagramatically in Fig. 3. A coherant light source is provided by a helium-cadmium laser (output=14mW, λ=441.6 nm), which illuminates the sample cell held in a thermostatically controlled water bath ($25\pm0.05^{\circ}$C). The resulting scattered light is detected at 90° by a variable angle photomultiplier assembly and the signal passed to a Malvern K7025, 64-channel, multibit autocorrelator via an amplifier/discriminator. The autocorrelator is interfaced with a microprocessor which analyses the data and also controls the system. Data are given as the translational diffusion coefficient of the colloid and the corresponding hydrodynamic diameter, together with the polydispersity index.

EXPERIMENTAL CONSIDERATIONS

Details regarding the design and optimal use of PCS systems have been described elsewhere.[8,9,16] Many of these factors should not concern the user simply interested in obtaining particle size data from a correctly adjusted system. However, there are certain procedures which should be understood to ensure the acquisition of accurate data:

a) Sample preparation. Stringent measures should be taken to exclude sample contamination by 'dust', since this can lead to eroneously large values for the particle size and polydispersity index. All glassware should be thoroughly clean and sample preparation preferably carried out in a laminar air flow cabinet. The sample should be diluted with pre-filtered solvent and then passed through a membrane filter that will remove large particles without affecting the particle size distribution of the colloid. Sample concentration should be as low as possible to avoid particle interactions, without interfering with the photocount efficiency.

b) Photomultiplier detection angle. Equation [3] shows the scattering vector, K, to be dependent on the measurement angle, θ. For very small particles $g_2(\tau)$ is independent of θ, but for large and/or polydisperse systems in the Mie scattering region $g_2(\tau)$ and the calculated particle size are a complex function of θ.[1] For smaller particles a measuring angle near 180° can avoid problems due to dust contamination and increases precision.[18] However, for larger polydisperse systems backscattering of light is a major problem. Generally, measurements are taken at 90° as a compromise or several readings taken at a series of angles and the size calculated from a plot of $D(\theta)$ vs $\sin^2(\theta/2)$ at $\sin^2(\theta/2)\rightarrow0$.[19] Correlation measurements at different angles can also give information on particle shape and in the case of large scatters, on size and polydispersity.[20]

c) Correlator sample time. The relationship between correlator sample time T, and the measured particle size and polydispersity index is shown graphically in Fig. 4. Similar results have been reported elsewhere.[18] The correct particle size is taken when Q is at a minimum and the measured size is in the plateau region. For accurate work the optimum value of T should be determined for each sample by taking results at a series of sample times. However, it is possible to determine the correct sample time from the shape of the correlation function as shown in Fig. 5.

d) Experimental duration. The experimental duration, NT, has little effect on the calculated particle size. However, in order to obtain a statistically reliable result, a sufficiently large number of samples, N, should be taken such that $N \not< 10^6$.

e) Precision of the data. Although the particle size can be found from a single experimental run, it is normal to take an average result obtained

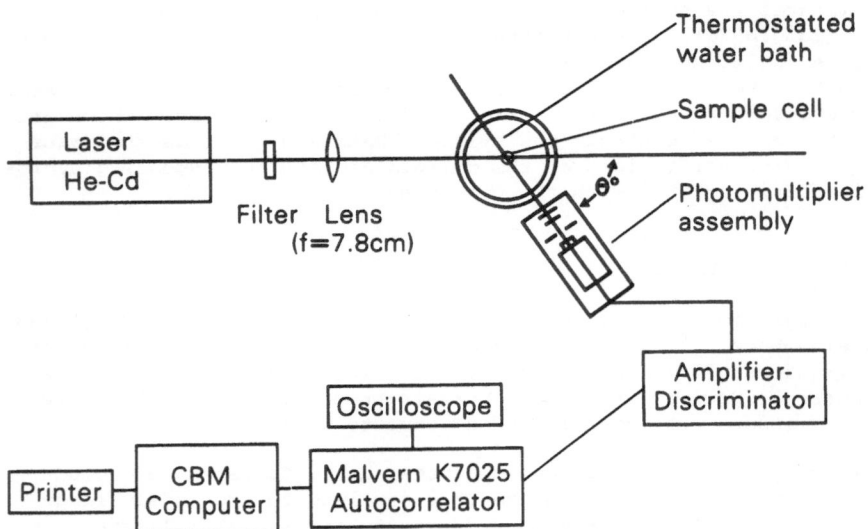

Fig. 3. Diagramatic representation of the PCS apparatus

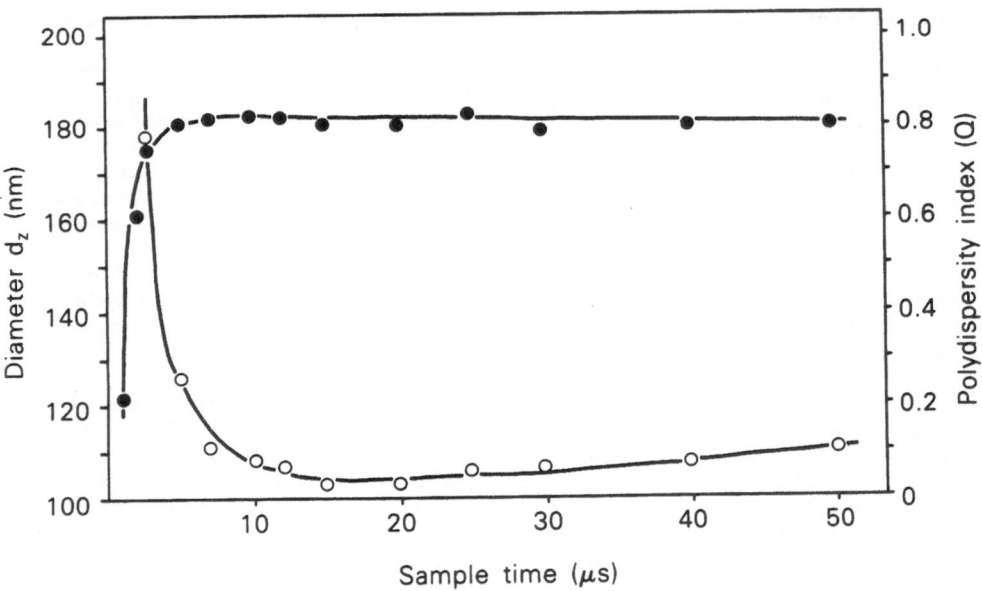

Fig. 4. Variation of PCS measured particle size (●) and
polydispersity index (o) as a function of correlator
sample time.

from several repeats. We normally analyse each sample a total of 15 times which gives a coefficient of variation for the diameter measurements of less than 3% and a measurement standard deviation for the polydispersity index of typically 0.02.

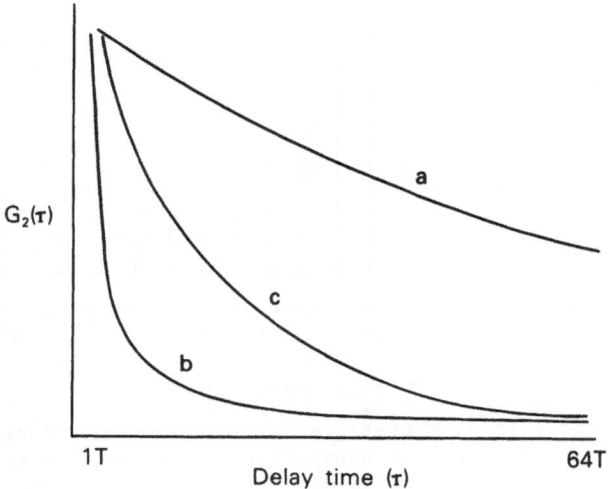

Fig. 5. Shapes of correlation functions obtained at various sample times; (a) Sample time is too low such that all delay times show high correlation. Values obtained for particle size would be too low and Q too high; (b) Sample time is too high, such that only the first few delay times show any degree of correlation. Values for both particle size and Q would be too high; (c) Sample time correct.

RESULTS WITH POLY (ALKYL 2-CYANOACRYLATE) NANOPARTICLES

PCS has been extensively used for the particle size analysis of poly-(butyl 2-cyanoacrylate) nanoparticles.[17,21] These particles are log-normally distributed with essentially a single mode as shown by the histogram in Fig. 6 which, when transferred to a log-probability plot, gives a straight line (r = 0.9977). It is therefore possible to apply Pearce's method[16] for determining the theoretical size distribution and this was found to be in good agreement with the size distribution obtained by electron microscopy.[17]

When determining the average diameter of nanoparticles it is necessary to consider the polydispersity index. For example, when investigating the effect of stirrer speed on nanoparticle formation[17] d_z was found to increase from 126 to 161 nm (Table 1). However, this was accompanied by an increase in Q such that d_n was relatively unchanged (Table 1). This indicates that

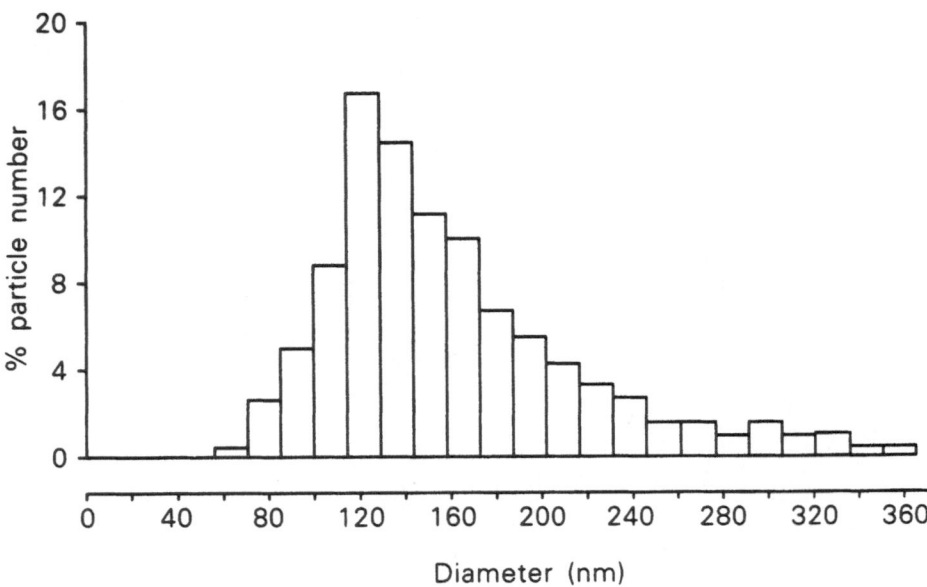

Fig. 6. Particle size histogram of a typical poly(butyl 2-cyanoacrylate) nanoparticle sample. The system is log-normally distributed with a single mode.

Table 1. Variation in d_z and Q of poly(butyl 2-cyanoacrylate) nanoparticles with stirrer speed.

Stirrer Speed (rpm)	Q	d_z (nm)	d_n (nm)
600	0.070	126	110
750	0.087	131	111
1000	0.099	136	113
1500	0.071	137	119
2000	0.140	142	109
2500	0.121	148	118
3000	0.149	161	122

The value of d_n shows no significant change with speed.[17]

stirrer speed was affecting the width of the particle size distribution (PSD) without having any significant effect on the mean diameter.

The PSD is also important in considering the extravasation of colloidal drug carriers. In terms of particle number the theoretical PSD for a poly (butyl 2-cyanoacrylate) system (d_z = 126 nm, Q = 0.13) is given in Fig. 7. If 100 nm is taken as an arbitary diameter below which particles can theoretically escape the vascular circulation via fenestrae, then integration of the particle number curve indicates that only 44% of the carrier could extravasate despite the average size (d_n = 98.6 nm) being approximately 100 nm. In terms of drug delivery even less of the drug would be expected to reach the target site, since the payload capacity of these nanoparticles is mainly dependent on surface sorption of the drug.[22] Thus, if the PSD is plotted in terms of particle surface area and integrated, only 21% of the therapeutic payload has the possibility of extravasation. For carrier systems where drug loading is a function of particle volume, such as liposomes, an even smaller percentage of the total payload would be contained in carriers with diameter less than 100 nm. Therefore, it can be seen that the particle size distribution could, theoretically, be a major factor in determining the targeting efficiency of colloidal carriers.

Particle diameter as measured by PCS is not only affected by changes in polydispersity but is also influenced by variations in the thickness of any tightly bound surface layer, since PCS yields a hydrodynamic diameter. This is well illustrated with nanoparticles stabilised by diethylaminoethyl (DEAE)–dextran.[23] This highly charged cationic polymer changes its conformation with variation in pH and ionic strength.[24] Thus, the PCS measured diameter of these DEAE-dextran coated nanoparticles is seen to be

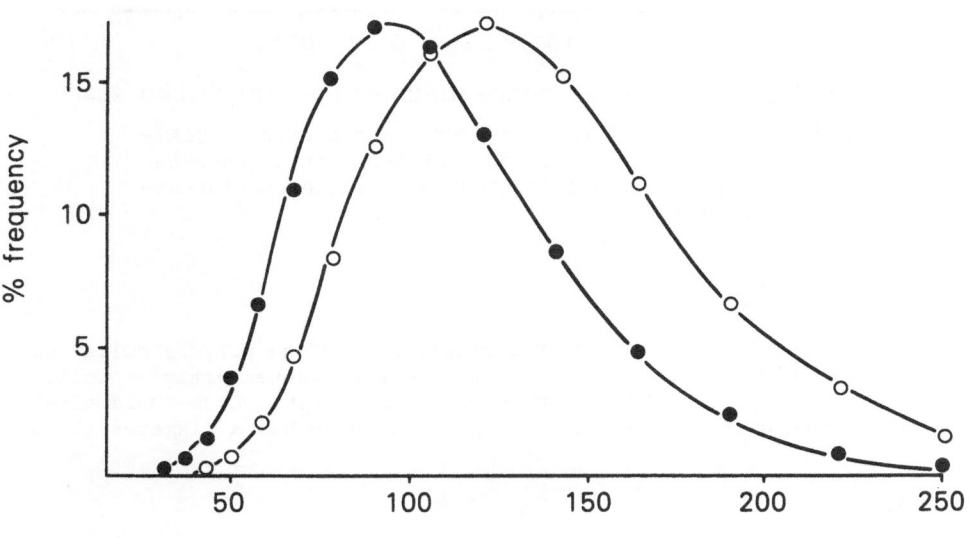

Fig. 7. Theoretical particle size distributions for a nanoparticle sample (d_z = 126nm; Q = 0.13) based on particle number (●) and particle surface area (o).

highly dependent on the pH and ionic strength of the dispersion medium (Fig. 8). As pH is increased the protonated amino groups lose their charge and the polymer adopts a less extended conformation due to the decrease in electrostatic repulsion. Similarly, increasing ionic strength decreases

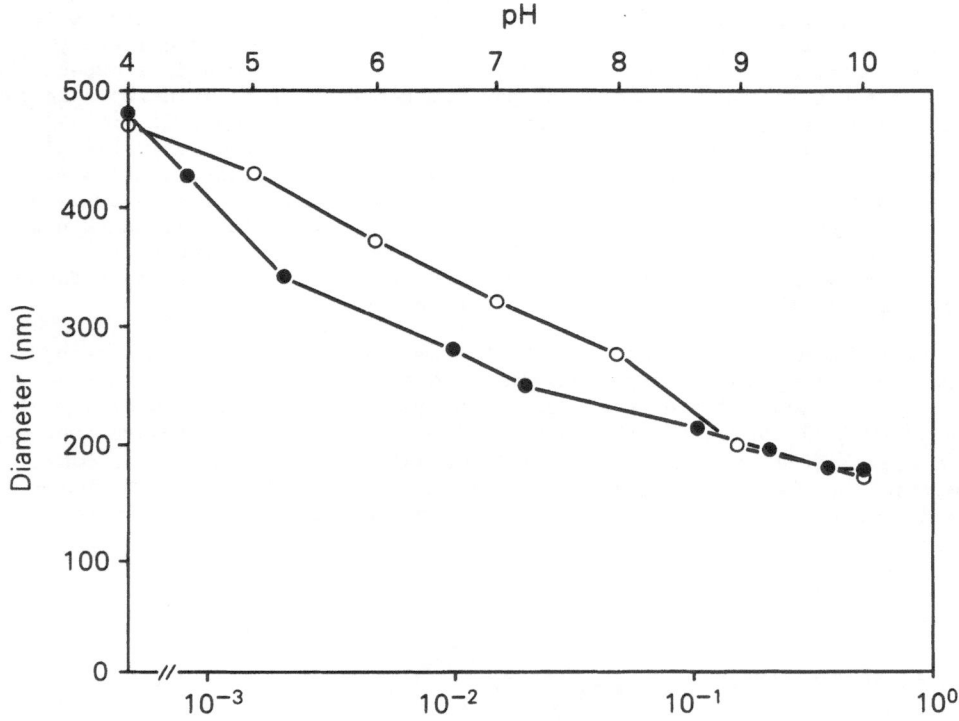

Fig. 8.   Variation of the measured PCS diameter of DEAE-dextran stabilised nanoparticles as a function of pH (o) and added sodium chloride concentration (•).

electrostatic repulsion due to charge shielding and the polymer collapses as depicted in Fig. 9. In terms of drug targeting these particles would be expected to have a low hydrodynamic diameter <u>in vivo</u> which could affect their ability to extravasate since the particle core has a diameter 100 nm.

CONCLUSIONS

PCS is a highly versatile instrument that can provide information on particle size, size distribution and surface characteristics. All of these properties are important considerations in the field of drug targeting. The ability of PCS to yield a hydrodynamic diameter is particularly important since this is more relevant to the in-vivo situation in considering the potential ability of a colloidal carrier to reach extravascular target sites.

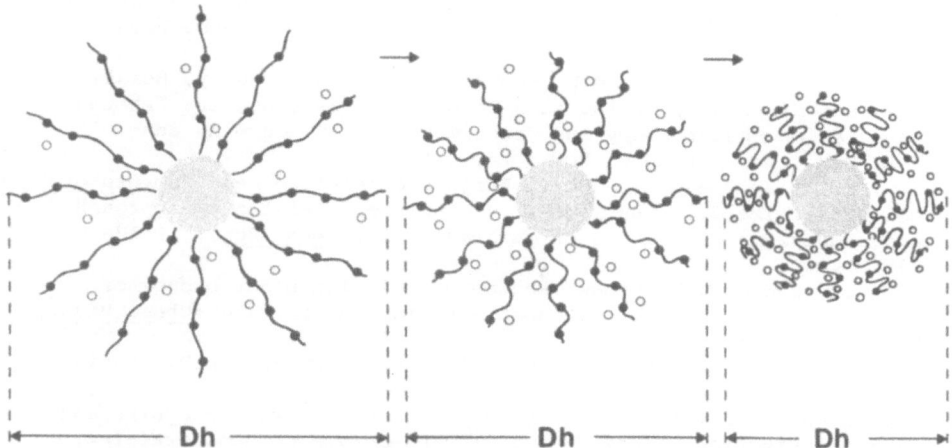

Fig. 9. Schematic representation of the decrease in hydrodynamic diameter (Dh) of DEAE-dextran stabilised
nanoparticles as ionic strength of the dispersion
medium is increased. Particle core is represented
by the shaded area. ∿ denotes charged polyer chain;
o denotes counterion in solution.

REFERENCES

1. M.L. McConnell, Particle size determination by quasielastic light
    scattering, Anal. Chem., 53:1007A (1981).
2. G.B. Benedek, Biological and medical applications of light scattering
    spectroscopy, in: "Photon Correlation Spectroscopy and Velocimetry",
    H.Z. Cummins and E.R. Pike, eds., Plenum, New York (1977).
3. J.W.S. Goossens and A. Zembrod, Characterization of the surface of
    polymer latexes by photon correlation spectroscopy, Colloid Polym.
    Sci., 257:437 (1979).
4. A.R. Goodall, K.J. Randle and M.C. Wilkinson, A study of the emulsif-
    ier-free polymerisation of styrene by laser light scattering tech-
    niques, J. Colloid Interface Sci., 75:493 (1980).
5. G.K. Von Schulthess, R.J. Cohen, N. Sakato and G.B. Benedek, Laser
    light scattering spectroscopic immunoassay for mouse IgA, Immuno-
    chem., 13:955 (1976).
6. H.Z. Cummins and E.R. Pike, (Eds.), "Photon correlation and Light
    Beating Spectroscopy", Plenum, New York, (1974).
7. H.Z. Cummins and E.R. Pike, (Eds.), "Photon Correlation Spectroscopy
    and Velocimetry", Plenum, New York, (1977).
8. B. Chu, "Laser Light Scattering", Academic Press, New York, (1974).
9. B.J. Berne, and R. Pecora, "Dynamic Light Scattering", John Wiley and
    Sons, New York, (1976).
10. P.N. Pusey, D.E. Koppel, D.W. Schaeffer, R.D. Camerini-Otero, and
    S.H. Koenig, Intensity fluctuation spectroscopy of laser light
    scattered by solutions of spherical viruses, R17, $Q\beta$, BSV, PM2 and
    T7. I. Light-scattering technique, Biochemistry, 13:952 (1974).
11. D.E. Koppell, Analysis of macromolecular polydispersity in intensity
    correlation spectroscopy. Method of cummulants, J. Chem. Phys.,
    (1972).

12. J.C. Brown, P.N. Pusey and R. Dietz, Photon correlation study of poly-disperse samples of polystyrene in cyclohexane, J. Chem. Phys., 62:1136 (1975).

13. D.J. Green, D.B. Sattelle, D.W. Westhead, and K.H. Langley, Relative size and dispersity of isolated chromassin granules, in: "Photon Correlation Spectroscopy and Velocimetry", H.Z. Cummins and E.R. Pike, eds., Plenum, New York, (1977).

14. B. Chu, E. Gulari, and E. Gulari, Photon correlation measurements of colloidal size distributions II. Details of a histogram approach and comparison of methods of data analysis, Phys. Ser., 19:476 (1979).

15. A.N. Lavery and J.C. Earnshaw, Photon correlation spectroscopy of particle polydispersity: a cubic B-spline analysis, J. Chem. Phys., 80:5438 (1984).

16. A.J. Pearce, A kinetic study of emulstion coalescence, Ph.D. Thesis, University of Nottingham (1984).

17. S.J. Douglas, L. Illum, S.S. Davis and J. Kreuter, Particle size and size distribution of poly(butyl 2-cyano-acrylate) nanoparticles. I. Influence of physicochemical factors, J. Colloid Interface Sci., 101:149 (1984).

18. E.J. Derderian and T.B. MacRury, Quasielastic light scattering on standard polystyrene lactices, J. Dispersion Sci. Tech., 2:345 (1981).

19. A.A. Al-Saden, A.T. Florence, T.L. Whateley, F. Puisieux and C. Vaution, Characterization of mixed nonionic surfactant micelles by photon correlation spectroscopy and viscosity, J. Colloid Interface Sci., 86:51 (1982).

20. P.N. Pusey and W. van Megan, Detection of small polydispersities by photon correlation spectroscopy, J. Chem. Phys., 80:3513 (1984).

21. S.J. Douglas, L. Illum and S.S. Davis, Particle size and size distrib-ution of poly (butyl 2-cyanoacrylate) nanoparticles. II. Influence of stabilizers, J. Colloid Interface Sci., 103:154 (1985).

22. M. El-Samaligy and P. Rohdewald, Triamcinolone diacetate nanoparticles, a sustained release drug delivery system suitable for parenteral administration, Pharm. Acta. Helv., 57:201 (1982).

23. S.J. Douglas, L. Illum and S.S. Davis, Poly (butyl 2-cyanoacrylate) nanoparticles with differing surface charges, J. Controlled Release, in press, (1986).

24. F. Gubensek and S. Laparije, Potentiometric titration studies of diethylaminoethyl dextran, J. Macromol. Sci. Chem., A2:1045 (1968).

STABILITY OF LIPOSOMES ON STORAGE

D.J.A. Crommelin,* G.J. Fransen,* and P.J.M. Salemink**

Dept. of Pharmaceutics,* University of Utrecht
Catharijnesingel 60, 3511 GH Utrecht, and Organon
International** B.V., P.O. Box 20, 5340 BH Oss
The Netherlands

INTRODUCTION

Liposomes that are developed to be used as drug carriers in therapy have to be sufficiently stable on storage. There is no consensus on the exact requirements, but shelf lives of over one year are certainly preferable. Liposomes can be unstable for a number of reasons. The liposome structure can change because of aggregation or fusion processes (physical stability), the associated drug can leak out of the vesicles and the phospholipids or the associated drug might be chemically unstable (hydrolysis, oxidation).[1,2] All these changes might have an impact on the therapeutic effect of the product. Therefore, concepts were developed to improve liposome stability on storage. These concepts can be classified into three classes:

1) Storage of liposomes as aqueous dispersions.
2) Storage of liposomes in frozen condition or as a freeze dried product (removal of water by sublimation).
3) Storage of liposomes as concentrated dispersions or after removal of all "free" water via evaporation.

The "state of the art" and the pros and cons of these three approaches will be discussed in this contribution.

The discussion will be limited to two aspects of the stability problem of liposomes: the physical stability against aggregation or fusion of the dispersion and the drug retention (or latency), defined as the percentage of drug that is liposome-associated. We will focus on two compounds: 5,6-carboxyfluorescein (CF) and doxorubicin (DXR). CF is an hydrophilic compound that is assumed to have a low tendency to interact with negatively charged bilayer structures;[3] DXR is an amphiphilic drug that interacts with the bilayer, mainly via an electrostatic interaction with the negative charge inducing phospholipids.[4,5]

For highly lipophilic compounds that strongly interact with the bilayer (e.g. cholesterol) leakage from the bilayer does not occur on storage. For these compounds only physical and chemical stability problems have to be addressed. In aqueous dispersions the proper choice of the bilayer constituents will prevent aggregation or fusion. If this system is chemically

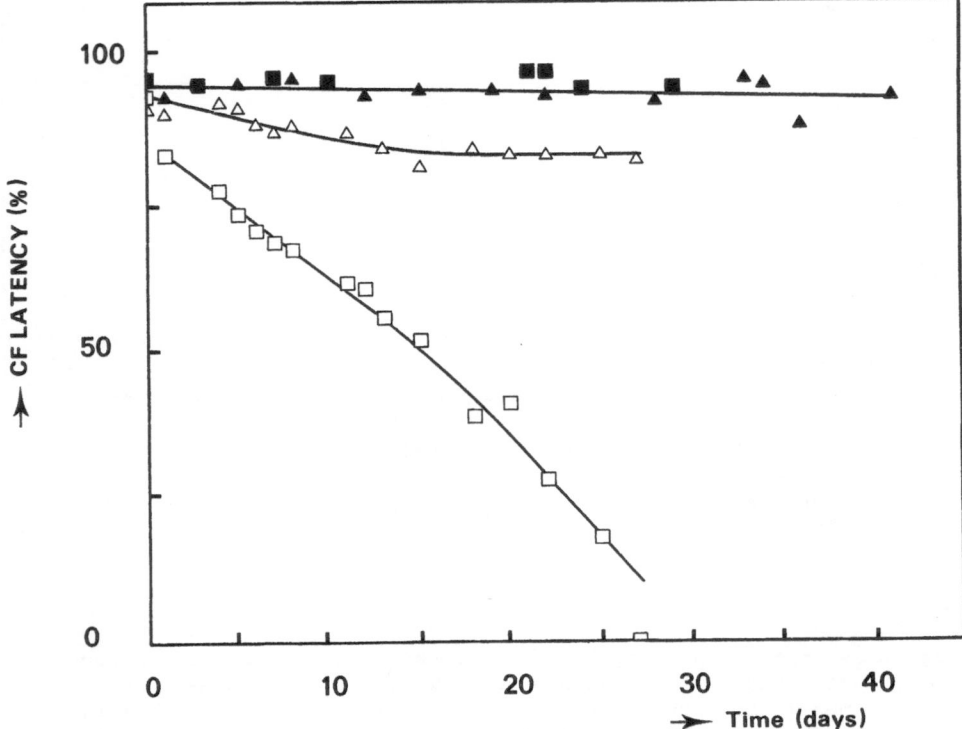

Fig. 1. CF retention (latency) in REV stored at 4-6°C.
Iso-osmotic sodium chloride/10 mM Tris solutions,
pH 7.4. Lipid composition of liposomes was
PC/PS (□) (9:1)(0.21 μm); PC/PS/chol (△)
(9:1:4)(0.23 μm); DSPC/DPPG (■)(10:1)(0.22 μm);
DSPC/DPPG/chol (▲)(10:1:5)(0.25 μm mean diameter);
PC, egg yolk phosphatidylcholine; PS, phosphatidylserine; DPPC, dipalmitoylphosphatidylcholine; DPPG,
dipalmitoylphosphatidylglycerol; chol, cholesterol;
all from Sigma Chemicals, St. Louis, MO; data taken
from Ref. 17.

unstable in water then, as will be shown later, the proper selection of
cryoprotectants offers the opportunity to lyophilize this dispersion with
the lipophilic drug successfully.

STORAGE OF LIPOSOMES AS AQUEOUS DISPERSIONS

Technically the simplest way to store liposome dispersions is to leave
the dispersions just as they are after finishing the hydration, sizing, free
drug removal and sterilization process. In this section the influence of
bilayer composition and type of liposome (multilamellar vesicles, MLV;
small unilamellar vesicles, SUV) on the retention of CF (hydrophilic, no
bilayer interaction) and DXR (amphiphilic, bilayer interacting) will be
discussed.

CF containing liposomes (reverse phase evaporation vesicles, REV) were
prepared.[6,7] Phosphilipid combinations that formed either fluid or gel

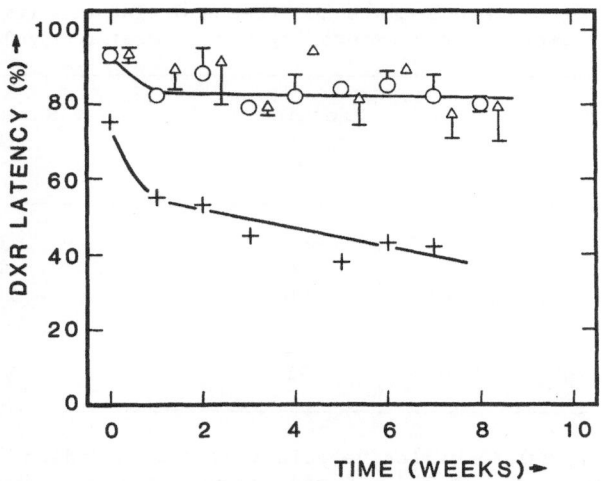

Fig. 2. DXR retention (latency) in extruded (0.2 μm pores)
or sonicated dispersions at 4-6°C.
○ and △ , PC/PS/chol liposomes (MLV) without
and with alpha-tocopherolacetate (10:1:4:0.1, molar
ratio). Vertical bars indicate standard deviations.
Phospholipid concentration was about 12 μmol/ml;
DXR concentration was about 0.4 μmol/ml. +, sonicated liposomes (SUV) composed of PC/PS/chol, 10:1:4
molar ratio. Phospholipid concentration was about
12 μmol/ml; DXR concentration was about 0.1 μmol/ml;
data taken from Ref. 8.

state bilayer structures at the experimental temperature were chosen. Not
only bilayers consisting of pure phospholipids were used. The effect of
cholesterol inclusion in the bilayer on the stability of these vesicles
was also investigated. The results are shown in Fig. 1. PC/PS liposomes
without cholesterol lose CF relatively fast. Inclusion of cholesterol reduced the release rate dramatically. DSPC/DPPG liposomes (with and without
cholesterol) lost less than 10% of CF over 6 months. These results clearly
demonstrate the impact of the bilayer structure on the transport rate of a
hydrophilic, non-bilayer interacting compound. If there is the option to
use a negatively charged liposome with the bilayer in the gel state, then
a shelf life of months is feasible as far as drug retention is concerned.
None of the dispersions mentioned in Fig. 1 showed a tendency to aggregate
or fuse during the observed time interval.

The stability of DXR-containing liposomes was also investigated.[8] DXR
is an amphiphilic, bilayer-interacting cytostatic.[5] One of the variables
under investigation was the type of liposome, e.g. SUV obtained by ultra-
sonication or MLV after a sizing procedure via extrusion through polycarbonate membrane filters. In Fig. 2 the results are presented. SUV lost DXR
faster and to a larger extent than MLV. In the case of MLV, DXR release
stopped after a loss of about 15% of the encapsulated drug. MLV did not
aggregate or fuse on storage; in SUV dispersions the mean particle size
tended to increase.

Table 1. Effect of cryoprotectants and freeze drying conditions on CF retention after a freeze drying/rehydration cycle.

Additive	Freezing Temperature (°C)	Retention (%)
no additive	-50	0
mannitol 2.5%	-50	0
mannitol 2.5% + glycerol 10%	-50	20
no additive	-21	0
mannitol 2.5% + glycerol 10%	-21	37

CF (50 mM)-containing multilamellar vesicles; hydration medium, 10 mM Tris and 100 mM sodium chloride, pH 7.4. Lipid composition: PC-H (phospholipon-100 H, Nattermann GmbH, Koln, FRG) and DCP (dicetylphosphate), 10:1 (molar ratio), extruded through 0.6 µm pore membrane filters; concentration of PC-H: around 5 µ.mol/ml. Freeze drying equipment: Microprocessor-controlled Lyovac GT 20, Leybold Heraeus.

For CF and DXR the liposome structure (rigidity of the bilayer, size and number of bilayers) turned out to be extremely important as far as shelf life is concerned. In this study no attention was paid to chemical degradation reactions of the liposome-associated drug or the bilayer structure. By proper selection of the bilayer structure and preparation conditions (pH, oxygen removal, antioxidants in bilayer or aqueous phase, phospholipids with saturated alkyl chains and storage at 4-6°C) sufficient reduction in the chemical decomposition rates might be achieved.[1,2,9-11]

STORAGE OF LIPOSOMES AS FREEZE-DRIED PRODUCT: CONCENTRATION OF THE LIPOSOME DISPERSIONS VIA SUBLIMATION

A number of articles on freeze drying of liposomes have been published.[12-19] However, hardly any systematic work has been done in this field. Basic insights into the processes that occur during freezing of the liposomes and the subsequent sublimation and secondary drying step are still not available. The nature of the encapsulated compound (bilayer/no bilayer interaction) and the experimental conditions appeared to be critical to obtain a product with high drug retention and no fusion and aggregation after rehydration. In Tables 1, 2 and 3 experimental data on drug retention are presented to illustrate this statement.

In these experiments mannitol and glycerol were used. Mannitol was added to provide a proper matrix and glycerol to protect the liposomes from disintegration. Other saccharides (glucose, trehalose) were also effective as cryoprotectant.[16,21] Very lipophilic compounds (e.g. cholesterol) remain in the bilayer structure after a freeze drying/rehydration cycle. As cryoprotectants adequately protect the liposomes from fusion and aggregation, freeze drying might be the preferred technique for storage of unstable liposome dispersions with lipophilic compounds. For hydrophilic compounds and weakly bilayer interacting compounds like DXR the freeze drying technique has to be further optimized. To improve freeze drying results by rationale instead of by "trial and error", a better insight into the fundamentals of the freeze drying process is required.

Table 2. Influence of the nature of the encapsulated compound on
          the retention after a freeze drying/rehydration cycle.

Compound	Retention (%)
CF*	37
DXR**	58

The liposomes (MLV) were composed of PC-H/DCP (10:1 molar ratio). For
further information see Table 1. Dispersions with DXR contained around
1.0 mg/ml liposome-associated DXR and 30 mol/ml PC-H; hydration medium:
10 mM Tris, pH 4; extrusion through 0.6 µm pore membrane filters.
Freezing temperature: -21°C. Cryoprotectants: mannitol/glycerol, 10%.
* Hydrophilic, non bilayer-interacting.
** Amphiphilic, bilayer-interacting.

Table 3. The influence of bilayer composition on DXR retention after
          a freeze drying/rehydration cycle.

Composition of bilayer (molar ratio)	Retention (%)
DPPC/DPPG 10/1	13
DPPC/DPPG/chol 10/1/4	58
PC-H/DCP 10/1	58

Multilamellar vesicles; hydration medium: 10 mM Tris, pH 4; extruded
through 0.6 µm pore membrane filters at elevated temperatures. Free DXR
removal by cation ion exchange resin.[20] DXR concentration: around 1 mg/ml;
lipid concentration: around 30 µ mol/ml. Freezing temperature: -21°C;
cryoprotectants: mannitol/glycerol, 10%. Freeze drying equipment: see
Table 1.

LIPOSOME DISPERSIONS DURING A FREEZING/THAWING CYCLE

     To gain more insight into the processes occurring during freeze drying
of liposome dispersions we paid special attention to one stage in the total
process: freezing the liposome dispersions. Cryobiology has a long trad-
ition in studying freezing/thawing of cellular material.[22,23] The experience gained in this field can be used to design research programs to unravel
the basics of freezing/thawing of liposome dispersions.

     We studied the behaviour of CF containing liposomes (MLV, PC-H/DCP
10/1, extruded through 0.6 µm pore membrane filters). Parameters that
turned out to have an impact on the drug retention and particle size after
a freezing/thawing cycle were: freezing time, freezing rate, freezing temperature and the presence of cryoprotectants.[21] During cooling down the
dispersion freezing starts in the external water phase. Freezing of the
water phase inside the liposome structure takes more time or requires lower
temperatures. As a rule, internal ice formation will destroy the liposome

Table 4. The influence of freezing temperature and freezing time
          on the CF retention.

Temperature (°C)	Retention (%)	
	after 20 min	after 5 h
−15	84	39
−25	91	56
−30	93	67
−40	24	23
−58	22	20

MLV consisting of PC-H/DCP 10:1 (molar ratio), extruded through 0.6 μm
pore membrane filters. Aqueous external phase: 10 mM Tris, pH 7.4,
100 mM sodium chloride; internal phase: 10 mM Tris, pH 7.4, 50 mM CF.
Freezing rate: 7°C min. -1. Data taken from Ref. 21.

Table 5. The influence of freezing temperature on CF retention
          after a freezing/thawing cycle with and without a cryo-
          protectant mixture (10% glycerol/10% mannitol).

Temperature (°C)	Retention (%)	
	− Cryoprotectant	+ Cryoprotectant
−30	93	96
−40	27	89
−58	22	83

The dispersions are briefly described in Table 4. Freezing time: 20 min;
freezing rate: 7°C min^{-1}. Data taken from Ref. 21.

structure. Another reason for losing drug from the liposomes is the
osmotic pressure difference over the bilayer as a result of increasing concentrations of dissolved material in the external phase when (pure) ice is
formed in the external phase. Tables 4 and 5 show a selection of the
results obtained with CF containing liposomes.

     For a high CF retention level both freezing temperature and freezing
time are important. Apparently the process of CF loss did not stop after
reaching the selected freezing temperature. It is an open question whether
this loss is induced by osmotic pressure differences over the bilayer alone
or in combination with internal ice formation.[21] The sharp drop in CF retention for temperatures below -30°C might be related to the eutectic temperature of the system. Above the eutectic temperature pockets of "fluid"
water are still present, only below this temperature the whole system should

be solid. Electrical conductivity data indicated the eutectic temperature to be around -33°C. In the presence of cryoprotectants (10% mannitol/10% glycerol) even at temperatures below the eutectic temperature (-48°C) a relatively high CF retention was found. In the absence of cryoprotectants the mean particle size of the systems tended to grow. The addition of a cryoprotectant (glycerol) prevented this size increase. Glycerol might protect the liposomes by a direct interaction with the bilayer structure, by a reduced ice crystallization rate because of an increased viscosity or by a reduction of the eutectic temperature compared to the system without cryoprotectant.

Although a complete insight into the processes involved is still not available these data can be used to optimize the experimental conditions to preserve liposome integrity during a freeze drying/rehydration cycle (see section: Storage of liposomes as freeze-dried product).

STORAGE AFTER CONCENTRATION OF THE AQUEOUS LIPOSOME DISPERSIONS VIA EVAPORATION

In this approach liposome dispersions are prepared at lipid concentrations that can be easily manipulated in the process of narrowing down the size distribution, removal of free drug or sterilization by filtration. After finishing the preparation procedure water is removed by evaporation under reduced pressure conditions in a vacuum chamber. The volume of the external aqueous phase is reduced to relatively small dimensions compared to the internal aqueous phase in the liposomes. In case of leakage of drug from the concentrated liposome dispersions on storage, the amount of drug in the external phase will be relatively small in the equilibrium state. Therefore, high levels of drug retention can be obtained in a stable system with no net transport between the internal and external aqueous phase.

The advantage of this approach compared to freezing or freeze drying the liposome dispersions is that problems concerning the physical stability (e.g. fusion and aggregation) are expected to be less likely to occur than with freeze drying. A strong reduction of the chemical degradation rate is not expected for degradation processes that depend on the presence of water. In these cases optimization of the experimental conditions (pH, the presence of additives in the bilayer and the aqueous phase, selection of phospholipids) might reduce the degradation rate to acceptable levels.

If all free water is removed from the dispersions by evaporation, then a "dry" product is obtained comparable to freeze-dried dispersions. During the freeze drying process, however, water is mainly removed at temperatures below 0°C. Thus, in the evaporation procedure damage to the liposomes originating from the freezing step (e.g. the passage of eutectic temperatures) is avoided. Besides, the use of relatively high temperatures (e.g. 20°C) to remove the water speeds up the process, requires a relatively simple technology and cuts down energy costs compared to freeze drying.

Here the results of preliminary experiments to evaluate the potential benefits of using the evaporation-concentration technique to increase shelf life of liposomes are presented. CF (no bilayer interaction) and DXR (amphiphilic, bilayer-interacting) were used as model compounds. Multilamellar vesicles were used. The details of the experimental conditions are given in tables 6 and 7.

From table 6 it is clear that already at relatively low levels of concentrating the dispersions (65% water removed) a substantial drop in CF retention was found. This drop might be the result of an increased osmotic pressure difference over the bilayer as a result of the concentration of

Table 6. CF retention and particle size after removing water by evaporation and readjustment to the initial volume.

Water removed (%)[a]	Retention (% ± S.D.)[b]	Diameter (μm)[c]
35	88 ± 9	0.51
65	71 ± 9	0.50
80	53 ± 14	n.d.
97	47 ± 19	0.77

CF-containing multilamellar vesicles; hydration medium: 5 mM Tris, pH 7.4, 2.5% glucose; PC-H/DCP 10:1 (molar ratio); extruded through 0.6 μm pore membrane filters; one ml samples. Evaporation was performed at room temperature in a vacuum chamber at pressures around 1 KPa.
(a) Reference point (0%) is initial water weight; 100% is theoretically calculated dry weight. Percentage water removed was calculated by monitoring sample weight.
(b) ratio of internal CF/phosphate related to initial ratio at t=0. n=3.
(c) measured by dynamic light scattering (Malvern type 7027, Malvern Instruments, U.K.)

Table 7. DXR retention after removing water by evaporation and readjustment to the initial volume.

Additive (2.5%)	Water removed (%)[a]	Retention (%)[b] 1 day	Retention (%)[b] 14 days
trehalose	88	87	92
trehalose	100	72(c)	72(c)
glucose	83	88	88
glucose	100	69(c)	64(c)
no additive	100	64(d)	61(d)

DXR retention was determined directly after readjustment of the volume. DXR-containing multilamellar vesicles; hydration medium: 5 mM sodium chloride, pH 4; 2.5% additive in the external phase; DPPC/DPPG/cholesterol 10:1:10 (molar ratio); 40 mMol lipids extruded through 0.6 μm pore membrane filters; one ml samples. Evaporation was performed in a vacuum chamber at room temperature at pressures around 1 KPa.

(a) See Table 6.
(b) Water was added after one or fourteen days.
(c) Mean diameter had a slight tendency to increase.
(d) Strong tendency to aggregate or fuse.

non-bilayer penetrating substances in the external phase. It was shown before that this type of liposome was sensitive to osmotic pressure differences.[21] Removal of 97% of the water did not induce extensive additional CF loss. Dynamic light scattering data indicated that these "dry" dispersions, in spite of the presence of a cryoprotectant, tended to aggregate or fuse upon rehydration with pure water.

DXR loss was less pronounced than CF loss at comparable levels of concentration. When 88% of the water was removed still around 90% of DXR was liposome-associated. During the process of the removal of the last 12% of water, substantial DXR loss occurred. The presence of the cryoprotectants (trehalose and glucose) did not reduce this drop of DXR retention as without the additives similar DXR retention data was found. However, after rehydration the additives protected the liposomes from fusion and aggregation. Two weeks of storage of the concentrated or dried dispersions did not change DXR retention levels after rehydration.

These preliminary experiments show that for DXR, a bilayer-interacting, amphiphilic compound, concentration of the aqueous phase offers perspectives. For non-bilayer interacting, hydrophylic compounds like CF additional work is necessary to increase drug retention. If all "free" water is removed, then additives are essential for protection of the vesicles against fusion and aggregation.

CONCLUDING REMARKS

There is still no standard "recipe" to stabilize all types of liposome-drug combinations. However, certain rules can be formulated. If the storage as an aqueous dispersion is preferred, then a proper selection of bilayer constituents (level of saturation and length of alkyl chains, charge inducing agents, cholesterol, antioxidants) and aqueous phase (additives, pH, ionic strength) can reduce the problems related to chemical and physical stability, and the release of the drugs from the liposomes to an acceptable level.

If the nature of the drug or the requirements for the bilayer structure do not permit optimum storage conditions as an aqueous dispersion, then freeze drying or concentration by evaporation might provide systems that are stable on storage. The results strongly depend on the experimental conditions. The processes controlling these techniques are still not well understood. In order to optimize the experimental conditions in a rational fashion, systematic work on the fundamentals of these processes should be carried out.

ACKNOWLEDGEMENTS

We acknowledge the help of Dr. Nassander and Mr. E. van Tuin for carrying out part of the experiments and Dr. H. Talsma for stimulating discussions.

REFERENCES

1. H.O. Hauser, 1971, The effect of ultrasonic irradiation on the chemical structure of egg lecithin, Biochem. Biophys. Res. Commun., 45:1049.
2. A.W.T. Konings, 1984, Lipid peroxidation in liposomes, in: Liposome Technology, G. Gregoriadis, ed., CRC Press, Inc., Boca Raton, Florida.
3. J.N. Weinstein, N. Yoshikami, P. Henkart, R. Blumenthal, and W.A. Hagins, 1977, Liposome-cell interaction. Transfer and intracellular release of a trapped fluorescent marker, Science, 195:489.
4. R. Goldman, T. Facchinetti, D. Bach, A. Raz, and M. Shinitzky, 1978, A differential interaction of daunomycin, adriamycin and their derivat-

ives with human erythrocytes and phospholipid bilayers, <u>Biochim. Biophys. Acta</u>, 512:254.

5. E. Goormaghtigh, P. Chatelain, J. Caspers, and J.M. Ruysschaert, 1980, Evidence of a specific complex between adriamycin and negatively charged phospholipids, <u>Biochim. Biophys. Acta</u>, 597:1.

6. J. Wilschut, 1982, Preparation and properties of phosphilipid vesicles, <u>in</u>: <u>Methodologie des Liposomes</u>, L.D. Leserman and J. Barbet, eds., Editions INSERM, Vol. 107, Paris.

7. F. Olson, C.A. Hunt, F. Szoka, W.J. Vail, and D. Papahadjopoulos, 1979, Preparation of liposomes of defined size distribution by extrusion through polycarbonate membranes, <u>Biochim. Biophys. Acta</u>, 557:9.

8. E.M.G. Van Bommel and D.J.A. Crommelin, 1984, Stability of doxorubicin-liposomes on storage: as an aqueous dispersion, frozen or freeze dried, <u>Int. J. Pharm.</u>, 22:299.

9. S. Frokjaer, E.L. Hjorth, and O. Worts, 1982, Stability and storage of liposomes, <u>in</u>: <u>Optimization of Drug Delivery</u>, H. Bundgaard, A. Bagger-Hansen, and H. Kofod, eds., Munkgaard, Copenhagen.

10. C.A. Hunt and S. Tsang, 1981, Alpha-Tocopherol retards autoxidation and prolongs the shelf-life of liposomes, <u>Int. J. Pharm.</u>, 8:101.

11. M.J.H. Janssen, D.J.A. Crommelin, G. Storm, and A. Hulshoff, 1985, Doxorubicin decomposition on storage. Effects of the pH, buffer components and liposome encapsulation, <u>Int. J. Pharm.</u>, 23:1.

12. J.R. Evans, F.J.Th. Fildes, and J.E. Oliver, 1978, Liposome, Deutsches Patentant 2818655.

13. G. Vanderberghe and R. Handjani, 1979, Improving the storage stability of aqueous dispersions of spherules, U.K. Patent Application GB 2013609 A.

14. R.E. Gordon, P.R. Mayer, and D.O. Kildsig, 1982, Lyophilization as a means of increasing shelf-life of phospholipid bilayer vesicles, <u>Drug Devel. and Ind. Pharm.</u>, 8:465.

15. S. Henry-Michelland, P. Poly, P. Puisieux, J. Delattre, and J. Likforman, 1983, Etude du compottement des liposomes lors d'experiments de congelation-decongelation et de lyophilisation: influence des cryoprotecteurs, C.R. 3eme Congres International de Technologie Pharmaceutique, APGI ed., Paris.

16. J.H. Crowe, L.M. Crowe, and R. Mouradian, 1983, Stabilization of biological membranes at low water activities, <u>Cryobiology</u>, 20:346.

17. D.J.A. Crommelin and E.M.G. Van Bommel, 1984, Stability of liposomes on storage: freeze dried, frozen or as an aqueous dispersion, <u>Pharm. Res.</u>, 1:159.

18. S.S. Abu-Zaid, M. Moril, and N. Takeguchi, 1984, Effects of freezing, freeze drying and cold storage on the size and membrane permeability of multilamellar liposomes, <u>Membrane</u> 9:43.

19. P.M. Shulkin, S.E. Seltzer, M.A. Davis, and D.F. Adams, 1984, Lyophilized liposomes: a new method for long term vesicular storage, <u>J. Microencap.</u>, 1:73.

20. G. Storm, L. Van Bloois, M. Brouwer, and D.J.A. Crommelin, The interaction of cytostatic drugs with adsorbents in aqueous media. The potential implications for liposome preparation, <u>Biochim. Biophys. Acta</u>, 818:343.

21. G.J. Fransen, P.J.M. Salemink, and D.J.A. Crommelin, 1986, Critical parameters in freezing of liposomes, <u>Biochim. Biophys. Acta</u>, submitted for publication.

22. P. Mazur, 1970, Cryobiology: the freezing of biological systems, <u>Science</u>, 168:939.

23. M.J. Ashwood-Smith, and J. Farrant, eds., 1980, <u>Low Temperature Preservation in Medicine and Biology</u>, Pitman Medical Ltd., Tunnbridge Wells, Kent.

Participants of the NATO Advanced Studies Institute "Targeting of Drugs with Synthetic Systems" held at Cape Sounion, Greece during 24 June–5 July 1985. The organizing committee included Gregory Gregoriadis (ASI Director and chairman), D. Chapman, J.B. Lloyd, G. Poste, J. Senior (Programme Co-ordinator) and Andre Trouet (ASI Co-director).

287

CONTRIBUTORS

Bakker-Woodenberg, I., Department of Clinical Microbiology, Erasmus
     University, Rotterdam, P.O. Box 1738, 3000 DR Rotterdam, The Nether-
     lands.

Barenholz, Y., Hebrew University-Hadassah Medical School, Jerusalem, Israel.

Brasseur, P., Laboratoire de Pharmacie Galenique, Universite Catholique
     de Louvain, 1200 Brussels, Belgium.

Couvreur, P., Laboratoire de Pharmacie Galenique, Universite de Parix XI,
     92290 Chatenay-Malabry, France.

Crommelin, D.J.A., Department of Pharmaceutics, University of Uttrecht,
     Catharijnesingel 60, 3511 GH Uttrecht, The Netherlands.

Daemen, T., Laboratory of Physiological Chemistry, University of Groningen,
     Bloemsingel 10, 9712 KZ Groningen, The Netherlands.

Davis, S.S., Pharmacy Department, University of Nottingham, Nottingham
     NG7 2RD, U.K.

Douglas, S.J., Pharmacy Department, University of Nottingham, Nottingham
     NG7 2RD, U.K.

Due, C., Cancer Biology laboratory, State University Hospital, (Rigshosp-
     italer), DK-2100 Copenhagen, Denmark.

Eppstein, D.A., Syntex Research, Palo Alto, CA 94394, U.S.A.

Ferruti, P., Facolta d'Ingegneria, Universita di Brescia, Viale Europa 39,
     25060 Brescia, Italy.

Fransen, G.J., Department of Pharmaceutics, University of Uttrecht,
     Catharijnesingel 60, 3511 GH Uttrecht, The Netherlands.

Gabizon, A., Cancer Research Institute, 1282-M, School of Medicine,
     University of California, San Francisco, CA 94143, U.S.A.

Goren, P., Hadassah University Hospital, Jerusalem, Israel.

Gregoriadis, G., Medical Research Council Group, Academic Department of
     Medicine, Royal Free Hospital School of Medicine, London NW3 2QG, U.K.

Grislain, L., Laboratoire de Pharmacie Galenique, Universite Catholique
     de Louvain, 1200 Brussels, Belgium.

289

Guiot, P., Laboratoire de Pharmacie Galenique, Universite Catholique de
    Louvain, 1200 Brussels, Belgium.

Hakomori, S.-I., Program of Biochemical Oncology and Membrane Research,
    Fred Hutchinson Cancer Research Centre, Departments of Pathobiology,
    Microbiology and Immunology, University of Washington, Seattle, WA
    98104, U.S.A.

Hopfer, R.L., Laboratory Medicine, The University of Texas M.D. Anderson
    Hospital and Tumor Institute at Houston, Houston, Texas, U.S.A.

Humphrey, M.J., Pfizer Central Research, Sandwich, Kent, U.K.

Ibrahim, A., Laboratoire de Pharmacie Galenique, Universite Catholique de
    Louvain, 1200 Brussels, Belgium.

Illum, L., Pharmaceutics Department, Royal Danish School of Pharmacy,
    2 Universitetsparken, Denmark.

Jones, P.D.E., Pharmacy Department, University of Nottingham, Nottingham
    NG7 2RD, U.K.

Juliano, R.L., Department of Pharmacology, The University of Texas Medical
    School at Houston, Houston, Texas, U.S.A.

Lenaerts, V., Laboratoire de Pharmacie Galenique, Universite Catholique
    de Louvain, 1200 Brussels, Belgium.

Lloyd, J.B., Biochemistry Research Laboratory, Department of Biological
    Sciences, University of Keele, Staffordshire, U.K.

Lopez-Berestein, G., Departments of Clinical Immunology and Biological
    Therapy, The University of Texas M.D. Anderson Hospital and Tumor
    Institute at Houston, Houston, Texas, U.S.A.

Mak, E., Pharmacy Department, University of Nottingham, Nottingham NG7 2RD
    U.K.

McIntyre, N., Academic Department of Medicine, Royal Free Hospital School
    of Medicine, Rowland Hill Street, London NW3 2PF, U.K.

McQueen, T., Departments of Clinical Immunology and Biological Therapy,
    The University of Texas M.D. Anderson Hospital and Tumor Institute
    at Houston, Houston, Texas, U.S.A.

Mehta, K., Departments of Clinical Immunology and Biological Therapy, The
    University of Texas M.D. Anderson Hospital and Tumor Institute at
    Houston, Houston, Texas, U.S.A.

Mehta, R., Department of Pharmacology, The University of Texas Medical
    School at Houston, Houston, Texas, U.S.A.

Muller, R.H., Pharmacy Department, University of Nottingham, Nottingham
    NG7 2RD, U.K.

Olsson, L., Cancer Biology Laboratory, State University Hospital (Rigs-
    hospitaler), DK-2100 Copenhagen, Denmark.

Pitha, J., Macromolecular Chemistry Section, National Institutes of Aging/
    GRC, NIH, 4940 Eastern Ave., Baltimore, Maryland, 21224, U.S.A.

Ramu, A., Hadassah University Hospital, Jerusalem, Israel.

Regts, J., Laboratory of Physiological Chemistry, University of Groningen, Bloemsingel 10, 9712 KZ Groningen, The Netherlands.

Ringrose, P.S., Pfizer Central Research, Sandwich, Kent, U.K.

Roerdink, F., Laboratory of Physiological Chemistry, University of Groningen, Bloemsingel 10, 9712 KZ Groningen, The Netherlands.

Ryser, H.J.-P., Department of Pathology, Boston University School of Medicine, 80 E. Concord Street, Boston, MASS 02118, U.S.A.

Salemink, P.J.M., Organon International BV., P.O. Box 20, 5340 BH Oss, The Netherlands.

Scherphof, G., Laboratory of Physiological Chemistry, University of Groningen, Bloemsingel 10, 9712 KZ Groningen, The Netherlands.

Senior, J., Medical Research Council Group, Academic Department of Medicine, Royal Free Hospital School of Medicine, London NW3 2QG, U.K.

Shen, W.-C., Department of Pathology, Boston University School of Medicine, 80 E. Concord Street, Boston, MASS 02118, U.S.A.

Summerfield, J., Academic Department of Medicine, Royal Free Hospital School of Medicine, Rowland Hill Street, London NW3 2PF, U.K.

Van Snick, L., Laboratoire de Pharmacie Galenique, Universite Catholique de Louvain, 1200 Brussels, Belgium.

Verdun, C., Laboratoire de Pharmacie Galenique, Universite Catholique de Louvain, 1200 Brussels, Belgium.